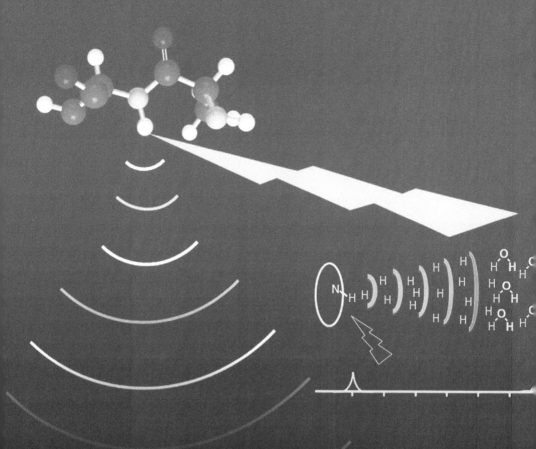

Chemical Exchange Saturation Transfer Imaging

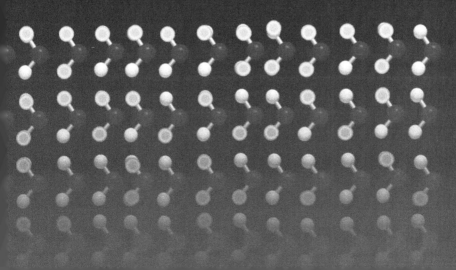

Chemical Exchange Saturation Transfer Imaging

Advances and Applications

edited by

Michael T. McMahon
Assaf A. Gilad
Jeff W. M. Bulte
Peter C. M. van Zijl

PAN STANFORD PUBLISHING

Published by

Pan Stanford Publishing Pte. Ltd.
Penthouse Level, Suntec Tower 3
8 Temasek Boulevard
Singapore 038988

Email: editorial@panstanford.com
Web: www.panstanford.com

British Library Cataloguing-in-Publication Data
A catalogue record for this book is available from the British Library.

Chemical Exchange Saturation Transfer Imaging: Advances and Applications

Copyright © 2017 Pan Stanford Publishing Pte. Ltd.

All rights reserved. This book, or parts thereof, may not be reproduced in any form or by any means, electronic or mechanical, including photocopying, recording or any information storage and retrieval system now known or to be invented, without written permission from the publisher.

For photocopying of material in this volume, please pay a copying fee through the Copyright Clearance Center, Inc., 222 Rosewood Drive, Danvers, MA 01923, USA. In this case permission to photocopy is not required from the publisher.

ISBN 978-981-4745-70-3 (Hardcover)
ISBN 978-1-315-36442-1 (eBook)

Printed in the USA

Contents

Preface   xiii

Section I From the 1960s to the 2010s: How Saturation Transfer Was First Discovered and Then Migrated Into Imaging

1 Discovery of the "Saturation Transfer" Method   3
Sture Forsén

2 Development of Chemical Exchange Saturation Transfer in Bethesda   9
Robert S. Balaban and Steven D. Wolff

3 History of In Vivo Exchange Transfer Spectroscopy and Imaging in Baltimore   17
Peter C. M. van Zijl
  3.1 Before There Was CEST   17
  3.2 Early CEST Experiments   20
  3.3 Amide Proton Transfer-Weighted MRI   24
  3.4 Expansion of the CEST Efforts   29
      3.4.1 Assaf Gilad's Recollections   29
      3.4.2 Mike McMahon's Recollections   31
  3.5 Translation to Human Scanners   32
  3.6 Active Growth in CEST   34

4 Early Discovery and Investigations of paraCEST Agents in Dallas   39
A. Dean Sherry

5 Birth of CEST Agents in Torino   47
   Silvio Aime

   SECTION II PULSE SEQUENCE, IMAGING, AND POST-PROCESSING
               SCHEMES FOR DETECTING CEST CONTRAST

6 General Theory of CEST Image Acquisition and
   Post-Processing   57
   Nirbhay N. Yadav, Jiadi Xu, Xiang Xu, Guanshu Liu,
   Michael T. McMahon, and Peter C. M. van Zijl
   6.1 Introduction   57
   6.2 Theory   59
        6.2.1 Low-Power Irradiation   60
        6.2.2 Sensitivity Enhancement   61
        6.2.3 High-Power Irradiation   63
        6.2.4 Pulsed-CEST   64
        6.2.5 Shape of RF Pulses   67
        6.2.6 Utilizing t_{exch} in LTMs to Extract CEST Contrast   68
        6.2.7 Utilizing Labeling Flip Angle to Filter Contrast   70
        6.2.8 OPARACHEE   70
        6.2.9 FLEX   71
        6.2.10 Alternative Ways for CEST Acquisition   72
               6.2.10.1 LOVARS   72
               6.2.10.2 Steady-state CEST   73
               6.2.10.3 SWIFT-CEST   74
               6.2.10.4 Ultrafast gradient-encoded
                        Z-spectroscopy   76
   6.3 Post-Processing   78
        6.3.1 B_0 Correction   79
               6.3.1.1 Using a pre-acquired B_0 map   79
               6.3.1.2 Fitting Z-spectral data   80
        6.3.2 Asymmetry Analysis   81
        6.3.3 Integration of CEST Effect over a Range of
               Offsets   82
        6.3.4 Non-MTR_{asym} Metrics   83
        6.3.5 Additional Image-Processing Steps   85
               6.3.5.1 SNR/CNR filtering   85
               6.3.5.2 Filters for image de-noising   85

		6.3.5.3	MTC-based image filtering	86
		6.3.5.4	B_1 correction	86
	6.4	Conclusion		87

7 Uniform-MT Method to Separate CEST Contrast from Asymmetric MT Effects — 97
Jae-Seung Lee, Ravinder R. Regatte, and Alexej Jerschow

7.1	Saturation of a Spin-1/2 System	98
7.2	Uniform Saturation of a Dipolar-Coupled Spin-1/2 System	102
7.3	Uniform-MT Methodology	104
7.4	Application to Brain MRI	110
7.5	Application to Knee MRI	114
7.6	Summary	117

8 HyperCEST Imaging — 121
Leif Schröder

8.1	HyperCEST in the Historic Context of CEST Development	122
8.2	Hyperpolarized Xenon NMR	126
	8.2.1 Xenon NMR Conditions Compared to Protons	128
	8.2.2 Production of Hyperpolarized Xenon	129
	8.2.3 Delivery of hp Xe and Optimized Use of Magnetization	131
	8.2.4 Fast Spectral Encoding (Gradient-Encoded CEST)	134
8.3	Xenon Host Structures	135
	8.3.1 Tailored Host Structures: Cryptophanes	135
	8.3.2 Compartmentalization of Xenon	137
	8.3.3 Targeted Hosts	138
	8.3.4 HyperCEST Modeling	140
8.4	Phospholipid Membrane Studies/Delta Spectroscopy	141
8.5	Live Cell NMR of Exchanging Xenon	143
8.6	Conclusion	147

SECTION III diaCEST/paraCEST/lipoCEST CONTRAST PROBES

9 Current Landscape of diaCEST Imaging Agents   161
Amnon Bar-Shir, Xing Yang, Xiaolei Song, Martin Pomper, and Michael T. McMahon
- 9.1 Introduction   161
- 9.2 Molecules with Alkyl Amines and Amides   165
- 9.3 Molecules with Alkyl Hydroxyls   166
- 9.4 N-H Containing Heterocyclic Compounds   168
- 9.5 Salicylic Acid and Anthranilic Acid Analogues   173
- 9.6 Macromolecules with Labile Protons   176
- 9.7 Fluorine and Chemical Exchange Saturation Transfer   177

10 Evolution of Genetically Encoded CEST MRI Reporters: Opportunities and Challenges   193
Ethel J. Ngen, Piotr Walczak, Jeff W. M. Bulte, and Assaf A. Gilad
- 10.1 Introduction   193
  - 10.1.1 Genetically Encoded Reporter Imaging   194
  - 10.1.2 Genetically Encoded MRI Reporters   196
- 10.2 CEST MRI Contrast Generation Mechanism   198
- 10.3 Genetically Encoded CEST MRI Reporters   200
  - 10.3.1 Genetically Encoded CEST-Responsive Protein-Based Reporters   203
    - 10.3.1.1 Lysine-rich protein (LRP)-based reporter genes   203
    - 10.3.1.2 Arginine-rich protein (ARP)-based reporter genes   204
    - 10.3.1.3 Superpositively charged green fluorescent proteins   204
  - 10.3.2 Genetically Encoded Enzyme/Probe CEST MRI Reporter Systems   205
    - 10.3.2.1 Protein kinase A   205
    - 10.3.2.2 Herpes simplex virus type 1 thymidine kinase   206
- 10.4 Genetically Encoded Hyperpolarized Xenon (^{129}Xe) CEST MRI Reporters   208

	10.5	Considerations in Developing CEST MRI Genetically Encoded Reporters	210
	10.6	Current Challenges and Future Directions	210
	10.7	Conclusion	211

11 ParaCEST Agents: Design, Discovery, and Implementation 219
Mark Milne, Yunkou Wu, and A. Dean Sherry

	11.1	Introduction	219
		11.1.1 History of paraCEST Agents	219
	11.2	Lanthanide-Induced Shifts	222
	11.3	T_1 and T_2 Considerations in the Design of paraCEST Agents	227
	11.4	Water Molecule Exchange, Proton Exchange, and CEST Contrast	235
	11.5	Modulation of Inner-Sphere Water Exchange Rates	238
	11.6	Techniques to Measure Exchange Rates	247
		11.6.1 Direct Measurement of ^{1}H NMR Resonance Line Widths	248
		11.6.2 Omega Plots	249
		11.6.3 Bloch Fitting	250
	11.7	Summary	252

12 Transition Metal paraCEST Probes as Alternatives to Lanthanides 257
Janet R. Morrow and Pavel B. Tsitovich

	12.1	Introduction	257
	12.2	Coordination Chemistry of Iron(II), Cobalt(II), and Nickel(II)	260
	12.3	NMR Spectra, CEST Spectra, and Imaging	263
		12.3.1 CEST Spectra	266
		12.3.2 CEST Imaging	269
	12.4	Responsive Agents	270
		12.4.1 pH-Responsive Agents	270
		12.4.2 Redox-Responsive Agents	271
		12.4.3 Temperature-Responsive Agents	274
	12.5	Toward In Vivo Studies	276
	12.6	Summary	277

13 Responsive paraCEST MRI Contrast Agents and Their Biomedical Applications 283
Iman Daryaei and Mark D. Pagel
- 13.1 Introduction 283
- 13.2 ParaCEST Agents That Detect Enzyme Activities 286
- 13.3 ParaCEST Agents That Detect Nucleic Acids 290
- 13.4 ParaCEST Agents That Detect Metabolites 292
- 13.5 ParaCEST Agents That Detect Ions 294
- 13.6 ParaCEST Agents That Detect Redox State 296
- 13.7 ParaCEST Agents That Measure pH 297
- 13.8 ParaCEST Agents That Measure Temperature 300
- 13.9 Future Directions for Clinical Translation of paraCEST Agents 301

14 Saturating Compartmentalized Water Protons: Liposome- and Cell-Based CEST Agents 311
Daniela Delli Castelli, Giuseppe Farrauto, Enzo Terreno, and Silvio Aime
- 14.1 Introduction 311
- 14.2 Basic Features of lipoCEST/cellCEST Agents 313
  - 14.2.1 Chemical Shift of Intravesicular Water Protons in Presence of Paramagnetic SR 313
  - 14.2.2 CEST Contrast in lipoCEST/cellCEST: Effect of Exchange Rate and Size 323
  - 14.2.3 Liposomes Loaded with CEST Agents 324
- 14.3 Applications 325
  - 14.3.1 LipoCEST Agents 325
  - 14.3.2 CellCEST Agents 330
  - 14.3.3 Liposomes Loaded with CEST Agents 336

SECTION IV EMERGING CLINICAL APPLICATIONS OF CEST IMAGING

15 Principles and Applications of Amide Proton Transfer Imaging 347
Jinyuan Zhou, Yi Zhang, Shanshan Jiang, Dong-Hoon Lee, Xuna Zhao, and Hye-Young Heo
- 15.1 Introduction 347
- 15.2 APT Imaging Principle and Theory 349

	15.3	APT Imaging of Stroke	352
	15.4	Differentiation between Ischemia and Hemorrhage	355
	15.5	APT Imaging of Brain Tumors	357
	15.6	Differentiation between Active Glioma and Radiation Necrosis	360
	15.7	Conclusions and Future Directions	362

16 Cartilage and Intervertebral Disc Imaging and Glycosaminoglycan Chemical Exchange Saturation Transfer (gagCEST) Experiment — 377

Joshua I. Friedman, Ravinder R. Regatte, Gil Navon, and Alexej Jerschow

	16.1	Introduction	377
	16.2	Composition and Organization of Cartilage	379
	16.3	Composition and Organization of Intervertebral Disc	381
	16.4	MRI Techniques for Measuring GAG (Other than CEST)	383
		16.4.1 Gadolinium-Enhanced Imaging	384
		16.4.2 Sodium Imaging	384
		16.4.3 $T_{1\rho}$ Contrast	385
	16.5	GagCEST	385
	16.6	Conclusion	393

17 GlucoCEST: Imaging Glucose in Tumors — 399

Francisco Torrealdea, Marilena Rega, and Xavier Golay

	17.1	Introduction	399
	17.2	Cancer Metabolism and the Warburg Effect	400
	17.3	Imaging Methods Targeting Metabolism	402
	17.4	GlucoCEST: The Concept	403
		17.4.1 Advantages	404
		17.4.2 Drawbacks	404
	17.5	GlucoCEST: State of the Art	405
		17.5.1 The Origins: GlycoCEST	406
		17.5.2 Cancer Studies	407
		17.5.3 Brain Studies	413
		17.5.4 Alternative Technique for Glucose Detection	417

	17.6	GlucoCEST: Good Practices	419
		17.6.1 Main Magnetic Field Drifts	419
		17.6.2 Timing of Frequency Offsets	420
		17.6.3 Offset and Integration Range	420
	17.7	Conclusion: Remaining Open Questions	421

18 Creatine Chemical Exchange Saturation Transfer Imaging — 427

Catherine DeBrosse, Feliks Kogan, Mohammad Haris, Kejia Cai, Anup Singh, Ravi P. R. Nanga, Mark Elliott, Hari Hariharan, and Ravinder Reddy

	18.1	Introduction	427
	18.2	Study of Energy Metabolism: ^{31}P MRS	428
		18.2.1 ^{31}P Magnetic Resonance Spectroscopy	428
		18.2.2 ^{31}P MRS versus CEST Imaging	429
	18.3	Development of Creatine CEST	430
		18.3.1 Definition of Exchangeable CK Amine Protons and Their Exchange Rates	430
		18.3.2 CrCEST Phantom Imaging	431
		18.3.3 In Vivo CrCEST Studies of Skeletal Muscle Exercise at Ultra-High Field	433
		18.3.4 Implementation of CrCEST at Clinical-Strength Field	436
		18.3.5 Application of CrCEST in Imaging of Myocardial Metabolism	438
		18.3.6 CrCEST Application in Brain Imaging	441
	18.4	Summary	442

19 Iodinated Contrast Media as pH-Responsive CEST Agents — 447

Dario Longo and Silvio Aime

	19.1	Iopamidol as a diaCEST Agent in Preclinical Studies	448
	19.2	Iopamidol as diaCEST Agent on a Clinical MRI Scanner (3 T)	454
	19.3	Iopromide as a diaCEST Agent in Preclinical Studies	456
	19.4	Iobitridol as a diaCEST Agent in Preclinical Studies	459
	19.5	Conclusion	463

Index — 467

Preface

Chemical exchange saturation transfer (CEST) imaging, a burgeoning field of magnetic resonance imaging (MRI), is only at the beginning of exploring some of the many possible applications and technology developments. The rapid expansion that CEST imaging has seen over the past 16 years since the seminal Ward and Balaban paper in 2000 has created a need for a graduate-level handbook that describes the wide assortment of issues that come into play when trying to apply CEST imaging to medicine. These issues include the fundamental principles of saturation transfer, the key features of the diaCEST, paraCEST, lipoCEST, and hyperCEST agents that enable the production of imaging contrast, and the practical aspects of preparing image acquisition and post-processing schemes suited for the application. An appreciation for the expansion of this field can be found in the CEST MRI publication statistics as follows (Fig. 1), obtained from the Web of Science in April 2016.

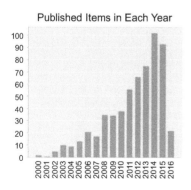
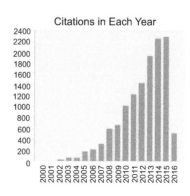

Figure 1 Publication statistics of CEST MRI.

Similarly, the first five CEST imaging workshops from 2010 to 2015 (see Figs. 2–6) saw an increasing number of participants from 49 to more than 200. The fifth CEST imaging workshop held in Philadelphia, PA, in 2015 (PennCEST) had three full days of presentations, displaying the tremendous breadth of the field. Included in the program were more than two sessions dedicated to presentations involving clinical applications, proving that CEST imaging is now indeed being considered by multiple sites as an imaging technology that is clinically translatable.

Figure 2   First CEST imaging workshop in Torino, Italy (January 2010).

Figure 3   Second CEST imaging workshop in Dallas, Texas (June 2011).

This textbook was developed after organizing the third CEST imaging workshop (OctoberCEST held in October 2012), which provided an outline for this textbook. While already incomplete due to the many novel methods and applications being developed continuously, it is meant to provide a current overview of the main

Figure 4 Third CEST imaging workshop (OctoberCEST) in Annapolis, Maryland (October 2012).

Figure 5 Fourth CEST imaging workshop in Torino, Italy (May 2014).

Figure 6 Fifth CEST imaging workshop (PennCEST), Philadelphia, Pennsylvania (October 2015).

aspects of CEST imaging technology, and as such we hope that this book provides guidance to both students and senior researchers with an interest in this topic. In order to provide a perspective on how the technology developed, the first five chapters describe the history of CEST imaging from the initial saturation transfer nuclear magnetic resonance experiments performed in the 1960s in Sweden by Ragnar Hoffman and Sture Forsén, to the pioneering work of Robert Balaban, Dean Sherry, Silvio Aime, and Peter van Zijl applying the principles of saturation transfer to MRI.

We would like to express gratitude to the many students, postdocs, and faculty members who have been important at Johns Hopkins University and the Kennedy Krieger Institute for developing the CEST imaging projects, whose names are in the many citations in the different chapters.

Michael T. McMahon
Assaf A. Gilad
Jeff W. M. Bulte
Peter C. M. van Zijl

Section I

From the 1960s to the 2010s: How Saturation Transfer Was First Discovered and Then Migrated Into Imaging

Sture Forsén, March 21, 2016

Chapter 1

Discovery of the "Saturation Transfer" Method

Sture Forsén

The Pufendorf Institute for Advanced Studies, Lund University, Lund, Sweden
sture.forsen@pi.lu.se

Let me take you back to around 1960, the early days of nuclear magnetic resonance (NMR) spectroscopy; some of you younger readers may think it was medieval times, but it was quite not that bad! At that time, there were essentially only two NMR instruments in entire Sweden: one at the Physics Department at the Uppsala University and one at the Royal Institute of Technology in Stockholm. This was, of course, in the pre-superconducting era, and at the core of these two Varian spectrometers were large heavy electromagnets with a field strength of 0.9 T (Uppsala) and 1.4 T (Stockholm) (40 resp. 60 MHz 1-H NMR frequency).

The instruments were real temperamental beasts that had to be domesticated every morning. The first problem was usually to get a good homogenous magnetic field over the sample probe. This entailed turning large nuts on the magnet yoke while simultaneously observing the shape of a water 1-H signal on an oscilloscope. To that end, we had several meters long monkey wrenches to turn the nuts ever so little. As a consequence of these demanding

---

Chemical Exchange Saturation Transfer Imaging: Advances and Applications
Edited by Michael T. McMahon, Assaf A. Gilad, Jeff W. M. Bulte, and Peter C. M. van Zijl
Copyright © 2017 Pan Stanford Publishing Pte. Ltd.
ISBN 978-981-4745-70-3 (Hardcover), 978-1-315-36442-1 (eBook)
www.panstanford.com

physical exercises, one developed a rather close relationship with the spectrometers!

The NMR instrument at Uppsala had been purchased by the head of the Physics Department, Kai Siegbahn (future Nobel Laureate in Physics), for precision measurements of nuclear magnetic moments. But since it turned out that chemical shifts had become the most irritating complication for this task, the instrument was eventually turned over to a young PhD student named Ragnar Hoffman, which allowed him to study chemical shifts and spectra of organic molecules. Ragnar indeed did these studies in collaboration with local organic chemists, and they mainly studied thiophene and furane derivatives that presented tractable 2-, 3-, and 4-proton spin systems.

The instrument at the Royal Institute of Technology in Stockholm was located at the Department of Physical Chemistry and had been acquired mainly for studies of water molecules in hydrogels. I had been persuaded to handle the spectrometer as part of my research work for a Doctor of Technology thesis, and this was a year before the instrument actually arrived in the late 1950s. This NMR instrument did not live up to the expectations of the department heads. But by reading the literature, I quickly realized that there were great opportunities in the studies of other physical chemical phenomena, and after some minor clashes with my supervisor, I was given the freedom to pursue them—in particular studies of inter- and intramolecular hydrogen bonds.

Ragnar and I had been aware of each other's existence for some time, but we had not really become personally acquainted until we had both passed our thesis defenses (quite demanding in those days). In 1962, both of us went to attend an NMR conference at Oxford, and as two of the few participants from Sweden, we were lodged together at Merton College—a great college, one of the oldest in Oxford, founded in 1264, and beautiful to look at. But after the founding—English readers, please forgive me—perhaps no major improvements in indoor housing and energy-saving measures seemed to have taken place. At least not where Ragnar and I were housed. We, pampered Swedes, were freezing like hell. We could not but laugh at our predicament and found that we shared a childish sense of humor. We decided right away that we must collaborate!

And when we were back in Sweden, we indeed began to plan what we should do.

The Stockholm instrument was equipped with a radio frequency (r.f.) modulation device that allowed us to conduct double resonance experiments with reasonable ease, irradiating one peak in a spin-coupled system while observing the effect on other peaks, or carry out Overhauser experiments and determine the sign of spin coupling constants and what not. These were families of experiments that could be performed with great ease years later with the advent of pulsed r.f. techniques and two- and three-dimensional NMR. So at full speed, we tried to use double resonance methods to study properties of spin systems of all kinds. We became increasingly familiar with magnetization transfer within complex systems of coupled spins. We, for one thing, developed a technique (transitory selective irradiation) whereby you transiently irradiate a peak in a 1-H NMR spectrum with a substantial r.f. field so fast that you force the population of the two involved levels to instantly become more or less equalized. Immediately afterward, if you recorded the whole 1-H NMR spectrum, before T_1 processes had a chance to restore the Boltzmann population, you could see intensity changes (increases or decreases) only in the lines that had an energy level in common with either of the two levels you had "equalized." Hereby the sign of spin coupling constants could be determined, a problem that occupied the minds of many NMR spectroscopists around 1960, and of course also later handled in a facile way with pulse techniques and two-dimensional NMR.

Meanwhile, I had simultaneously continued 1-H NMR studies of inter- and intramolecular hydrogen bonding in different organic systems. It was notoriously hard, if not outright impossible, to determine the chemical shift of OH signals of hydroxy compounds in aqueous media. It was obvious that the OH hydrogens were involved in chemical exchange with those of the solvent H_2O. It was a different situation with alcohols in inert organic solvents; the OH signal from ethanol in solvents like per-deutero cyclohexane or CS_2 was usually seen as a sharp triplet due to spin coupling with the neighboring CH_2 group. Signals from OH groups involved in intramolecular hydrogen bonds—for example those in salicylaldehyde, 2-hydroxy acetophenone, or enolic systems—were usually easy to study and the 1-H NMR chemical shift seemed to reflect the strength of the hydrogen bond. Determining the rate of hydrogen exchange between OH groups with NMR techniques in mixed systems was generally difficult. Through studies of line shapes, one could, in fortunate cases, obtain a value of the rate of hydrogen exchange for a particular OH group.

During the spring of 1963, I had prepared a CS_2 solution with an equimolar mix of a sterically crowded aliphatic alcohol (tert-butylalcohol) and a molecule with an intramolecular hydrogen bond (2-hydroxyacetophenone). In this solution, separate, well-resolved signals from the two OH groups were observed. Clearly, the hydrogen exchange between the OH groups was "slow": but how slow? I discussed the system with Ragnar on a visit to Uppsala University and more or less out of the blue asked him what would happen if we used double resonance to irradiate one of the OH signals. I do not think we really dared to predict the outcome; "let us take a look and see" was our joint decision!

And so we did. A few weeks later, we booked the 60 MHz DP 60 at Stockholm for 2 days. As I alluded to earlier, it generally took some time to get the "beast" in good working order, and it was quite late when we were finally in business. We repeatedly recorded the NMR signal of the tert-butyl OH groups and put our strong second r.f. field at the peak of the OH group from 2-hydroxyacetophenone in order to "saturate" it, i.e., to equalize the population of "up" and "down" spin levels.

With delight and amazement, we observed how the tert-butyl OH signal slowly decayed in what looked like an exponential process! Success was visible, but it was too early to get excited. Was the phenomenon reproducible or had we overlooked something? We repeated the experiment and got the same result. What would happen if we turn off the second r.f. field after the signal had decayed to zero? Yes, the obliterated peak gradually returned to normal intensity in an apparently exponential process.

What happened if we did our experiment but switched the OH molecules we irradiated? Similar results. Now we finally felt convinced that we had discovered a way to measure the rate of chemical exchange of OH hydrogens between two different sites, in our case two kinds of molecules! Upbeat late that night, we actually started dancing on the floor around the magnet. A security guard, probably alarmed by the noise we made, put his puzzled head in through the door and must have thought we had gone totally bananas. But we managed to put him at ease.

Exhausted, we finally departed to our respective homes realizing that we needed to put together a theoretical description of the method and its results as soon as possible. But our brains needed re-oxygenation and at least a night's rest before we took on this task.

Over the next few days, we put forward a theory—initially involving only two sites—starting with the Bloch equations, so brilliantly modified in 1958 by Harden M. McConnell, to include chemical exchange. It was fairly straightforward to show that the z-magnetization $[M_z(A)]$ in the observed site A, when the z-magnetization $[M_z(B)]$ of the irradiated, "saturated" site B was equal to zero, would decay with a time constant involving both the chemical exchange rate $1/\tau_A$ (the inverse lifetime of protons at site A) and the relaxation rate $1/T_1$ (A). This also indicated the range of exchange rates accessible with the "saturation transfer" method; the longer the T_1, the slower was the exchange rates that could be studied.

We rushed off an account of our findings to a local journal (*Acta Chemica Scandinavica*) mainly to ascertain priority so that we could have time to elaborate on the theory and perform a few more experiments. A full paper was submitted to the *Journal of Chemical Physics* in late June 1963 and was published in the December 1 issue the same year (Vol. **39**, pp. 2892–2901). This paper was soon followed by a second paper in the same journal, where the method was extended and applied to a system of three non-equivalent sites—a keto–enol equilibrium [Vol. **40**, pp. 1189–1196 (1964)].

I have often been asked why we called the method "saturation transfer" rather than "magnetization transfer" because transferring magnetization was really what we were doing. I think it simply has to do with the way we did the experiments in those days; remember this was before pulse techniques were commonly available, at least it was not to us. We used continuous r.f. irradiation to "saturate" an NMR signal, meaning that we forced an equalization of the spin population in the two energy levels involved. Had we done the experiments on a modern pulsed NMR spectrometer, we could have played around with the magnetization in the "irradiated" spin system and turned the initial z-magnetization vector 90° or 180° or whatever.

I cannot but find delight in the many subsequent ways saturation transfer methods have found applications in modern MRI studies, as so well documented in this volume. And I am indebted to Mike McMahon for inviting me to bring to life the events that led to our accounts of the method in 1963. I finally regret that Ragnar Hoffman is no longer with us to share the reminiscing he would have loved to be part of!

Chapter 2

Development of Chemical Exchange Saturation Transfer in Bethesda

Robert S. Balaban[a] and Steven D. Wolff[b]

[a]*Laboratory of Cardiac Energetics, National Heart Lung and Blood Institute, Bethesda, MD 20892, USA*
[b]*Carnegie Hill Radiology, New York, NY 10075, USA*
balabanr@nhlbi.nih.gov

The development of chemical exchange saturation transfer (CEST) imaging agents in the Laboratory of Cardiac Energetics began with our interest in the creatine kinase reaction kinetics in vivo using ^{31}P nuclear magnetic resonance (NMR) to monitor the exchange between creatine phosphate (CrP) and the γP of adenosine triphosphate (ATP). We had been examining the reaction kinetics of this system in different tissues but were concerned that the summing of the reaction kinetics over the large voxel of the classical spectroscopy methods might be averaging different rates across the tissues. Paul Hsieh was a Howard Hughes Medical Institute-NIH Research Scholar who had done an excellent job of characterizing the gross tissue exchange rate [1], and we decided that the 25 mM CrP signal was sufficiently large that we might be able to image the kinetics of the creatine kinase reaction by monitoring the effect of saturating the lower concentration γP-ATP and reading this out on the amplitude of CrP. We had previously

---

Chemical Exchange Saturation Transfer Imaging: Advances and Applications
Edited by Michael T. McMahon, Assaf A. Gilad, Jeff W. M. Bulte, and Peter C. M. van Zijl
Copyright © 2017 Pan Stanford Publishing Pte. Ltd.
ISBN 978-981-4745-70-3 (Hardcover), 978-1-315-36442-1 (eBook)
www.panstanford.com

established that the saturation transfer process is a great way to amplify the signal from low concentration exchange partners and is even capable of observing enzyme substrate complexes. This was recently reviewed by Alan Koretsky and Robert Balaban [2]. Important in this experiment was a control irradiation on the equal but opposite side of the CrP resonance to control for imperfections in the irradiation. The importance of this will become clear in the following paragraphs. Paul was very successful in this effort. We believe publishing the first image of a chemical exchange process in living tissue [3]. Unfortunately, this paper was published in the *Journal of Magnetic Resonance*, one of the best basic journals in the NMR field, in 1987 before it was covered in Pub Med and is thus not easy to find by the biological and medical community.

After completing these studies, we sat with Paul in our lab in the basement of Bldg 1 under the Director of NIH's office discussing the next steps. Steve was using ^{31}P spectroscopy and ^{1}H imaging at 4.7 T to monitor a group of organic solutes that are in relatively high concentration in the rabbit kidney, a project that was appropriate for the Laboratory of Kidney and Electrolyte Metabolism, which we were associated with at the time [4]. We hypothesized that the phenomenon that Paul exploited by imaging the relatively strong CrP signal could be applied analogously to imaging the strong water proton signal. In this case, the image would be sensitive to the chemical exchange of protons between water and metabolites that had chemically exchangeable protons. Steve Wolff took this project on with the idea of imaging the high concentration of exchangeable molecules in the kidney such as urea and ammonia. These molecules play a key role in the physiology of the kidney. Again we were in the Laboratory of Kidney and Electrolyte Metabolism, and earlier I had tried to image urea and other trimethylamines using ^{14}N with Mark Knepper without much success due to signal to noise issues [5]. Steve ran a large series of experiments using phantoms that demonstrated that the exchange process with urea could be easily detected via specific saturation transfer to water resulting in a large amplification of the urea signal. The results suggested that the technique should work in vivo at physiologic concentrations of urea and physiologic pH. We have included a page from Steve's notebook from one of these early studies in 1987 (Fig. 2.1). Please

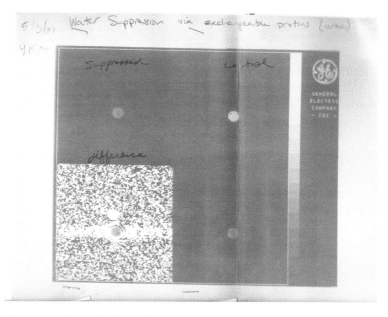

Figure 2.1 One of the early sets of images displaying CEST contrast collected on a 1 M urea phantom at NIH.

excuse the rough appearance of these images since we were quite new to this field at the time. The image presents a water proton cross section of a 15 ml test tube containing 1 M urea. After a control irradiation, discussed earlier, Steve irradiated the urea resonance and observed the predicted decrease in the water amplitude. These solution studies with urea and ammonia were the first to use CEST imaging for water, but they were not published for several years [6] for reasons that will become apparent in the following paragraphs.

After Steve demonstrated the utility of the approach, he immediately went to his kidney model and attempted to image the urea or ammonia in the kidney. The experimental idea was to irradiate the metabolite resonance and see a much larger drop in the water proton signal due to signal amplification by chemical exchange. Steve was initially excited by the huge drop in the water signal when he irradiated the metabolite position. However, when he conducted the control irradiation on the opposite side of the water signal, he got a similar but not identical effect (see Fig. 2.1).

This result was initially perplexing, and a number of possible explanations were explored, such as an instrument error (not uncommon in those days), or the presence of poorly shimmed water. A consistent observation was that properly controlled CEST imaging could be easily performed on the urine taken from the rabbit or numerous prepared solutions but was more difficult in tissues. A key experiment was when Steve demonstrated a change in the water T_1 with the control or metabolite-specific irradiation demonstrating that most of the water was experiencing the off-resonance energy from a broad underlying resonance from the macromolecules. This work resulted in the original publication on magnetization transfer contrast (MTC) in the kidney [7]. Thus, the interference of macromolecular magnetization transfer in attempting to use a CEST approach to study specific metabolites resulted in the discovery of MTC in our laboratory. MTC became a major effort in the lab for many years; however, we also appreciated the importance of the metabolite-specific amplification provided by the saturation transfer to the water resonance and retrospectively published Steve's control experiments in 1990 [6]. Again this paper was really the first published CEST study from our lab.

One arm of the laboratory began to explore the sources of the magnetization transfer between the spin-coupled macromolecular protons and the free water. This really began to help us understand the chemistry of the exchange process at the hydration layer of the macromolecule. Teresa Fralix and Toni Ceckler, post-docs, and Sid Simon, on sabbatical from Duke University, decided to focus on lipid bilayer preparations (a specific expertise of Dr. Simon's) where we could alter the chemistry of the hydrophilic region of the lipid bilayer quite easily using different types of lipids. This was a very revealing study for us since it showed that simply having a hydrophilic region, such as in pure phosphotidlylcholine (PC), did not result in an exchange process. You needed a specific "exchange antenna" to transfer the magnetization to the water. In the case of PC, we found that by adding a small of amount of cholesterol, with its hydroxyl group intact, we could then observe magnetization transfer to the bulk water [8, 9]. These studies illustrated to us that the chemistry of the water macromolecule interface was critical and likely chemical exchange was one of the plausible mechanisms at the interface rather than spin diffusion to highly restricted water, at least in this model system. Again, chemical exchange was the critical element in these studies. This model system also allowed us to explore different chemical groups at the surface of the lipids, which were valuable in our subsequent studies of developing exogenous CEST agents.

Shortly after these studies had completed, Steve and I wanted to return to the original experiment of detecting the metabolites that had originally started this line of research in the kidney. A new graduate student had joined the lab, Valeska Scharen-Guivel, and we thought that a complete analysis of how to extract this metabolite-specific information from the background MTC process would make a good thesis project. Returning to the kidney, we demonstrated that by using appropriate controls to compensate for the MTC effects, we could image the distribution of urea and other molecules in the kidney [10]. We also validated some of the assignments using enzymatic methods.

With the development of methods to compensate for the MTC "interference," we turned to attempting to design molecules that could serve as exogenous contrast agents using the intrinsic amplification of transferring magnetization to the water resonance through saturation transfer as originally described in Steve's earlier work [7]. Thus, taking the concept of chemical exchange sites from the lipid work and the methods of detecting individual exchange sites in the presence of the background MTC, we initiated studies on designing CEST-specific agents. We thought this was a laudable approach since this would be the only contrast agent available that does not involve a potentially toxic metal, could be turned off and on by the NMR experiment permitting many types of signal averaging and noise reduction schemes, could generate agents that might be able to enter the cells and report the intracellular milieu or reactions, and finally could even envision genetically programmed contrast agents to insert in the cell much like the green fluorescence protein that had recently been exploited. To help in the development of the MRI imaging approaches, we initiated a project with Kathleen Ward, a post-doctoral fellow in the lab, and Anthony Aletras, a staff scientist, to find appropriate molecules for this task. Steve had left the NIH by this time to continue his career in cardiac MRI, which he also initiated in our laboratory. By simple analysis, we wanted to find agents with the largest chemical shift to attain the highest exchange rate without approaching the fast exchange limit. The faster the exchange, the more saturation transfer could be realized to the water and thus the more sensitive the assay with regards to the concentration of the exogenous agent. In these studies that we published in 2000 [11], we found many plausible molecular classes that had the appropriate properties of large chemical shift with high proton exchange sites and demonstrated the feasibility of this approach.

Why did we call it chemical exchange saturation transfer? In water–macromolecule magnetization transfer, which we had been extensively studying, both chemical exchange (CE) and magnetization transfer (MT) (dipolar coupling, spin diffusion) play a role. CE occurs primarily at the water–macromolecule interface, while MT occurs within the macromolecule. Some MT may also occur at the interface under specialized conditions. Thus, we thought it was important to specify CE rather than MT in generating the saturation transfer effect since the agents we were developing are working through CE alone.

Most notable in these studies was the obvious dependence on pH, which we had actually characterized in our earliest studies. Thus, as a direct measure of tissue pH in targeted compartments, CEST was an attractive application, but one needed to have an appropriate reference to compensate for water T_1 and concentration of the CEST probe. Kathleen and I published a ratiometric approach in which by using two exchange sites, a measure of pH could be determined independent of the water T_1 or probe concentration [12].

Since these initial observations, many innovations and improvements in both the probes, metals or intramolecular hydrogen bonding to increase the chemical shift, multivalent molecules, hyperpolarization, etc. as well as methodology to acquire the data have been developed in active research groups outlined in the chapters in this monogram. The main advantages remain evident again: not requiring potential toxic metals, ease of cell entry, ability to dynamically modulate the effect with RF pulses (timing somewhat limited by T_1), intrinsic pH information, monitoring of enzymatic activity in cells, and potential genetic programming of probes in specific cell compartment. I am sure this field will continue to develop as the information content and versatility of the approach is extensive, and the progress in the last decade has greatly exceeded what we thought was possible.

References

1. Hsieh PS and Balaban RS. Saturation and inversion transfer studies of creatine kinase kinetics in rabbit skeletal muscle in vivo. *Mag Reson Med*, 1988; 7: 56–64.
2. Balaban RS and Koretsky AP. Interpretation of ^{31}P NMR saturation transfer experiments: What you can't see might confuse you. Focus on "Standard magnetic resonance-based measurements of the $P_i \rightarrow$ ATP rate do not index the rate of oxidative phosphorylation in cardiac and skeletal muscles." *Am J Physiol Cell Physiol*, 2011; 301(1): C12–15.
3. Hsieh PS and Balaban RS. ^{31}P imaging of in vivo creatine kinase reaction rates. *J Magn Res*, 1987; 74(3): 574–579.
4. Wolff SD and Balaban RS. Proton NMR-spectroscopy and imaging of the rabbit kidney, in vivo, at 4.7 tesla. *J Magn Res*, 1987; 75(1): 190–192.
5. Balaban RS and Knepper MA. Nitrogen-14 nuclear magnetic resonance spectroscopy of mammalian tissues. *Am J Physiol* 1983; 245(5 Pt 1): C439–C444.
6. Wolff SD and Balaban RS. NMR imaging of labile proton-exchange. *J Magn Reson*, 1990; 86(1): 164–169.
7. Wolff SD and Balaban RS. Magnetization transfer contrast (MTC) and tissue water proton relaxation, in vivo. *Mag Reson Med*, 1989; 10(1): 135–144.
8. Ceckler TL, Wolff SD, Yip V, Simon SA, and Balaban RS. Dynamic and chemical factors affecting water proton relaxation by macromolecules. *J Magn Reson*, 1992; 98(3): 637–645.
9. Fralix TA, Ceckler TL, Wolff SD, Simon SA, and Balaban RS. Lipid bilayer and water proton magnetization transfer: Effect of cholesterol. *Mag Reson Med*, 1991; 18(1): 214–223.
10. Scharen-Guival VLM, Sinnwell TM, Balaban RS, and Wolff SD. Imaging of proton-water chemical exchange in the kidney using saturation transfer. *Proc Int Soc Magn Res Med*, 1998; 1: 470.

11. Ward KM, Aletras AH, and Balaban RS. A new class of contrast agents for MRI based on proton chemical exchange dependent saturation transfer (CEST). *J Magn Reson*, 2000; 143(1): 79–87.
12. Ward KM and Balaban RS. Determination of pH using water protons and chemical exchange dependent saturation transfer (CEST). *Mag Reson Med*, 2000; 44(5): 799–802.

Chapter 3

History of In Vivo Exchange Transfer Spectroscopy and Imaging in Baltimore

Peter C. M. van Zijl

F. M. Kirby Research Center, Kennedy Krieger Institute, 707 N Broadway, Baltimore, MD 21205, USA
Department of Radiology, Johns Hopkins University School of Medicine, 707 N Broadway, Baltimore, MD 21205, USA
pvanzijl@mri.jhu.edu

3.1 Before There Was CEST

My early interest in chemical exchange saturation transfer (CEST) imaging was a natural progression from work on exchange-based nuclear magnetic resonance (NMR) during the 1990s, when Susumu Mori, still a graduate student at the time, and I were developing multi-dimensional ^{1}H NMR spectroscopy methods to study amide protons in proteins [1–6]. At that time, it was difficult to detect rapidly exchanging amide protons using proton NMR, because the water suppression that was essential to perform spectroscopy

---

Chemical Exchange Saturation Transfer Imaging: Advances and Applications
Edited by Michael T. McMahon, Assaf A. Gilad, Jeff W. M. Bulte, and Peter C. M. van Zijl
Copyright © 2017 Pan Stanford Publishing Pte. Ltd.
ISBN 978-981-4745-70-3 (Hardcover), 978-1-315-36442-1 (eBook)
www.panstanford.com

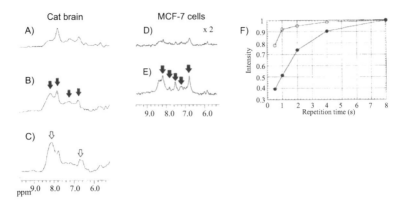

Figure 3.1 ^{1}H MRS spectra of exchangeable protons in cat brain (a–c) and perfused breast cancer cells (d, e). When comparing spectra with CHESS water suppression before the pulse sequence (a, d) to those without (b, e), it can be seen that much of the signal is lost due to exchange of saturated water protons to the protons of interest. In the postmortem case in the cat, exchange is reduced due to lower pH and even less signal is lost (c), which is especially clear at the composite amide resonance at 8.3 ppm. In (f) the large gain in signal when using a water flip-back pulse is shown for the cells. Reprinted with permission from Ref. [8], Copyright 2005, John Wiley and Sons.

with an acceptable dynamic range also caused suppression of the exchangeable proton signals due to the transfer of the water saturation during the measurement. For a single acquisition, this could be solved by applying suppression just before detection, but spectroscopy of millimolar concentration compounds required multiple scans, and many exchangeable protons remained invisible due to the steady-state saturation from the water suppression staying around between scans. So a long repetition time (TR) was required to detect exchangeable protons. Susumu resolved that issue by incorporating a so-called water flip-back pulse that returned the water magnetization to the z-axis before suppression, which allowed rapid scanning because the bulk of the water signal was no longer saturated after spectroscopy detection and fresh magnetization from the large water proton pool was now supplied quickly to fast exchanging protons [7].

As my lab was interested mainly in biomedical applications, we started to transfer these methods to the study of exchangeable protons in cells and tissues. Such protons are mainly in the low-field (high frequency) range of the spectrum with respect to the water signal (∼5–10 ppm) and usually not observed in ^{1}H MRS in vivo due to the typical water suppression approaches used. The goal was to measure pH changes during ischemia [8], which we expected to be possible based on the known pH dependence of amide proton exchange from high-resolution NMR studies [9, 10]. In addition, we studied perfused MCF-7 cancer cells, where we found many exchangeable protons, including a large combined amide resonance around 8.3 ppm in the proton spectrum. Similar to the high-resolution NMR studies on proteins, we found the signal from exchangeable protons in vivo and in perfused cancer cells to be low when using conventional CHEmical Shift Selective (CHESS) water suppression before the pulse sequence (Fig. 3.1a,d), which we again attributed to saturation transfer from water. We were able to reduce this signal loss by using the so-called WATERGATE sequence in which binomial excitation is used to suppress the water signal just before detection, leading to strongly increased signals of amide protons (Fig. 3.1b,e). In vivo, due to the need for spatial localization, it was hard to do a proper water flip-back pulse similar to the high-resolution experiment as not all effects of saturation could be removed. As a consequence, when studying the brain postmortem (i.e., at reduced pH and thus reduced exchange rate), an additional strong increase in signal was seen in particular at the composite amide proton resonance (Fig. 3.1c). In perfused cells, on the other hand, we were able to use a water flip-back pulse to minimize the presence of saturated water protons during the recovery period, which allowed much improved detection of the exchangeable protons (see Fig. 3.1e,f), especially at short repetition times.

In the methods section of the paper describing this first work [8] in cell systems, submitted in early 1997, we have the following passage: "*In fact, the relative intensity of solvent-exchangeable and exchange-relayed peaks to nonexchangeable peaks increases with rapid scanning because, during the short recovery period, magnetization of exchangeable peaks is replenished from the large* *reservoir of nonsaturated water magnetization by physical exchange, whereas nonexchanging peaks recover as a function of T_1. Without the water flipback scheme, water is saturated and recovery of magnetization at exchanging sites is determined by the T_1 of water."* This principle, which was also used in the earlier high-resolution NMR studies mentioned earlier, is actually the inverse of the CEST effect being exploited to enhance the sensitivity of exchangeable protons in ^{1}H NMR spectra.

3.2 Early CEST Experiments

However, it was not until later in 1997 that I realized the possibility of exploiting the inverse pathway of exchange transfer to water for imaging of molecules containing exchangeable protons. This occurred when I read a poster by Bob Balaban's group at the 10th anniversary of the in vivo NMR Center at NIH (October 7, 1997). Neither Bob, nor I remember exactly what the poster was about, but it probably was the early work of Guivel-Sharen [11]. The potential to get large signal enhancements for the detection of low-concentration molecules was so exciting that I immediately asked Tsang-Lin Hwang, a postdoctoral fellow, to start working on this. Based on my vague familiarity with earlier magnetization transfer contrast (MTC) work on semi-solid tissues, I had somehow understood from the Balaban poster that it was necessary to have large immobile molecules to saturate the molecular backbone protons and then have a fast nuclear overhauser enhancement (NOE) transfer through the backbone and to the exchangeable protons, followed by exchange to water. So I asked Tsang-Lin to look at glycogens of large molecular weight, which should have fast spin diffusion and backbone protons in close contact with a large number of exchangeable groups. Interestingly, however, we were not able to find much of a transfer effect when irradiating outside the typical high-resolution NMR proton spectral range, presumably into the broad background signal, and we only found an effect when directly saturating the OH resonance. This was somewhat confusing at first, and I probably should have read the Balaban poster better. The only valid excuse for this poor interpretation is that the poster session was during a

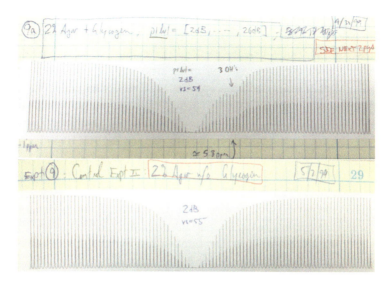

Figure 3.2 Some of our first data on glycogen in 1999. On top glycogen in 2% agar, on the bottom just 2% agar from a reference experiment performed a few days later. Note that the frequency scale is with high frequency on the right and low on the left, contrary to the convention used for Z-spectra today.

wine and beer reception following the workshop. In 1998, Tsang-Lin left the laboratory and the work was taken over by Nick Goffeney, a graduate student. Since I was interested in the NOE transfer to OH, I asked him to immobilize the glycogen with agar and we found the semi-solid effect using a higher B1 strength for saturation. However, at lower B1, we still saw mainly the OH groups and direct water saturation. Figure 3.2 shows some old data from my summary notebook in 1999. The confusion delayed publication of any of the results. In addition, the years 1998 and 1999 were a busy and scientifically exciting period, in which I was occupied with establishing the F. M. Kirby Research Center for Functional Brain Imaging at the Kennedy Krieger Research Institute and, together with Risto Kauppinen and his group, thinking about the origin of fMRI signals [12]. In addition, Susumu Mori and I spent a lot of time on the first work on fiber tracking with DTI [13]. So the inverse exchange work stayed a bit on the backburner until I got reminded by a fabulous presentation of Bob Balaban at the ISMRM in Denver in 2000, where he showed the first CEST agents, and this revamped my interest in the topic. At the meeting, I discussed this with Jeff Bulte and Jeff Duyn (both at NIH at the time), and we decided we should publish something quickly with amide-containing molecules. Jeff Bulte suggested to look at dendrimers and poly-L-lysine (PLL), which he was using for his research. So Nick Goffeney started to work on that, and at the ISMRM in Glasgow in 2001, we presented two CEST abstracts, one on the aforementioned polymers [14] and one on the pH dependence of the CEST contrast produced by amide protons in dendrimers [15]. Our first paper appeared in the *Journal of the American Chemical Society* later in 2001 [16] and also contained the first more detailed analytical CEST theory, including back exchange effects that I wrote up quickly based on general principles and intuition. This theory used the simple assumptions of fast exchange with respect to longitudinal relaxation of the NH protons but sufficiently slow exchange on the NMR timescale to allow at least partial saturation of the proton before exchange. The theory fitted the amide proton data very well and was confirmed a few years later by Jinyuan Zhou with a more precise derivation of an analytical solution for CEST contrast based on the Bloch equations [17]. When I finally realized that the saturation of the exchangeable groups was transferred directly and not via backbone-based NOEs, it immediately occurred to me that saturation was not needed and that, due to the finite line width of the exchangeable protons, one should also be able to do these kinds of experiments using repeated excitation, i.e., using pulsed approaches. Figure 3.3 shows the first page of a disclosure I wrote for this in 2000, but I never filed it as we did not have data. Over the following decade, I asked several students and postdoctoral fellows to explore this somewhat complicated approach, but nobody was able to make it work until 2009, when Josh Friedman, a graduate student in pharmacology, succeeded. We published the results as a communication in the *Journal of the American Chemical Society* in 2010 and named it the frequency-labeled exchange (FLEX) transfer method [18]. A patent was finally filed too. The glycogen work was also published very slowly and only as a consequence of me talking to Dean Sherry about it around 2006. He insisted we work on it more with him and Craig Malloy. As a consequence, Craig Jones (from Hopkins) and

JOHNS HOPKINS UNIVERSITY
SCHOOL OF MEDICINE
and
THE JOHNS HOPKINS HOSPITAL

DEPARTMENT OF RADIOLOGY
AND
RADIOLOGICAL SCIENCES
DR. PETER C.M. VAN ZIJL
TEL (410) 614-2733 or 955-4221
FAX (410) 614-1948
e-mail: pvanzij@mri.jhu.edu

Mailing Address:
THE JOHNS HOPKINS HOSPITAL
DEPT OF RADIOLOGY/217 TRAYLOR
720 RUTLAND AVE, BALTIMORE, MD 21205
U.S.A

July 10, 2000

Disclosure of Invention: Strongly Enhanced Sensitivity in MRI and NMR using Selective Transfer of Exchangeable protons in combination with Chemical Shift labeling, Scalar Coupling labeling and/or Gradient Evolution labeling.

Standard MR experiments suffer from low sensitivity and signal enhancement methods would be welcome. Most enhancement methods are based on relaxation enhancement using paramagnetic contrast agents or on hyperpolarized samples. Recently, Balaban et al. (J. Magn. Reson. **143**, 79-87, 2000) showed that enhancements by a factor up to 1000 can be achieved by selectively irradiating exchangeable protons, which subsequently exchange with water. The effect of this becomes visible as a signal attenuation of the water resonance. The enhancement depends on the exchange rate of the particular protons (NH, OH or SH), the number of protons per molecule, and the pH (which influences the exchange rate, Liepinsh & Otting, Magn. Reson. Med. **35**, 30-42 1996, Mori et al., Magn. Reson. Med. **40**, 36-42, 1998). Performing proton saturation transfer experiments on polylysine and dendrimers, we recently obtained enhancement factors of 10^3-10^6 (data enclosed). For instance we can detect 100 micromolar polylysine (a delivery molecule for gene transfer) with a 50% change in the water signal. However, the proton saturation transfer approach is severely hampered by dynamic range, as lower concentrations would show only small changes on the large water resonance. For instance, a 1 micromolar polylysine sample would only give an 0.5% change on water, which is borderline for detection. If this dynamic range problem could be alleviated, an additional factor of 1000 could be gained in the absolute concentration detection limit (not the sensitivity enhancement), extending it to the nanomolar range and allow some receptor binding studies. The reason is that the sensitivity of the MRI scanners is sufficiently good to detect signals that are 10,000-100,000 times smaller than water, i.e. as is done in spectroscopy, where optimum dynamic range can be used. In the present context, this dynamic range problem is an issue of selectivity, in which we only want to detect protons that were originally located on the exchangeable sites of the molecule of interest. Thus these protons should be labeled (not saturated) in a selective manner by an MRI approach. We propose several novel methodologies here to achieve that, which can be summarized in the following pulse sequence approach:

{selective labeling – wait for exchange -}$_n$ excite – detect   n is integer; n>= 1

In which the exchange enhancement comes from the n times that the labeling is repeated after the first labeled protons (n = 1) have exchanged with the much larger water proton pool and the fact that there are multiple exchangeable protons on the molecule of interest. I propose that this type of imaging can be achieved by three types of selective labeling and/or combinations thereof, namely:

1) labeling the exchangeable protons with their chemical shift frequency evolution/phase
2) labeling the exchangeable protons with a scalar coupling frequency evolution/phase, e.g. when coupled to a neighboring MRI/NMR sensitive nucleus such as 15N.
3) labeling the exchangeable protons with a magnetic-field-gradient-induced phase after selectively exciting the exchangeable nucleus,

and subsequently transferring this information via proton exchange to the water.

Figure 3.3 Page 1 of a disclosure for the repeated selective excitation with exchange transfer that I wrote in 2000. It was never filed due to lack of data and some notes are visible that we added later. Part of this was published in 2010 as the FLEX method.

Jimin Ren (from UTSW, Dallas) performed successful experiments leading to a collaborative paper in 2007 with the UTSW group [19].

3.3 Amide Proton Transfer–Weighted MRI

One of the ultimate goals of our work always is application in vivo, and as soon as I became aware of the CEST phenomenon, I wanted to apply it to the study of amide protons in proteins and peptides in situ. There were several exciting possibilities for important applications based on the earlier work with Susumu Mori. First, our spectroscopy data showed that the combined signal of multiple amide protons around 8.3 ppm (3.5 ppm from the water line, which is used as reference in CEST spectra) was larger at low pH due to a reduced exchange rate (Fig. 3.1c). This offered the possibility to now image pH changes using the tissue water signal. In addition, the large amide proton signals in tumor cells indicated potential for imaging of proteins and peptides in cancer. We still had some unpublished data that indicated that such CEST MRI of amide protons should be possible. These were so-called water-exchange (WEX) spectra based on a pulse sequence for detecting amide protons by labeling the water signal with an inversion pulse [1] and detecting the increase in exchange-based signals in the proton spectrum. Figure 3.4 shows results for this type of acquisition in RIF-1 cancer cells (older data acquired by Susumu Mori) and in vivo in rat brain, which we ultimately published in 2003 [20]. In these spectra, acquired as a function of time after the inversion labeling of water, one can see the exchange effects appear first in the amide proton region and subsequently as exchange-relayed NOEs in the aliphatic region. These cell and in vivo data looked very similar to the protein NMR data Susumu and I had acquired in solution [2, 3, 21] in the 1990s, and I had become convinced that we should be able to detect proteins and peptides in vivo.

With this in mind, I asked Jinyuan Zhou to start studying properties of the amide protons using CEST in vivo early in the summer of 2000. The progress at the beginning was slow because a long saturation pulse was not available in our old animal scanners. Thus, we spent much time to design and optimize a pulse train

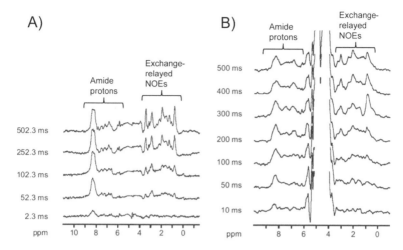

Figure 3.4 Water exchange (WEX) spectra of perfused RIF-1 cancer cells (A) and in vivo rat brain (B). The WEX pulse sequence consist of selective magnetic labeling (inversion) of bulk water (inverse of CEST/MT approaches) followed by a waiting period. Chemical exchange and cross relaxation occur, the latter either exchange-relayed or through direct excitation of C(a) protons. The data show fast buildup of exchangeable proton signals (especially amide protons around 8.3 ppm) as a function of time after inversion followed by gradual transfer to aliphatic protons through intramolecular NOEs. Notice that rapidly exchangeable amine and hydroxyl groups are not visible as they remain in the baseline due to their large linewidths relative to the amide protons. Reprinted with permission from Ref. [20], Copyright 2003, John Wiley and Sons.

saturation scheme. This work was stimulated by the arrival in August 2000 of Dr. Jean-Francois Payen, a brain surgeon from the University of Grenoble, France, who performed a 1-year sabbatical in my lab, during which Jinyuan and he demonstrated the possibility of performing amide proton transfer (APT) MRI in vivo in the brain. Based on our old paper showing the amide proton increase postmortem (Fig. 3.1) and my interest in acute stroke, I suggested they study ischemia and other pH effects (using CO_2 gas) as it should be possible for the first time to perform pH imaging using the water signal. Despite many roadblocks, including small effect size and a 20 min walk through the tunnels of Johns Hopkins University with the anesthetized animals from the lab to the scanner, they were able to convincingly demonstrate the APT effect at the correct frequency. We also benefitted greatly from the expertise for animal ischemia studies of David Wilson and Dick Traystman. The paper describing these results was published in 2003 [22]. We were also able to generate a pH image and show changes during early ischemia and used WEX spectroscopy to confirm that the signal change was due to amide proton exchange and not a consequence of a change in protein concentration, a potential alternative explanation. A few years later, middle cerebral artery occlusion experiments on rats by Philip Sun in our lab convincingly demonstrated the ability of APT imaging to detect ischemia before changes are visible in diffusion MRI [23], which is expected based on the fact that aerobic metabolism becomes impaired at a higher cerebral blood flow threshold than the one at which cellular depolarization occurs.

As can be seen in Fig. 3.4, the WEX spectral data on perfused RIF-1 cancer cells [20] showed a very large composite amide proton signal around 8.3 ppm (~3.5 ppm from water), opening up the exciting possibility to image tumors based on CEST principles. The older MCF-7 data in Fig. 3.1 had also pointed in that direction. I, therefore, asked Jinyuan to look at cancer using APT imaging, and we started a collaboration with John Laterra, an expert on brain tumors. Again we were fortunate and found a very large APT effect at the composite amide proton offset. In addition, the images did not show a CEST effect in areas with edema and clearly contrasted the tumor from normal brain and edematous regions.

This early work in vivo clearly showed that when looking at CEST/APT images, it is important to understand that there are multiple contributions, the size of which may depend on the way the data are recorded (B_1, B_0, pulse timing) and analyzed. In order to illustrate the shape and complexity of the CEST data, Fig. 3.5 shows Z-spectra (Fig. 3.5a,d), magnetization transfer ratio asymmetry spectra (MTR_{asym}, Fig. 3.5b,e), and proton transfer difference spectra (Fig. 3.5c,f) between lesion and contralateral brain for the cases of glioma (Fig. 3.5a–c) and ischemia (Fig. 3.5d–f) from our studies on rats in early 2000. The Z-spectra contain contributions from multiple origins, namely, (i) exchangeable protons of many metabolites and proteins, peptides, and sugars at positive frequencies with respect to the water resonance; (ii)

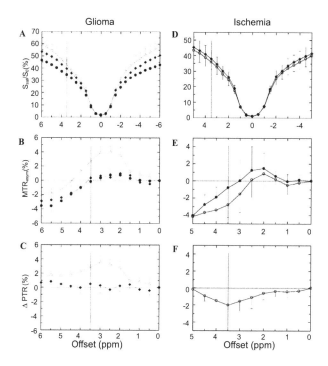

Figure 3.5 CEST data for 9L glioma (A–C) and ischemia (D–F) in rat brain, showing Z-spectra (A,D), MTR_{asym} spectra (B,E) and proton transfer ratio (PTR) difference spectra (C,F) in which DPTR = MTR_{asym} (ROI) - MTR_{asym} (contralateral brain). Solid circle: contralateral brain; diamond: peritumoral tissue (A–C); open circle: tumor (A–C) or early ischemia (D–F). Signal attenuation in Z-spectra (A, D) is due mainly to direct water saturation close to the water frequency and a large MTC effect (~40–60%) over the whole spectral range. The MTC contribution is reduced for edema and tumor, indicating a higher water content for these tissues. The Z-spectra are slightly asymmetric, which becomes visible when performing MTR asymmetry analysis (B, E). Both exchange effects and asymmetry in the MTC contribute to the residual curve, with the latter reflected in a relatively constant negative signal (2–3%) at offsets above 5 ppm (B,E). The PTR for tissue changes can be estimated by comparing tumor and edema with normal brain, removing some but not all of the MTC effects. This shows that edematous regions can be separated from tumor (C) by analyzing the PTR difference with normal brain. When applying asymmetry analysis (E) and a PTR difference analysis between ischemic and normal brain, the PTR difference is negative (F). Reprinted with permission from Ref. 20, Copyright 2003, John Wiley and Sons (A–C) and from Ref. [22], Copyright 2003, Nature Publishing Group (D–F).

NOE-relayed effects of mobile proteins at negative frequencies with respect to water (and possibly some aromatic ones at positive offsets); (iii) a symmetric signal reduction due to direct water saturation (DS), the width of which depends on B_1; (iv) a very broad asymmetric signal reduction due to NOEs in the semi-solid lattice, better known as the conventional MTC. Thus, the signal reduction is more at high field (lower frequency) than at low field with respect to water. More specific, in the tumor data, it can be seen that there is a contrast between tumor and normal brain (Fig. 3.5c) not only at the amide proton frequency around 3.5 ppm but also between 3.5 and 1.5 ppm from water (probably due to amide and amine protons from such proteins and possibly other metabolites) and above 5 ppm from water. The latter can only be due to an asymmetry in the conventional MTC based on semi-solid protons. In the equations describing the contrast, we, therefore, added a correction term MTR'_{asym} to account for these and other potentially confounding contributions, which are still under discussion in the current literature. Based on these issues, it is best to call the clinical measure of MTR_{asym} (3.5 ppm) the APT-weighted intensity. However, from a clinical point of view, we were extremely fortunate as all contribution factors (asymmetric MTC, NOE, and APT) add up constructively. This APT-weighted contrast has been shown to be usable for separating tumor from edema, high-grade from low-grade glioma, and recurrent tumor from treatment necrosis. In recent years, Jinyuan Zhou (currently associate professor of radiology) has taken the lead in APT-based MRI studies and has expended this application tremendously. In addition, Philips has licensed the patent [24] and Jochen Keupp (Philips Hamburg) is working on the first automated approach for push-button use in the clinic [25].

Interestingly, these tumor MTR_{asym} and PTR difference spectra differ in appearance from those for ischemia (Fig. 3.5e,f) in that there is a clean maximum effect at the amide proton main frequency around 3.5 ppm and thus a cleaner APT-based difference. Of course, the original Z-spectra and MTR_{asym} spectra still have the contributions mentioned above, but the ischemic lesion contrast is mainly APT, which was confirmed via WEX spectra and pH-dependent ^{1}H and ^{31}P measurements in vivo [22].

3.4 Expansion of the CEST Efforts

In 2001, I had met Maurice Guéron at the International Society for Magnetic Resonance (ISMAR) meeting in Rhodos (Greece), a world-renowned expert in basic NMR and its application to nucleic acids. During a small session, I presented the pH-dependent PLL and dendrimer studies, which raised his interest. We talked a bit after the session and agreed that the imino protons in nucleic acids (large chemical shift offset of about 6 ppm from water and fast exchange rate of ~5800 Hz under physiological conditions) should be an excellent source of CEST contrast. He sent over a very talented postdoctoral fellow named Karim Snoussi, who within a short period of 4 months worked day and night to perform an impressive series of experiments on polyuridilic acid (poly(rU)) and its binding to dendrimers (gene carriers). A paper on this was published in 2003 [26].

In 2003, I was lucky to have two excellent applicants for postdoctoral positions, namely, Mike McMahon and Assaf Gilad. I hired Mike, while Jeff Bulte (new Hopkins Faculty at that time) and I supported Assaf together. They formed a great team, and the CEST studies expanded to the quantification of CEST contrast and the rational design of new contrast agents (Mike) and the cellular production of such agents (Assaf). They are still at Johns Hopkins, now associate professors with their own research groups, and I asked each of them to summarize their recollections.

3.4.1 *Assaf Gilad's Recollections*

In the spring of 2003, I was looking for a position as a postdoc. I applied to work with Dr. Peter van Zijl, and when he replied within 30 min, I knew that this would be the place for me. I met with Peter for the first time for dinner at the ISMRM in Toronto in July 2003. During dinner, Peter described a new contrast mechanism that would allow to image peptides with specific properties such as PLL [16]. Peter also described a new idea he had for targeted imaging that involved labeling of antibodies against specific cell membrane antigens with PLL, dendrimers, and ribonucleic acid (RNA). He had just published a paper describing the CEST properties of RNA

[26] and proposed the targeted CEST imaging as part of Dr. Zaver Bhujwalla's P50 center grant as a pilot project.

I went back to the Weizmann Institute to complete my PhD studies where I was working with Dr. Michal Neeman, who was at that time working on developing ferritin as a reporter gene for MRI. This study was published in 2005 in the journal *Neoplasia* [27]. Ferritin protein synthesis is based on oligomerization of 24 subunits and requires the presence of high intracellular iron concentrations to generate T_2 contrast. It had occurred to me that we could use CEST to develop an alternative reporter gene based on CEST, which is metal free. Since I could not find any protein in nature that is similar enough to PLL, I thought of making a synthetic gene that would encode to such protein. I started to work on the technical details and was happy to learn from Peter that he wanted me to work with a new faculty member in his center, Dr. Jeff Bulte.

I presented the idea of a synthetic CEST reporter to Peter and Jeff a few weeks after arriving to Baltimore, and luckily both agreed that this was a risky idea but worth trying. I teamed with Dr. Venu Raman and Paul Winnard to perform the molecular cloning. That was before the era of synthetic biology, and unlike today when it is possible to have a company synthesize a whole gene for you. We, therefore, had to "stitch" it together from 88 base-pair long DNA oligomers. Another lucky coincidence was that Peter assigned me to share an office with Dr. Michael McMahon, another new hire in the center. Together Mike and I started to set up the CEST imaging of cell extracts expressing the new synthetic protein that we termed lysine-rich protein (LRP). When it came to performing animal studies, we got great help from Dr. Piotr Walczak who had just started as a new postdoc in Jeff's lab. He helped us set up the mouse studies on the old 500 MHz vertical bore instrument. We ran into many technical hurdles as the scanner was not really equipped for animal imaging, but with the help of Piotr and lots of plastic tubing and masking tape, we managed to overcome all of these, eventually publishing the first paper describing the LRP [28]. A second paper that demonstrated the applicability of the LRP in monitoring oncoviral therapy has just came out in the journal *Radiology* [29]. But in between, we have expanded the scope of this research line and developed several other reporter genes based on CEST [30–32].

3.4.2 Mike McMahon's Recollections

During my final year of postdoctoral studies in solid state NMR research in Robert Griffin's lab at MIT, I became interested in moving into MRI research based on having a number of nice conversations with colleagues. I contacted a few people who were in the area between NMR and MRI, with Peter van Zijl's group capturing my eye as an excellent place to transition from chemistry to radiology. In particular, Peter seemed to straddle the two worlds, for example showing the utility of field gradients widely used in MRI for coherence selection in high-resolution NMR experiments [33] and transferring pulse sequences for highlighting WEX into imaging [20]. I started in the summer of 2003. To get my feet wet, I decided to assist Seth Smith and Xavier Golay with some of their magnetization transfer imaging studies, writing some code to numerically solve the Bloch equations and testing this code on 2% and 4% agar phantoms. I also met Jeff Bulte and became quite interested in his work using mostly iron oxides to track cells after transplantation, attending his group meetings regularly. While discussing these results with Peter over a beer at the weekly happy hour at the 9th floor of the Traylor Building, we came to the conclusion that I should try and finish up a project of Nick Goffeney's, an old student of his, which involved investigating the pH dependence of the saturation transfer contrast of PLL and poly-amino-amine dendrimers. This sounded quite attractive to me as I could use my background in NMR spectroscopy and modeling. As part of this, I started using the old 500 MHz scanner and coded up the WEX filter that Susumu Mori had developed with Peter and tested this out on the polymers, and compared this with simple variation of the saturation parameters (B_1-field and saturation time) in a systematic fashion for exchange rate estimation, which ultimately became the QUEST and QUESP experiments [34]. Assaf Gilad also joined our center at that time, who became my officemate and good friend very quickly. Since he was particularly interested in Nick's result that PLL generated excellent CEST contrast, and I was working on PLL and becoming interested in finding ways to model CEST contrast, we started to scheme how to combine our expertise. I started to try and find a way to use Walter Englander's work on protein folding to really tune up a peptide to be much better than LRP, first through ordering a wide range of poly-amino acids from Sigma and then after this by obtaining quotes for a set of peptides that I estimated would produce better contrast according to Englander's empirical expressions. This led to the discovery that the protamine family of peptides would produce the largest in vitro CEST contrast to date, which we published in *Magnetic Resonance in Medicine* and also to developing a method to highlight lysine-rich peptides separately from arginine-rich peptides, which we decided to call "multi-color imaging" based on assigning colors to different frequencies (chemical shifts) in a way similar to optical imaging. Based on this initial work and the potential improvements I felt could be made, I decided to focus most of my time on developing organic compounds as CEST imaging agents, with a particular interest in the discovery of aromatic compounds that have very nice exchange properties for use on clinical scanners.

3.5 Translation to Human Scanners

The ultimate goal of developing new MRI methods is to make them clinically useful, and we quickly started to implement our methods on the human scanner. This is not straightforward due to the practical limitations when performing radiofrequency (RF) irradiation. Originally, we were worried about SAR limitations, especially at 3 Tesla, as the power deposition goes with the square of the magnetic field strength. However, a bigger problem was the restriction imposed by the manufacturer on the length of the RF pulses and the duty cycle of the amplifier, especially when using body coil transmission, which had become standard after the emergence of parallel imaging methods. Most of the work for the first implementation was performed by Craig Jones, a postdoctoral fellow at the time and an expert on programming clinical scanners, together with Jinyuan Zhou. To simplify matters, we used a transmit head coil and only acquired a single slice. It was very gratifying to see that this initial work on glioma patients showed that, similar to the animal experiments in 2003, we could clearly separate tumor from edema and normal brain (Fig. 3.6) and distinguish high-grade and low-grade tumors [35]. We were quite excited about

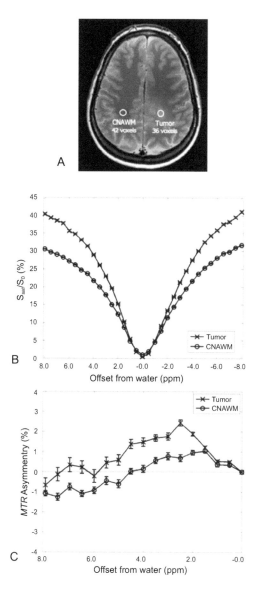

Figure 3.6 CEST data experimental results for a patient with an astrocytoma: (A) regions (circles) of the tumor and contralateral normal appearing white matter (CNAWM) used to create the MT curves, and (B) Z-spectra and (C) MTR_{asym} spectra. Saturation power 3 µT, duration 3 s. The maximum APT appears in an offset range between 2 and 4 ppm relative to water. Reprinted with permission from Ref. [35], Copyright 2006, John Wiley and Sons.

this success, which was published in 2006 and confirmed in later studies [36, 37]. However, the neurosurgeons told us that they were still not that interested because most patients undergo surgery anyway. The neuroradiologist (Dr. Doris Lin) indicated that the most important application would be to separate out recurrent tumor from treatment necrosis and to monitor whether the tumor responds to treatment. In a recent study, Jinyuan Zhou has been able to prove just that and early patient studies now ongoing are confirming his animal studies.

3.6 Active Growth in CEST

The aforementioned sections summarize the early development of diamagnetic CEST (diaCEST) contrast at Johns Hopkins University and Kennedy Krieger Institute. We focused mainly on diaCEST compounds as these relate to non-metallic contrast agents or endogenous substances that have the potential for faster translation to human scanners. In the last decade, the CEST field has grown rapidly, and we now have a large group of young and talented researchers working on pulse sequence development and applications. Several of them are contributing chapters in this book. The field looks promising and the number of applications seems unlimited. Looking on PubMed in July 2016, there are already 60 papers found for this year alone when using "CEST, MRI" as key words. This number was 66 for the whole of 2014. Many groups at multiple universities are now working in this field and new pulse sequences (both excitation and saturation based) and applications are being presented. Whole-organ MRI sequences that are sufficiently fast for clinical studies are starting to become available. I foresee that, in addition to APT-weighted MRI for the study of tumors, many important human applications will soon be apparent.

References

1. Mori S, Johnson MO, Berg JM, and van Zijl PCM. Water exchange filter (WEX filter) for nuclear-magnetic-resonance studies of macromolecules. *J. Am. Chem. Soc.*, 1994; 116(26): 11982–11984.

2. Mori S, Abeygunawardana C, van Zijl PCM, and Berg JM. Water exchange filter with improved sensitivity (WEX II) to study solvent-exchangeable protons. Application to the consensus zinc finger peptide CP-I. *J. Magn. Reson. Ser. B.*, 1996; 110(1): 96–101.

3. Mori S, Berg JM, and van Zijl PCM. Separation of intramolecular NOE and exchange peaks in water exchange spectroscopy using spin-echo filters. *J. Biomol. NMR.*, 1996; 7(1): 77–82.

4. Mori S, van Zijl PCM, and Shortle D. Measurement of water-amide proton exchange rates in the denatured state of staphylococcal nuclease by a magnetization transfer technique. *Proteins: Struct. Funct. Genet.*, 1997; 28(3): 325–332.

5. Mori S, Abeygunawardana C, Berg JM, and van Zijl PCM. NMR study of rapidly exchanging backbone amide protons in staphylococcal nuclease and the correlation with structural and dynamic properties. *J. Am. Chem. Soc.*, 1997; 119(29): 6844–6852.

6. Hwang TL, Mori S, Shaka AJ, and van Zijl PCM. Application of phase-modulated CLEAN chemical EXchange spectroscopy (CLEANEX-PM) to detect water-protein proton exchange and intermolecular NOEs. *J. Am. Chem. Soc.*, 1997; 119(26): 6203–6204.

7. Mori S, Abeygunawardana C, Johnson MO, and van Zijl PCM. Improved sensitivity of HSQC spectra of exchanging protons at short interscan delays using a new fast HSQC (FHSQC) detection scheme that avoids water saturation. *J. Magn. Reson. Ser. B.*, 1995; 108(1): 94–98.

8. Mori S, Eleff SM, Pilatus U, Mori N, and van Zijl PCM. Proton NMR spectroscopy of solvent-saturable resonances: A new approach to study pH effects in situ. *Magn. Reson. Med.*, 1998; 40(1): 36–42.

9. Woodward CK and Hilton BD. Hydrogen-exchange kinetics and internal motions in proteins and nucleic-acids. *Ann. Rev. Biophys. Bioeng.*, 1979; 8: 99–127.

10. Englander SW, Downer NW, and Teitelba H. Hydrogen-exchange. *Annu. Rev. Biochem.*, 1972; 41: 903.

11. Guivel-Scharen V, Sinnwell T, Wolff SD, and Balaban RS. Detection of proton chemical exchange between metabolites and water in biological tissues. *J. Magn. Reson.*, 1998; 133(1): 36–45.

12. van Zijl PCM, Eleff SM, Ulatowski JA, et al. Quantitative assessment of blood flow, blood volume and blood oxygenation effects in functional magnetic resonance imaging. *Nat. Med.*, 1998; 4(2): 159–167.

13. Mori S, Crain BJ, Chacko VP, and van Zijl PCM. Three-dimensional tracking of axonal projections in the brain by magnetic resonance imaging. *Ann. Neurol.*, 1999; 45(2): 265–269.

14. Goffeney N, Bulte JWM, Duyn JH, Bryant LH, and van Zijl PC. Detection of cationic-polymer-based gene delivery systems without paramagnetic metals. Paper presented at: *9th International Society for Magnetic Resonance in Medicine*; April 21–27, 2001; Glasgow, Scottland.
15. van Zijl PCM, Goffeney N, Duyn JH, Bryant LH, and Bulte JWM. The use of starburst dendrimers as pH contrast agents. Paper presented at: *9th International Society for Magnetic Resonance in Medicine*; April 21–27, 2001; Glasgow, Scottland.
16. Goffeney N, Bulte JWM, Duyn J, Bryant L.H, Jr., and van Zijl PCM. Sensitive NMR detection of cationic-polymer-based gene delivery systems using saturation transfer via proton exchange. *J. Am. Chem. Soc.*, 2001; 123(35): 8628–8629.
17. Zhou J, Wilson DA, Sun PZ, Klaus JA, and van Zijl PCM. Quantitative description of proton exchange processes between water and endogenous and exogenous agents for WEX, CEST, and APT experiments. *Magn. Reson. Med.*, 2004; 51(5): 945–952.
18. Friedman JI, McMahon MT, Stivers JT, and van Zijl PC. Indirect detection of labile solute proton spectra via the water signal using frequency-labeled exchange (FLEX) transfer. *J. Am. Chem. Soc.*, 2010; 132(6): 1813–1815.
19. van Zijl PCM, Jones CK, Ren J, Malloy CR, and Sherry AD. MRI detection of glycogen in vivo by using chemical exchange saturation transfer imaging (glycoCEST). *Proc. Natl. Acad. Sci. U.S.A.*, 2007; 104(11): 4359–4364.
20. van Zijl PCM, Zhou J, Mori N, Payen JF, Wilson D, and Mori S. Mechanism of magnetization transfer during on-resonance water saturation. A new approach to detect mobile proteins, peptides, and lipids. *Magn. Reson. Med.*, 2003; 49(3): 440–449.
21. Hwang TL, van Zijl PCM, and Mori S. Accurate quantitation of water-amide proton exchange rates using the Phase-Modulated CLEAN chemical EXchange (CLEANEX-PM) approach with a Fast-HSQC (FHSQC) detection scheme. *J. Biomol. NMR.*, 1998; 11(2): 221–226.
22. Zhou J, Payen JF, Wilson DA, Traystman RJ, and van Zijl PCM. Using the amide proton signals of intracellular proteins and peptides to detect pH effects in MRI. *Nat. Med.*, 2003; 9(8): 1085–1090.
23. Sun PZ, Zhou JY, Sun WY, Huang J, and van Zijl PCM. Detection of the ischemic penumbra using pH-weighted MRI. *J. Cereb. Blood Flow Metab.*, 2007; 27(6): 1129–1136.
24. van Zijl PCM, Zhou J, Inventors. Magnetic resonance method for assessing amide proton transfer effects between amide protons of

endogenous mobile proteins and peptides and tissue water in situ and its use for imaging ph and mobile protein/peptide content. US patent US 6,943,033 B2; Provisional filed: Dec. 13, 2001; patent filed: Dec 13, 2002.

25. Hye-Young H, Zhang Y, Keupp J, et al. Towards an optimized and standardized amide proton transfer (APT) MRI sequence and protocol for clinical applications. *Int. Soc. Magn. Reson. Med.*, Toronto, Ontario 2015: 0016.

26. Snoussi K, Bulte JWM, Guéron M, and van Zijl PC. Sensitive CEST agents based on nucleic acid imino proton exchange: Detection of poly(rU) and of a dendrimer-poly(rU) model for nucleic acid delivery and pharmacology. *Magn. Reson. Med.*, 2003; 49(6): 998–1005.

27. Cohen B, Dafni H, Meir G, Harmelin A, and Neeman M. Ferritin as an endogenous MRI reporter for noninvasive imaging of gene expression in C6 glioma tumors. *Neoplasia*, 2005; 7(2): 109–117.

28. Gilad AA, McMahon MT, Walczak P, et al. Artificial reporter gene providing MRI contrast based on proton exchange. *Nat. Biotechnol.*, 2007; 25(2): 217–219.

29. Farrar CT, Buhrman JS, Liu GS, et al. Establishing the lysine-rich protein CEST reporter gene as a CEST MR imaging detector for oncolytic virotherapy. *Radiology*, 2015; 275(3): 746–754.

30. Bar-Shir A, Liu GS, Liang YJ, et al. Transforming thymidine into a magnetic resonance imaging probe for monitoring gene expression. *J. Am. Chem. Soc.*, 2013; 135(4): 1617–1624.

31. Bar-Shir A, Liang YJ, Chan KWY, Gilad AA, and Bulte JWM. Supercharged green fluorescent proteins as bimodal reporter genes for CEST MRI and optical imaging. *Chem. Commun.*, 2015; 51(23): 4869–4871.

32. Bar-Shir A, Liu GS, Chan KWY, et al. Human protamine-1 as an MRI reporter gene based on chemical exchange. *ACS Chem. Biol.*, 2014; 9(1): 134–138.

33. Ruiz-Cabello J, Vuister GW, Moonen CTW, Vangelderen P, Cohen JS, and van Zijl PCM. Gradient-enhanced heteronuclear correlation spectroscopy: Theory and experimental aspects. *J. Magn. Reson.*, 1992; 100(2): 282–302.

34. McMahon MT, Gilad AA, Zhou J, Sun PZ, Bulte JWM, and van Zijl PC. Quantifying exchange rates in chemical exchange saturation transfer agents using the saturation time and saturation power dependencies of the magnetization transfer effect on the magnetic resonance imaging signal (QUEST and QUESP): Ph calibration for poly-L-lysine and a starburst dendrimer. *Magn. Reson. Med.*, 2006; 55(4): 836–847.

35. Jones CK, Schlosser MJ, van Zijl PCM, Pomper MG, Golay X, and Zhou JY. Amide proton transfer imaging of human brain tumors at 3T. *Magn. Reson. Med.*, 2006; 56(3): 585–592.
36. Salhotra A, Lal B, Laterra J, Sun PZ, van Zijl PCM, and Zhou JY. Amide proton transfer imaging of 9L gliosarcoma and human glioblastoma xenografts. *NMR Biomed.*, 2008; 21(5): 489–497.
37. Zhou JY, Zhu H, Lim M, et al. Three-dimensional amide proton transfer MR imaging of gliomas: Initial experience and comparison with gadolinium enhancement. *J. Magn. Reson. Imaging*, 2013; 38(5): 1119–1128.

Chapter 4

Early Discovery and Investigations of paraCEST Agents in Dallas

A. Dean Sherry

The University of Texas at Dallas, Richardson, Texas 75080-3021, USA
University of Texas Southwestern Medical Center, Dallas, TX 75390, USA
dean.sherry@utsouthwestern.edu

Our first observation of a chemical exchange saturation transfer (CEST) signal from a paramagnetic lanthanide complex came when one of my chemistry PhD students, Kuangcong Wu, collected a high-resolution nuclear magnetic resonance (NMR) spectrum of the Eu^{3+} complex, $EuDOTA\text{-}(gly\text{-}OEt)_4^{3+}$, in a mixture of water/acetonitrile (the ligand $DOTA\text{-}(gly\text{-}OEt)_4$ is a tetraamide derivative of DOTA). Much to our surprise, the ^1H NMR spectrum of this complex showed eight hyperfine shifted ligand proton resonances over the chemical shift range covering $+24$ to -18 ppm (as expected) plus an unexpected resonance near $+50$ ppm with an integrated intensity of two protons. No one at the time would have considered the possibility that this extra resonance could reflect a water molecule coordinated directly to the Eu^{3+} in the complex because water exchange in all known lanthanide complexes was thought to be much too fast for a separate resonance to appear. A water resonance at $+50$ ppm in a ^1H 500 MHz NMR spectrum would dictate the rate of water exchange to be $<<1.6 \times 10^5$ s^{-1}, about 10-fold

---

Chemical Exchange Saturation Transfer Imaging: Advances and Applications
Edited by Michael T. McMahon, Assaf A. Gilad, Jeff W. M. Bulte, and Peter C. M. van Zijl
Copyright © 2017 Pan Stanford Publishing Pte. Ltd.
ISBN 978-981-4745-70-3 (Hardcover), 978-1-315-36442-1 (eBook)
www.panstanford.com

slower than any other known lanthanide complex at that time. Yet, further investigations proved that the resonance at +50 ppm did indeed reflect a single Eu^{3+}-bound water molecule. One experiment Kuangcong did in an effort to assign the peak at +50 ppm was to apply a long frequency-selective RF pulse on the shifted water resonance to see whether it was spin-coupled to another resonance. Saturation of this peak did not alter the intensities of the other ligand peaks but did result in substantial reduction in the intensity of the solvent water resonance; hence, this was the first CEST experiment performed in Dallas! The details of this experiment were described in our first paraCEST paper in 2001 [1]. We quickly realized the beauty of having an exchanging proton peak +50 ppm away from the water signal was that it could be saturated with negligible off-resonance saturation of water so that any decrease in water signal could be directly ascribed to CEST. This suggested that one should be able to create responsive paraCEST agents and detect them easily by simple "off" minus "on" image intensity differences. We immediately set out to measure Z-spectra for each of the other paramagnetic lanthanide ion complexes with DOTA-(gly-OEt)$_4$ hoping that they would all display equally slow water exchange kinetics favorable for CEST. This would provide us with a family of paraCEST agents with water exchange peaks suitable for CEST activation over an incredibly wide frequency range (± 600–700 ppm). Although we succeeded in detecting a water exchange CEST peak for the entire series of lanthanide complexes (except the Yb^{3+} complex, which as we learned later does not have an inner-sphere water molecule), the lanthanide complexes that produce the largest hyperfine shifts (Tb, Dy, Ho, Tm) all displayed extraordinarily broad water exchange peaks and required extremely high-power B_1 pulses for detection [2]. This was the first indication that water exchange in these complexes must be highly variable and in some cases quite fast. A later more complete study of the water exchange kinetics of the entire lanthanide series of complexes revealed that the Eu^{3+} complex surprisingly displayed the slowest rate of water molecule exchange of all the lanthanide complexes [3]. The origins of this effect have not been fully proven, but based on the elegant work of Merbach et al. [4–6] on thermodynamic measures of water exchange, it is reasonable to conclude that water exchange is slowest for Eu^{3+}

because this ion is near the optimum size for the DOTA-(gly-OEt)$_4$ macrocyclic cavity and this size match forces the mechanism of water exchange to switch from an associative mechanism for the larger lanthanide ion (La^{3+}, Ce^{3+}, Pr^{3+}, and Nd^{3+}) complexes to a dissociative mechanism at Eu^{3+}. This switch in water exchange mechanism could, of course, differ for other types of CEST ligands having different metal ion binding cavity sizes, but this has yet to be explored in sufficient detail for other paraCEST systems. In general, it appears to hold that water exchange is slowest for all Eu-based CEST complexes reported so far (compared to other lanthanide complexes with the same ligand), while exchange appears to be generally faster for the smaller lanthanide ion complexes (Tb^{3+}, Dy^{3+}, Ho^{3+}, Er^{3+}, Tm^{3+}, and Yb^{3+}).

We were also fortunate at the time that Don Woessner was working in our NMR group as an Adjunct Professor. Don had retired from Mobil Oil Field Research Labs after 34 years of service and wanted to remain active in NMR research. Don was a superb NMR spectroscopist and had deep insights into virtually all aspects NMR. Paul Lauterbur once wrote in a supporting letter for Don, "Nobody knows more about nuclear relaxation than Don Woessner." Don was particularly well versed in chemical exchange theory, and he loved the Bloch equations. So it was not too difficult to get Don hooked on CEST agents. It was Don who suggested that we call our newly discovered agents paraCEST agents. He would sit in his office day after day doing calculations and thinking about chemical exchange and the idea of CEST in general. His thoughts and calculations were summarized in a highly cited *Magnetic Resonance in Medicine* paper [7], which quickly became the "bible" for all students and postdocs working in my lab on new paraCEST designs.

An important additional question about EuDOTA-(gly-OEt)$_4$$^{3+}$ concerning the origin of the CEST signal is: does it reflect exchange of the entire water *molecule* or exchange of *protons* from a Eu^{3+}-bound water molecule with protons in bulk water. This question was answered by comparing the rate of water exchange in this complex as measured by ^{17}O NMR versus ^{1}H NMR [8]. Those studies demonstrated quite clearly that the CEST mechanism involved exchange of the entire water molecule from the inner coordination sphere of the Eu^{3+} in the complex into bulk solvent.

That observation led us to think about how one might be able to modify water molecule exchange rates in these complexes.

Once we had proven that CEST in these complexes was due to water molecule exchange and not proton exchange, we set out to investigate whether we might be able to modulate the rate of water molecule exchange using ligand design concepts. Our goal was to use this chemical feature to create paraCEST agents that respond to changes in various physiologic or metabolic parameters. It did not take long for my students and postdocs to come up with paraCEST agents that responded to the presence of Zn^{2+} ions [9], glucose [10], singlet oxygen [11], hydrogen peroxide [12], or redox reactions involving NADH [13]. Given that these systems are weakly paramagnetic and produce large hyperfine NMR shifts, it was also quickly shown that they act as highly sensitive temperature probes [14]. Further investigations revealed that the chemical nature of the amide side-chains in DOTA-tetraamide ligands played a major role in determining the water exchange rates in these complexes, with negatively charged carboxylates tending to slow water exchange, while more hydrophobic side-chains tending to speed up water exchange. We concluded from many studies that water exchange between the Eu^{3+}-bound water molecule and bulk solvent is controlled by the structure and organization of water in the second and third water layers that surround the complex. It was later shown that even subtle differences in ligand structure could not only alter the water exchange rate but also affect the frequency of the bound water molecule [15, 16]. This fundamentally important observation stimulated Yunkou Wu to design a pH-sensitive paraCEST agent that changes CEST frequency rather than CEST intensity with changes in pH [17]. We have since tested this pH sensor in vivo and found that it is reasonably non-toxic and that it responds to alterations in tissue pH much like it does in vitro [18]. We are now focused on improving the CEST sensitivity of this pH sensor and hope to use this agent or a derivative to obtain reliable CEST images of tissue pH after a single injection of a non-toxic dose of the sensor.

Translation of paraCEST sensors from in vitro NMR tube experiments to in vivo imaging experiments poses other questions yet to be answered. First, is it reasonable to assume that water exchange rates and the exchange mechanism in vivo are identical to those measured in vitro? The answer is not likely because other factors such as the presence of biological anions such as phosphate, carbonate, or other charged species are known to catalyze proton exchange. Again, we were fortunate to have an active collaboration with another NMR legend Tom Dixon who derived a relationship between CEST intensity and applied B_1 power that provides the rate of water exchange without knowing the exact concentration of the paraCEST agent [19]. This method, commonly referred to as the Omega plot method, allowed for the first time the possibility of measuring water exchange rates in paraCEST complexes in vivo. A second related question is one of sensitivity. Will it be possible to image biologically important parameters such as tissue pH using acceptable amounts of exogenous paraCEST sensors? Theoretically, the answer is yes, but it will require agents with optimal water exchange rates that can be fully activated using B_1's within acceptable SAR limits. There are also alternative imaging sequences being developed by others in the field that separate out faster exchanging CEST processes from more slowly exchanging CEST processes. So if one considers the advances the entire CEST community has made in these agents since they were first introduced in 2001, I remain optimistic that this community will discover a clear path forward to allow translation of these interesting agents into clinical practice.

References

1. Zhang S, Winter P, Wu K, and Sherry AD. A novel europium(III)-based MRI contrast agent. *J Am Chem Soc*, 2001; 123(7): 1517–1518.
2. Zhang S, Wu K, and Sherry AD. Unusually sharp dependence of water exchange rate versus lanthanide ionic radii for a series of tetraamide complexes. *J Am Chem Soc*, 2002; 124(16): 4226–4227.
3. Zhang S and Sherry AD. Physical characteristics of lanthanide complexes that act as magnetization transfer (MT) contrast agents. *J Solid State Chem*, 2003; 171: 38–43.
4. Merbach AE and Tóth É (eds). *The Chemistry of Contrast Agents in Medical Magnetic Resonance Imaging*: Wiley; 2001.

5. Micskei K, Helm L, Brucher E, and Merbach AE. Oxygen-17 NMR study of water exchange on gadolinium polyaminopolyacetates [Gd(DTPA)(H$_2$O)]$^{2-}$ and [Gd(DOTA)(H$_2$O)]$^-$ related to NMR imaging. *Inorg Chem*, 1993; 32(18): 3844–3850.
6. Micskei K, Powell DH, Helm L, Brucher E, and Merbach AE. Water exchange on gadolinium (aqua)(propylenediamine tetraacetate) complexes [Gd(H$_2$O)$_8$]$^{3+}$ and [Gd(PDTA)(H$_2$O)$_2$]$^-$ in aqueous solution: A variable-pressure, -temperature and -magnetic field oxygen-17 NMR study. *Magn. Reson. Chem*, 1993; 31(11): 1011–1020.
7. Woessner DE, Zhang S, Merritt ME, and Sherry AD. Numerical solution of the Bloch equations provides insights into the optimum design of PARACEST agents for MRI. *Magn Reson Med*, 2005; 53(4): 790–799.
8. Zhang S, Wu K, Biewer MC, and Sherry AD. ^{1}H and ^{17}O NMR detection of a lanthanide-bound water molecule at ambient temperatures in pure water as solvent. *Inorg Chem*, 2001; 40(17): 4284–4290.
9. Trokowski R, Ren J, Kálmán FK, and Sherry AD. Selective sensing of zinc ions with a PARACEST contrast agent. *Angew Chem Int Ed*, 2005; 44(42): 6920–6923.
10. Ren J, Trokowski R, Zhang S, Malloy CR, and Sherry AD. Imaging the tissue distribution of glucose in livers using a PARACEST sensor. *Magn Reson Med*, 2008; 60(5): 1047–1055.
11. Song B, Wu Y, Yu M, et al. A europium(iii)-based PARACEST agent for sensing singlet oxygen by MRI. *Dalton Trans*, 2013; 42(22): 8066–8069.
12. Fidelino LC. *A paraCEST-Based Hydrogen Peroxide Sensor* [dissertation]: PhD dissertation, University of Texas at Dallas; 2013.
13. Ratnakar SJ, Viswanathan S, Kovacs Z, Jindal AK, Green KN, and Sherry AD. Europium(III) DOTA-tetraamide complexes as redox-active MRI sensors. *J Am Chem Soc*, 2012; 134(13): 5798–5800.
14. Zhang S, Malloy CR, and Sherry AD. MRI thermometry based on PARACEST agents. *J Am Chem Soc*, 2005; 127(50): 17572–17573.
15. Green KN, Viswanathan S, Rojas-Quijano FA, Kovacs Z, and Sherry AD. Europium(III) DOTA-derivatives having ketone donor pendant arms display dramatically slower water exchange. *Inorg Chem*, 2011; 50(5): 1648–1655.
16. Viswanathan S, Ratnakar SJ, Green KN, Kovacs Z, De León-Rodríguez LM, and Sherry AD. Multi-frequency PARACEST agents based on europium(III)-DOTA-tetraamide ligands. *Angew Chem Int Ed*, 2009; 48(49): 9330–9333.

17. Wu Y, Soesbe T, Kiefer GE, Zhao P, and Sherry AD. A responsive europium(III) chelate that provides a direct readout of pH by MRI. *J Am Chem Soc*, 2010; 132: 14002–14003.
18. Wu Y, Zhang S, Soesbe TC, et al. pH imaging of mouse kidneys in vivo using a frequency-dependent paraCEST agent. *Magn Reson Med*, 2016; 75(6): 2432–2441.
19. Dixon WT, Ren J, Lubag AJ, et al. A concentration-independent method to measure exchange rates in PARACEST agents. *Magn Reson Med*, 2010; 63(3): 625–632.

Chapter 5

Birth of CEST Agents in Torino

Silvio Aime

Molecular & Preclinical Imaging Centers, Department of Molecular Biotechnology and Health Sciences, University of Torino, Torino 10126, Italy
silvio.aime@unito.it

Our work, as magnetic resonance imaging (MRI) probe developers, was always frustrated by the drawbacks that relaxation enhancers have with respect to the colored reporters biologists dealt with in their confocal images. In fact, our beloved paramagnetic agents fail when two of them are targeted to different epitopes in the same anatomical region. Their presence is indistinguishable in an MR image, as they simply add their effect on water proton relaxation rates. Moreover, also for the class of Gd-based responsive agents (where chemists have identified a number of modes to make the observed relaxivity responsive to specific physicochemical and biological parameters), there were problems because their in vivo exploitation required the knowledge of the local concentration in order to monitor the variable of interest from the relaxation enhancement values.

Altogether, the limitations of Gd-based agents prompted chemists to look at other routes to affect the contrast in MR images. Basically, it was clear that one has to look at frequency-encoding agents in order to recover the full potential of the nuclear magnetic

---

Chemical Exchange Saturation Transfer Imaging: Advances and Applications
Edited by Michael T. McMahon, Assaf A. Gilad, Jeff W. M. Bulte, and Peter C. M. van Zijl
Copyright © 2017 Pan Stanford Publishing Pte. Ltd.
ISBN 978-981-4745-70-3 (Hardcover), 978-1-315-36442-1 (eBook)
www.panstanford.com

resonance (NMR) experiments. Attention was first devoted to the class of paramagnetic shift reagents as they may show highly shifted ^{1}H resonances characterized by very short relaxation times. For instance, we showed that chemical shift images can be obtained, in a relatively short time, from the four equivalent methyl groups of paramagnetic Ln-DOTMA complexes (Ln = Dy, Tm, Yb) [1]. Later, with the introduction of improved pulse sequences, it was shown that this approach might be of interest, and in vivo proofs of concept studies were reported [2]. Obviously, frequency-encoding agents for multitasking images can be enormously fuelled by designing molecules in which the imaging reporting unit is represented by an NMR-active heteronuclear species. This direction is currently under intense scrutiny with regard to the field of ^{19}F-containing agents and, even more intriguing, the class of hyperpolarized molecules labelled with ^{13}C, ^{15}N, etc.

However, the idea of coupling the advantages associated with the use of frequency-encoding agents to the established modality of creating contrast through the topological representation of water signal intensities was considered mature in many labs. I remember an interesting discussion I had with Ivano Bertini and his colleagues in the late 1990s on the possibility of affecting water signal intensity by exploiting magnetization transfer from N-H peptide moieties, in analogy to what was routinely done by NMR-structural biologists to determine the access to water solvent of the different types of amide functionalities in the protein backbone. The idea of "pumping-out" upon irradiating a specific site and looking at the effect of the chemical exchange at another one was well established in NMR spectroscopy since the seminal work of Forsen and Hoffman in 1963 as is described eloquently in Chapter 1 [3, 4].

Actually, at the time we entered the field of MRI contrast agents, we surmised that a simple method to affect the T_2 of water protons would be through the chemical exchange of mobile protons. Years later, we extended this method to Eu(III) chelates using the exchange between the coordinated and *bulk* water as a means to reduce the T_2 of the solvent water resonance and suppress it in ^{1}H-NMR spectrum by applying a proper T_2-weighted sequence. However, it was only in the late 1990s that we started to perform double-resonance experiments with the aim of developing chemical exchange saturation transfer (CEST) experiments. First, we started with the diamagnetic compound Iopamidol (which we and others had already reported for possible use in MRI as a T_2-agent, thanks to the prototropic exchange from OH and NH moieties). The experiments were done (erroneously!) on an NMR spectrometer operating at 2.1 T, and the observed saturation transfer (ST) effect upon irradiating the two proton amide resonances was not considered a significant progress with respect to the "spontaneously occurring" T_2-effect. Having in mind the sensitivity issue of Gd-based CAs, we were a little biased by small ST effects that could have been generated by Iopamidol in spite of the high quantities that are usually given in vivo. We thought that paramagnetic systems containing highly shifted exchangeable protons could provide more room for enhanced effects, thanks to the possibility of using much higher prototropic exchange rates. We selected the Yb(III) complex of a DOTA tetraamide ligand DOTAM-Gly with the aim of detecting a CEST contrast upon saturating the four equivalent amide protons (by *Enzo Terreno* and *Daniela Delli Castelli*).

During this study, we became aware of the founding paper of CEST agents by Ward and Balaban [5, 6] who proposed a ratiometric method that allows one to disregard the concentration in CEST systems having two magnetically non-equivalent pools of mobile protons. As the CEST effect generated by the saturation of the amide protons of (Yb-DOTAM-Gly)$^-$ displayed a good pH responsiveness, we designed the first concentration-independent pH sensor based on paraCEST agents that exploited the peculiarity of Ln-complexes of the same ligand to have similar structure, but different magnetic properties. The system was represented by a mixture of Yb- and (Eu-DOTAMGly)$^-$ complexes, whose amide protons displayed different chemical shift and pH responsiveness. This method was submitted to *Magnetic Resonance in Medicine* in July 2001 and accepted 4 months later [7]. This was the debut of the Torino group in the CEST field.

Keeping focused on pH sensors, the successive year we published in *Angewandte Chemie* a series of (Ln-DOTAM-Gly)$^-$ complexes (Ln = Nd, Pr, and Eu) able to act as individual paraCEST pH reporters by virtue of the presence of two CEST active pools in their structure (amide + metal-coordinated water protons) [8].

These studies highlighted the sensitivity issues suffered by these systems (especially if compared to Gd-based T_1-agents) and prompted us to find approaches to overcome this limitation. As a first approach, we exploited our experience on non-covalent interactions between Ln-complexes and macromolecules to design the first supramolecular CEST adduct [9]. Here a paramagnetic shift reagent for cations (Tm-HDOTP)$^{4-}$ was used to shift the resonance of the mobile spins of the positively charged poly-arginine away from the diamagnetic region, thus improving the sensitivity in the contrast detection of approximately three order of magnitude.

However, a real breakthrough in the way of designing highly sensitive CEST agents was attained in 2005, when the story on lipoCEST started in our lab. At that time, we were interested in using liposomes as biocompatible carriers for Gd-complexes (especially for targeting experiments), but we realized soon that exploiting the huge number of exchangeable solvent water protons in the inner core of the nanovesicle, not achievable by any other classical paraCEST-based system, would have brought the sensitivity of CEST agents down to sub-nM detection!

Again, our experiences on paramagnetic Ln-complexes were essential to find a solution for inducing a sufficient chemical shift separation between intraliposomal and bulk protons: to encapsulate in the nanovesicles an efficient paramagnetic shift reagent for solvent protons (by *Daniela Delli Castelli* and *Enzo Terreno*). Using (Tm-DOTMA)$^-$, we succeeded in detecting CEST contrast at a frequency offset of 3.1 ppm at a particle concentration of 90 pM [10]. In the same period, we also provided the first proof-of-concept of the feasibility to detect and track different cell populations with paraCEST agents [10].

Going back to the lipoCEST story, we realized that a frequency offset of a few ppm would have significantly limited the in vivo potential of such systems. Hence, we started thinking of possible approaches to fix this. Important hints were provided by some papers from the group of Charles Springer Jr., who (15 years earlier) studied bulk magnetic susceptibility shift (BMS) effects of compartmentalized paramagnetic species. A BMS shift contribution occurs when the shift reagent is confined in a non-spherical compartment, and its magnitude and sign depend on the paramagnetism of the metal center as well as the orientation of the compartment in the B_0 field.

Subjecting spherical liposomes to hyper-osmotic conditions, the high softness of liposomes, associated with their semipermeable bilayer, allowed the preparation of lens-shaped non-spherical lipoCEST agents. Besides the BMS shift contribution from the encapsulated complex, we further extended the frequency offset range of non-spherical liposomes (to approximately 50 ppm!) through the incorporation of paramagnetic complexes in the liposome's bilayer that controlled the orientation of the vesicles in the B_0 field [11]. We achieved an additional increase in the offset by encapsulating non-ionic poly-metallic shift reagents [12]. The extension in the range of frequency offset allowed the first detection of different lipoCEST agents co-localized in the same image voxels [13]. We demonstrated that non-spherical lipoCEST probes improved the potential of liposomes in diagnostic/theranostic field, especially in MRI multi-contrast procedures. In fact, in addition to CEST contrast, liposomes loaded with paramagnetic complexes can simultaneously generate T_1 and T_2 contrast. Very interestingly, the encapsulation of a Gd-complex in non-spherical liposomes allowed the design of a three-modal (T_1, T_2, and CEST) contrast agent [14].

The outstanding results obtained with lipoCEST agents prompted us to test cells as vesicular containers (by *Giuseppe Ferrauto and Daniela Delli Castelli in collaboration with researchers at Philips*). The analogy with lipoCEST is striking: Both systems can be loaded with shift reagents and the permeability of cellular membranes to water molecules is well established. Moreover, cells (in particular RBCs) are not spherical and, as such, they naturally provide access to the BMS contribution to the chemical shift of the intracellular water protons. We proved the feasibility of this approach with RBCs (cellCEST by *Giuseppe Ferrauto*), showing that their detection required the presence of only 5% of labelled cells with respect to the total RBCs [15]. More recently, we proposed a new system that combines lipo- and cellCEST. Cationic lipoCEST agents were formulated and stuck to the negatively charged surface of RBCs where they can affect the frequency offset of the intracellular water protons by the BMS effect [16]. Such adducts might provide novel insights for the development of innovative theranostic agents.

In addition to vesicle-based CEST agents, in the last years my group has focused its attention on the possible clinical translation of CEST agents. This purpose was accomplished by investigating the properties of two molecules: one already used in the clinic (the CT agent Iopamidol by *Dario Longo*), and the other with the same bio-physicochemical properties of an approved MRI agent (Yb-HPDO3A, the Yb analogue of Gadoteridol by *Daniela Delli Castelli* and *Giuseppe Ferrauto*). Both these systems have two pools of magnetically non-equivalent spins and have been demonstrated to enable the monitoring of pH in vivo in different pathologies [17, 18].

Another interesting result recently obtained in our laboratory deals with the design of the first CEST agent based on silica particles (by *Giuseppe Ferrauto* in collaboration with *Mauro Botta*'s group). By binding a coordinatively unsaturated Ln-DO3A (Ln = Eu, Tb, Tm) complex on the silica surface, it has been found that the neighboring –OH moieties reversibly interact with the paramagnetic shift reagent. The net result is the appearance in the Z-spectrum of a strong absorption due to the ST originated by the shift exchangeable hydroxyl protons [19].

Hence, the marriage between Ln-based shift reagents and CEST properties continued to yield beautiful fruits and mentioning the word "fruits" why not thinking of CEST of fruit!

Recently, we showed that CEST contrast could be very useful to characterize specific molecules containing exchangeable protons when they are constituents of a given fruit (by *Daniela Delli Castelli*) [20]. For instance, plums are very rich in aspargines whose amide protons represent the exchanging pool for a CEST experiment. Asparagine (that is an important marker of the ripening process) can be quantitatively determined in fruits. Sugars are obvious candidates to assess the sweetness of a fruit; maybe the CEST label will arrive at the market desk.

References

1. Aime S, Botta M, Fasano M, et al. A new ytterbium chelate as contrast agent in chemical shift imaging and temperature sensitive probe for MR spectroscopy. *Magn. Reson. Med.*, 1996; 35(5): 648–651.

2. Schmidt R, Nippe N, Strobel K, et al. Highly shifted proton MR imaging: Cell tracking by using direct detection of paramagnetic compounds. *Radiology*, 2014; 272(3): 785–795.
3. Forsen S and Hoffman RA. A new method for study of moderately rapid chemical exchange rates employing nuclear magnetic double resonance. *Acta Chem. Scand.*, 1963; 17(6): 1787.
4. Forsen S and Hoffman RA. Study of moderately rapid chemical exchange reactions by means of nuclear magnetic double resonance. *J. Chem. Phys.*, 1963; 39(11): 2892.
5. Ward KM, Aletras AH, and Balaban RS. A new class of contrast agents for MRI based on proton chemical exchange dependent saturation transfer (CEST). *J. Magn. Reson.*, 2000; 143(1): 79–87.
6. Ward KM and Balaban RS. Determination of pH using water protons and chemical exchange dependent saturation transfer (CEST). *Magn. Reson. Med.*, 2000; 44(5): 799–802.
7. Aime S, Barge A, Delli Castelli D, et al. Paramagnetic lanthanide(III) complexes as pH-sensitive chemical exchange saturation transfer (CEST) contrast agents for MRI applications. *Magn. Reson. Med.*, 2002; 47(4): 639–648.
8. Aime S, Delli Castelli D, and Terreno E. Novel pH-reporter MRI contrast agents. *Angew. Chem.-Int. Edit.*, 2002; 41(22): 4334–4336.
9. Aime S, Delli Castelli D, and Terreno E. Supramolecular adducts between poly-L-arginine and Tm(III)dotp: A route to sensitivity-enhanced magnetic resonance imaging-chemical exchange saturation transfer agents. *Angew. Chem.-Int. Edit.*, 2003; 42(37): 4527–4529.
10. Aime S, Carrera C, Delli Castelli D, Crich SG, and Terreno E. Tunable imaging of cells labeled with MRI-PARACEST agents. *Angew. Chem.-Int. Edit.*, 2005; 44(12): 1813–1815.
11. Terreno E, Cabella C, Carrera C, et al. From spherical to osmotically shrunken paramagnetic liposomes: An improved generation of LIPOCEST MRI agents with highly shifted water protons. *Angew. Chem.-Int. Edit.*, 2007; 46(6): 966–968.
12. Terreno E, Barge A, Beltrami L, et al. Highly shifted LIPOCEST agents based on the encapsulation of neutral polynuclear paramagnetic shift reagents. *Chem. Commun.*, 2008; (5): 600–602.
13. Terreno E, Delli Castelli D, Milone L, et al. First ex-vivo MRI co-localization of two LIPOCEST agents. *Contrast Media Mol. Imaging.*, 2008; 3(1): 38–43.

14. Aime S, Delli Castelli D, Lawson D, and Terreno E. Gd-loaded liposomes as T-1, susceptibility, and CEST agents, all in one. *J. Am. Chem. Soc.*, 2007; 129(9): 2430.
15. Ferrauto G, Delli Castelli D, Di Gregorio E, et al. Lanthanide-loaded erythrocytes as highly sensitive chemical exchange saturation transfer MRI contrast agents. *J. Am. Chem. Soc.*, 2014; 136(2): 638–641.
16. Ferrauto G, Di Gregorio E, Baroni S, and Aime S. Frequency-encoded MRI-CEST agents based on paramagnetic liposomes/RBC aggregates. *Nano Lett.*, 2014; 14(12): 6857–6862.
17. Longo DL, Busato A, Lanzardo S, Antico F, and Aime S. Imaging the pH evolution of an acute kidney injury model by means of iopamidol, a MRI-CEST pH-responsive contrast agent. *Magn. Reson. Med.*, 2013; 70(3): 859–864.
18. Delli Castelli D, Terreno E, and Aime S. Yb-III-HPDO3A: A dual pH- and temperature-responsive CEST agent. *Angew. Chem.-Int. Edit.*, 2011; 50(8): 1798–1800.
19. Ferrauto G, Carniato F, Tei L, Hu H, Aime S, and Botta M. MRI nanoprobes based on chemical exchange saturation transfer: Ln(III) chelates anchored on the surface of mesoporous silica nanoparticles. *Nanoscale*, 2014; 6(16): 9604–9607.
20. Podda R, Delli Castelli D, Digilio G, Gullino ML, and Aime S. Asparagine in plums detected by CEST-MRI. *Food Chem.*, 2015; 169: 1–4.

Section II

Pulse Sequence, Imaging, and Post-processing Schemes for Detecting CEST Contrast

Chapter 6

General Theory of CEST Image Acquisition and Post-Processing

Nirbhay N. Yadav, Jiadi Xu, Xiang Xu, Guanshu Liu, Michael T. McMahon, and Peter C. M. van Zijl

F. M. Kirby Research Center, Kennedy Krieger Institute, 707 N Broadway, Baltimore, MD 21205, USA
Department of Radiology, Johns Hopkins University School of Medicine, 707 N Broadway, Baltimore, MD 21218, USA
yadav@kennedykrieger.org, pvanzijl@mri.jhu.edu

6.1 Introduction

Nuclear magnetic resonance (NMR) is an extremely powerful method for detecting and quantifying both the structure of molecules and dynamic processes that are occurring involving these molecules. An example of a dynamic process is chemical exchange, which is special in that this process can be used to impart large signal contrast in NMR imaging (MRI), as we will describe in detail in this textbook. Herein, we define chemical exchange as a nuclear spin experiencing a change in its molecular environment with these environmental changes imparting a perturbation in the chemical shift of the nuclear spin. NMR spectra are sensitive to many forms

---

Chemical Exchange Saturation Transfer Imaging: Advances and Applications
Edited by Michael T. McMahon, Assaf A. Gilad, Jeff W. M. Bulte, and Peter C. M. van Zijl
Copyright © 2017 Pan Stanford Publishing Pte. Ltd.
ISBN 978-981-4745-70-3 (Hardcover), 978-1-315-36442-1 (eBook)
www.panstanford.com

of chemical exchange via the parameters of chemical shift, scalar coupling, dipolar coupling, and relaxation. Further, NMR can be used to study chemical exchange without the use of dyes or tracers and thus can measure chemical exchange dynamics without perturbing the physical system and under equilibrium conditions. As such NMR is a leading research and clinical tool for measuring exchange phenomena, including investigations of protein dynamics, substrate binding, and since NMR is the foundation of MRI, for generating MRI contrast related to these phenomena.

As mentioned earlier, NMR is sensitive to chemical exchange via changes in chemical shift, scalar coupling, dipolar coupling, and relaxation times produced by environmental differences. The extent to which these changes perturb the NMR lineshapes is determined by the timescale over which the exchange occurs (Larmor, spectral, or relaxation timescales). For instance, exchange on the spectral timescale (measured by the inverse width of the NMR spectrum) causes noticeable changes to NMR lineshapes. For chemical exchange saturation transfer (CEST) experiments, we are primarily concerned with measuring signal changes from exchange occurring over the spectral timescale. Exchange rates can be further subdivided into regimes (slow, intermediate, fast) based on the spectral distance between the exchanging sites, and this has important implications for CEST experimental parameters (see below).

Most CEST experiments are performed in a similar manner to conventional magnetization transfer (MT) experiments where off-resonance radiofrequency (RF) pulses label solute molecules, after which this labeling is transferred to the solvent. Here we focus on bulk water as the solvent pool, but this is not always the case (e.g., ^{19}F-CEST, hyperCEST, see the diamagnetic CEST agents and hyperCEST imaging chapters). The change in longitudinal magnetization (z) of the water pool is then measured to characterize the effect. There are several pathways that exist for transferring magnetization to water, including intermolecular dipolar coupling, chemical exchange, and exchange-relayed intramolecular dipolar coupling. In a historical sense, CEST aims to separate out only the exchanging component from other MT pathways, but more recently, several studies have been able to measure interesting phenomena in experiments where chemical exchange is but one of a multi-step process [1–3].

6.2 Theory

Since the description of intermolecular saturation transfer via proton exchange and dipolar cross-relaxation is theoretically equivalent [4], we here describe only chemical exchange and note that the theory below can also provide insight into simple cross-relaxation processes. If we take a simple two-pool system, the exchange process can be described by the modified Bloch equations with exchange terms [5]:

$$\frac{dM_{xs}}{dt} = -\Delta\omega_s M_{ys} - R_{2s} M_{xs} - \kappa_{sw} M_{xs} + \kappa_{ws} M_{xw}$$

$$\frac{dM_{ys}}{dt} = \Delta\omega_s M_{xs} + \omega_1 M_{zs} - R_{2s} M_{ys} - \kappa_{sw} M_{ys} + \kappa_{ws} M_{yw}$$

$$\frac{dM_{zs}}{dt} = -\omega_1 M_{ys} - R_{1s}(M_{zs} - M_{0s}) - \kappa_{sw} M_{zs} + \kappa_{ws} M_{zw}$$

$$\frac{dM_{xw}}{dt} = -\Delta\omega_w M_{yw} - R_{2w} M_{xw} + \kappa_{sw} M_{xs} - \kappa_{ws} M_{xw}$$

$$\frac{dM_{yw}}{dt} = \Delta\omega_w M_{xw} + \omega_1 M_{zw} - R_{2w} M_{yw} + \kappa_{sw} M_{ys} - \kappa_{ws} M_{yw}$$

$$\frac{dM_{zw}}{dt} = -\omega_1 M_{yw} - R_{1w}(M_{zw} - M_{0w}) + \kappa_{sw} M_{zs} - \kappa_{ws} M_{zw} \quad (6.1)$$

where R_1 and R_2 are the spin–lattice and spin–spin relaxation rates, respectively; M_0 is the equilibrium magnetization; and k_{sw} and k_{ws} are the exchange rates of protons from pool s (solute protons) to pool w (water protons), and vice versa. The MRI parameters are defined as $\omega_0 = \gamma B_0$, $\omega_1 = \gamma B_1$ (where γ is the nuclear gyromagnetic ratio, B_0 is the static magnetic field strength, and B_1 is the applied RF field strength), $\Delta\omega = \omega - \omega_0$, and the RF field is applied along the x direction in the rotating frame.

It is mathematically intractable to solve Eq. 6.1 exactly, even in the case of two-site exchange. Therefore, an approximate analytical solution [6, 7] must be used or the equations must be numerically solved [8–10].

6.2.1 Low-Power Irradiation

To obtain an analytical solution, we can assume that the irradiation is applied on the solute resonance ($\Delta\omega_s = 0$), the system is at equilibrium ($dM_{0s}/dt = 0$), and $k_{sw}M_{0s} = k_{ws}M_{0w}$. Furthermore, we can make simplifying assumptions that the B_1 field does not perturb the water pool ($\Delta\omega_w \to \infty$); thus, for the case of complete saturation (cs), Eq. 6.1 becomes

$$M_{zw}^{cs} = \frac{R_{1w}}{R_{1w} + \kappa_{ws}} M_{0w} \tag{6.2}$$

The reduction in the water signal intensity (proton transfer ratio, PTR) can then be written as

$$\text{PTR}^{cs} = \frac{M_{0w} - M_{zw}^{cs}}{M_{0w}} = \frac{\kappa_{ws}}{R_{1w} + \kappa_{ws}} \tag{6.3}$$

Note that it is impossible to achieve complete saturation for a solute pool using a selective irradiation pulse (i.e., this requires a very high B_1); thus, a steady-state solution [6] can be obtained by modifying Eq. 6.1 to

$$\frac{dM_{ys}}{dt} = \omega_1 m_{zs} - r_{2s} M_{ys} + \kappa_{ws} M_{yw} + \omega_1 M_{0s}$$

$$\frac{dM_{zs}}{dt} = -\omega_1 m_{ys} - r_{1s} m_{zs} + \kappa_{ws} m_{zw}$$

$$\frac{dM_{yw}}{dt} = -r_{2w} M_{yw} + \kappa_{sw} M_{ys}$$

$$\frac{dM_{zw}}{dt} = -r_{1w} m_{zw} + \kappa_{sw} m_{zs} \tag{6.4}$$

where $m_{zs} = M_{zs} - M_{0s}$, $m_{zw} = M_{zw} - M_{0w}$, $r_{1s} = R_{1s} + k_{sw}$, $r_{1w} = R_{1w} + k_{ws}$, $r_{2s} = R_{2s} + k_{sw}$, and $r_{2w} + k_{ws}$.

Assuming steady state (i.e., $dM/dt = 0$), Eq. 6.4 can be solved to give

$$m_{zs}^{ss} = -\frac{\omega_1^2 M_{0s}}{\omega_1^2 + pq}$$

$$M_{ys}^{ss} = -\frac{\omega_1 M_{0s}}{\omega_1^2 + pq}$$

$$m_{zs}^{ss} = -\frac{\kappa_{sw}}{r_{1w}} \frac{\omega_1^2 M_{0s}}{(\omega_1^2 + pq)} = \frac{\kappa_{ws}}{r_{1w}} \frac{\omega_1^2 M_{0w}}{(\omega_1^2 + pq)}$$

$$m_{yw}^{ss} = -\frac{\kappa_{sw}}{r_{2w}} \frac{\omega_1 M_{0s}}{(\omega_1^2 + pq)} = \frac{\kappa_{ws}}{r_{2w}} \frac{\omega_1 M_{0w}}{(\omega_1^2 + pq)} \tag{6.5}$$

where $p = r_{2s} - (k_{sw}k_{ws}/r_{2w})$ and $q = r_{1s} - (k_{sw}k_{ws}/r_{1w})$. Often, particularly in MRI experiments, it is impractical to use long continuous wave (CW) pulses for saturation (see Fig. 6.1). Thus, the magnetization for the water pool will not be in a steady state. To obtain an analytical solution, we can assume that the solute pool approaches a steady state $(m_{zs}^{ss} = M_{zs}^{ss} - M_{0s})$ instantly; thus, the longitudinal magnetization (z) of the water pool (w) can be described by

$$\frac{dm_{zw}}{dt} = -r_{1w}m_{zw} + \kappa_{sw}m_{zs}^{ss} \tag{6.6}$$

which has a solution

$$m_{zw}(t) = \frac{\kappa_{sw}m_{zs}^{ss}}{r_{1w}}\left[1 - e^{-r_{1w}(t-t_0)}\right] + m_{zw}(t_0)e^{-r_{1w}(t-t_0)} \tag{6.7}$$

The second term in Eq. 6.7 can be eliminated if we assume the experiment begins from equilibrium ($M_{zw}(t_0) = M_{0w}$, thus $m_{zw}(t_0) = 0$). The PTR obtained from Eq. 6.7 is

$$\text{PTR} = \frac{M_{0w} - M_{zw}(t_{sat})}{M_{0w}} = \frac{\kappa_{sw}\alpha M_{0s}}{(R_{1w} + \kappa_{ws})M_{0w}}\left[1 - e^{-(R_{1w}+\kappa_{ws})t_{sat}}\right] \tag{6.8}$$

where $t_{sat} = t - t_0$ and α is defined as the saturation efficiency (α)

$$\alpha = -m_{zs}^{ss}/M_{0s} = \omega_1^2/(\omega_1^2 + pq) \tag{6.9}$$

6.2.2 Sensitivity Enhancement

One of the great advantages of CEST is that the solute signal need not be visible in the NMR spectrum to be detected using CEST [11]. From Eqs. 6.3 and 6.9, we can see that prolonged irradiation from a CW pulse results in an amplification of the solute signal. This sensitivity enhancement was noted in early CEST studies [13–16] and became more clear for large macromolecular [14, 17–20] and lipoCEST [21–23] agents, where enhancements of up to six orders of magnitude are possible [12]. The greatest enhancements are achieved when the solute concentration is much smaller than the solvent concentration. This ensures a high probability that saturated solute protons are replaced by unsaturated protons from the much larger water/solvent pool. Prolonged irradiation leads to several iterations of the saturation transfer effect on water leading to the water pool having a larger

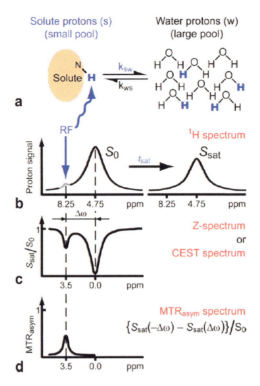

Figure 6.1 Chemical exchange saturation transfer (CEST): principles and measurement approach for pure exchange effects. (a, b) Solute protons (blue) are saturated at their specific resonance frequency in the proton spectrum (here 8.25 ppm for amide protons). This saturation is transferred to water (4.75 ppm) at exchange rate k_{sw} and non-saturated protons (black) return. After a period (t_{sat}), this effect becomes visible on the water signal (b, right). (c) Measurement of normalized water saturation (S_{sat}/S_0) as a function of irradiation frequency, generating a so-called Z-spectrum (or CEST spectrum or MT spectrum). When water protons are irradiated at 4.75 ppm, the signal disappears due to direct (water) saturation (DS). This frequency is assigned to 0 ppm in Z-spectra. At short saturation times, only this direct saturation is apparent. At longer t_{sat}, the CEST effect becomes visible at the frequency of the low-concentration exchangeable solute protons, now assigned to 8.25−4.75 = 3.5 ppm in the Z-spectrum. (d) Result of magnetization transfer ratio ($MTR = 1 - S_{sat}/S_0$) asymmetry analysis of the Z-spectrum with respect to the water frequency to remove the effect of direct saturation. In the remainder of this chapter, we will use the standard NMR chemical shift assignment for water at 4.75 ppm in ^{1}H spectra, while the 0 ppm assignment will be used in Z-spectra. Reprinted with permission from Ref. [12], Copyright 2004, John Wiley and Sons.

concentration of saturated protons compared to the original solute concentration.

From Eq. 6.8, we can obtain a proton transfer enhancement (PTE)

$$PTE = \frac{2[H_2O]PTR}{[solute]}$$

$$= \frac{\kappa_{sw}\alpha}{R_{1w} + x_{CA}\kappa_{sw}}\left[1 - e^{-(R_{1w}+\kappa_{sw}x_{CA})t_{sat}}\right] \quad (6.10)$$

in which $x_{CA} = (M_{0s}/M_{0w})$ is the exchangeable proton fraction of the contrast agent relative to the water proton concentration.

The analytical equations listed above are useful for understanding important parameters for CEST experiments and, in ideal systems (see assumptions above), can be used for interpreting CEST data, especially for exchange rates on the order of tens to hundreds of Hz [10]. However, more commonly, the assumptions made above are invalid (e.g., intermediate-fast exchange regime, partial irradiation of water pool, multi-pool exchange), and we must resort to numerical solutions of Eq. 6.1 for interpreting the data (see Section 6.3).

6.2.3 High-Power Irradiation

To derive the PTR values above, we made the important assumption that the B_1 field does not perturb the water pool ($\Delta\omega_w \to \infty$). However, there are several cases where this assumption is not valid (e.g., applying large amplitude B_1 fields close to the water resonance frequency). Partial irradiation of the water pool by the B_1 field results in a reduction of the water signal intensity. This reduction is in addition to the reduction of the water signal due to the CEST mechanism. Several groups [24–26] have derived CEST expressions in this case based on the enhanced relaxation rate R_{ex} of water. Since the partial saturation of the water pool is due to off-resonance effects of the saturation pulse, and thus the effective B_1 is tilted away from the x-y plane towards the z-axis, the RF pre-pulse can be used to lock the magnetization along the effective B_1 and thus measure $R_{1\rho}$. The observed water signal reduction with respect to the total saturation time t_{sat} can be described using

$$PTR = \frac{\cos^2\theta \cdot R_{1w}}{R_{1\rho}}(1 - e^{-R_{1\rho}t_{sat}}) \quad (6.11)$$

In the two-pool model, the measured rotating frame relaxation is the combined effect between rotating frame relaxation R_{eff} due to water direct saturation and the CEST effect R_{cest}:

$$R_{1\rho} = R_{\text{eff}} + R_{\text{cest}} \tag{6.12}$$

where

$$R_{\text{eff}} = \sin^2\theta\, R_{1w} + \cos^2\theta\, R_{2w} \tag{6.13}$$

$$\sin^2\theta = \frac{\Delta\omega^2}{\omega_1^2 + \Delta\omega^2};\ \cos^2\theta = \frac{\omega_1^2}{\omega_1^2 + \Delta\omega^2} \tag{6.14}$$

The CEST introduced rotation frame relaxation is given by

$$R_{\text{cest}} = \frac{x_s k_{sw} \omega_1^2}{\omega_1^2 + k_{ws}(k_{sw} + R_{2s})} \tag{6.15}$$

The enhanced relaxation from chemical exchange is useful for generating contrast when the exchange rate falls within the intermediate exchange regime. However, this experiment is limited on clinical scanners due to the high-power deposition produced when using appropriate pulses based on specific absorption rate (SAR) and hardware duty cycle considerations.

6.2.4 Pulsed-CEST

Classical CW saturation is used to equilibrate the populations of the two quantum states of the solute protons. While doing so, partial saturation is transferred continuously to bulk water, which can result in large water signal changes. However, it is often not possible to run CW-CEST based on amplifier, transmission coil, and SAR considerations, especially on human MRI scanners. Consequently, most CEST experiments on human subjects have used pulsed-CEST approaches [27–32]. Pulsed-CEST approaches use shorter but higher field strength (B_1) frequency-selective pulses for labeling. This labeling pulse is followed by a delay that allows labeled protons to exchange with bulk water. This whole process can be repeated many times to build up the fraction of labeled protons in bulk water.

For pulsed-CEST using rectangular shaped pulses, we can obtain an expression for longitudinal magnetization under steady-state conditions starting from Eq. 6.6, but now the initial condition ($t = 0$)

is non-zero for pulses after the first one ($m_{zw}{}^1$). The longitudinal magnetization for subsequent pulses ($m_{zw}{}^{n+1}$) is given by

$$m_{zw}^{n+1} = m_{zw}^n e^{-r_{1w}t_p} + \frac{k_{sw}m_{zs}^{ss}}{r_{1s}}(1 - e^{-r_{1w}t_p}) \qquad (6.16)$$

where t_p is the pulse duration. If there are no delays between the pulses, the solution will be similar to Eq. 6.7 where $t_{sat} = n \cdot t_p$. In the case of delays between the pulses (t_{mix}), we can assume that the magnetization of the solute pool reaches the same steady state (m_{zs}^{ss}) at the end of each pulse. This is relatively accurate in cases of a large number of pulses or in systems with short T_2. If we assume that the solute magnetization remains in steady state during t_{mix} ($m_{zs} = m_{zs}^{ss}$) and $m_{zw} = m_{zw}^n$, the longitudinal water magnetization for a pulsed-CEST experiment is given by

$$m_{zw} = \frac{k_{sw}m_{zs}^{ss}}{r_{1w}} \frac{\left(1 + e^{-r_{1w}(t_p+t_{mix})}\right)}{\left(1 - e^{r_{1w}(t_p+t_{mix})}\right)} \qquad (6.17)$$

In addition to the lower SAR and duty cycle for pulsed-CEST experiments, another significant advantage is their versatility compared to CW-CEST sequences. CEST compounds typically have much longer transverse relaxation times compared to compounds that generate conventional MT contrast (MTC) ($T_{2,mobile}^* \gg T_{2,solid}^*$). The longer relaxation times allow the use of more sophisticated labeling schemes to detect CEST agents and suppress confounding signal sources (e.g., MTC). Some examples include inversion [33], rotation [34], and frequency encoding [35], but there are many other approaches that have yet to be explored. Inversion and dephasing (shown in Fig. 6.2c,d) are saturation-like schemes that lead to a reduction in the bulk water signal intensity within a label-transfer module (LTM). In contrast, frequency labeling or "FLEX" modulates the water signal intensity depending on the exchangeable groups by encoding the chemical shift evolution of exchanging protons (Fig. 6.2e). For FLEX, each LTM consists of (i) a selective φ_x RF excitation pulse that excites protons over a range of frequencies, (ii) a delay (t_{evol}) during which excited protons undergo chemical shift evolution, (iii) a selective φ_{-x} RF pulse that flips the magnetization back to the longitudinal axis. This labeling period is followed by a delay (t_{exch}) that allows time for labeled protons to exchange with bulk water and can be used as an exchange rate filter. Each

66 | General Theory of CEST Image Acquisition and Post-Processing

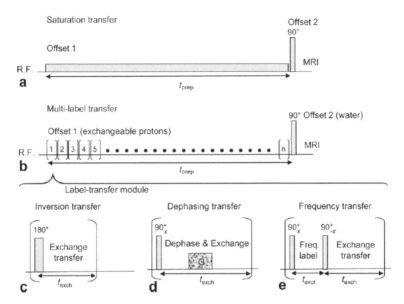

Figure 6.2 Possible schemes for exchange transfer MRI. (a) Standard CW-CEST: Protons are labeled through continuous wave saturation and transfer signal loss continuously during labeling. (b) Exchange transfer using label-transfer modules (LTMs): Protons are rapidly labeled through either selective inversion (c) or selective excitation followed by a magnetic manipulation (d, e). This can be gradient dephasing (d) or frequency labeling during an evolution time t_{evol} followed by selective flipback to the z-axis (e). After labeling, exchange transfer to water protons occurs during t_{exch}. The LTMs (n total) are repeated continuously during preparation period t_{prep} to enhance the effect on bulk water. The water-labeling efficiency depends on the exchange rate, which, together with the power deposition limits, determines the number of modules that can be used. Reprinted with permission from Ref. [12], Copyright 2004, John Wiley and Sons.

LTM can be repeated several times (n) to build up the fraction of labeled protons in the water pool. Since excitation is used for labeling, the labeling period can be made as short as hardware and SAR limitations allow, and thus it is possible to label rapidly exchanging protons or components [36] with extremely short T_{2*} [37]. Several pulsed methods have also shown the capability to filter CEST contrast based on exchange rate using parameters such as the pulse flip angle [38, 39], t^1_{exch} or t_{evol} [40, 41].

6.2.5 Shape of RF Pulses

One very important consideration for pulsed saturation transfer is the shape of the RF pulses. To date, a number of waveforms have been tested, including Fermi [42, 43], e-burp [44, 45], Gaussian [46, 47], SEDUCE [48], d-SNOB [48], and Blackman-shaped inversion pulses [49]. However, a systematic comparison of the efficiency for different waveforms has not been performed, and as a result, there is no well-established optimal waveform. The Z-spectra produced by a particular waveform can be simulated either using Bloch equation [50] or density matrix based approaches [51]. Compared to simulations of CW saturation transfer, simulating shaped pulse saturation transfer requires a number of additional parameters: pulse shape and duration as defined by the time-dependent function $B_1(t)$, t_{exch}, and number of pulses (n). Based on this, considerations are: flip angle of saturation pulse (θ), which is defined by the net effect of RF pulse on a spin over the period of t_p, and duty cycle, which is defined by $t_p/(t_p + t_{mix})$.

Similar to CW experiments, the B_1 of shaped pulses also requires optimization based on the exchange rate of the protons of interest. A simple way to estimate the optimal B_1 for shaped pulses is to calculate their CW-equivalent average B_1 [30, 39]. For example, Zu et al. have derived the average B_1 power ($B_{\text{avg\_power}}$) in their recent work [39] as:

$$B_{\text{ave\_power}} = \sqrt{\frac{1}{t_p + t_{mix}} \int_0^{t_p} B_1^2(t) dt} = \frac{\pi \theta}{180 \cdot \gamma p_1} \sqrt{\frac{p_2}{(t_p + t_{mix})t_p}}$$

(6.18)

where p_1 is the ratio of the average amplitude to the maximum amplitude and p_2 is the ratio of the average of the square of the amplitude to the square of the maximum amplitude, respectively, for the shaped pulse. p_1 and p_2 are well defined for a specific shape, for example, $p_1 = 0.416$, $p_2 = 0.295$ for a Gaussian pulse.[39] The parameters then can be adjusted interactively to achieve the optimal average B_1 power ($B_{\text{avg\_power}}$) for a given $B_1(t)$ function.

More exotic pulse shapes, which involve more than just changes in amplitude, can also be exploited. One nice example is two-frequency irradiation pulse trains. Irradiating at two separate frequencies simultaneously can be useful to detect CEST contrast [52, 53]. Two-frequency irradiation can be accomplished directly by alternating every other saturation pulse in a pulse train between two frequencies on opposite sides of the water resonance to obtain a signal of $S_{sat}(\omega_+ + \omega_-)$ [52]. Another method to accomplish this is by using phase-modulated pulses to generate multi-frequency excitation or inversion pulses [53, 54]. A detailed treatment of considerations when using this irradiation scheme, including how to post-process the resulting Z-spectra, is included in Chapter 7.

6.2.6 Utilizing t_{exch} in LTMs to Extract CEST Contrast

Saturation transfer is not dependent only on the shape of the labeling pulses. As a result, delay times such as t_{exch} in pulsed-CEST methods can be varied in a systematic fashion to manipulate the contrast in images. As mentioned above, the primary purpose of t_{exch} is to allow labeled spins to exchange with the bulk water pool. McMahon et al. [55] showed that CEST contrast could be enhanced for longer CW pulses through the increased number of labeled spins that exchange with water. The buildup curves in that particular study (QUEST) were used to determine exchange rates of the solute protons; however, as revealed in Xu et al. [1], collecting CEST contrast buildup curves for a couple of t_{exch} enables filtration of CEST contrast data, providing a relatively simple and rapid method for obtaining CEST images. This was termed the variable delay multi-pulse (VDMP) technique, with the buildup curves different from the QUEST buildup curves in that the number of pulses, and hence the total saturation time, does not change (although the length of the saturation module does). Hence, for slow-exchanging solute protons, the CEST signal buildup curves increase initially due to requiring a longer time to exchange with water and then gradually decay due to the effects of longitudinal relaxation and back exchange. For fast-exchanging protons, saturated protons are almost completely transferred during the labeling pulse and no additional saturation is transferred during t_{exch}. Thus, for rapidly exchanging protons, the VDMP-CEST signal starts at a maximum value and then drops with increasing t_{exch} due to longitudinal relaxation and back exchange, and for sufficiently fast exchange, this signal intensity

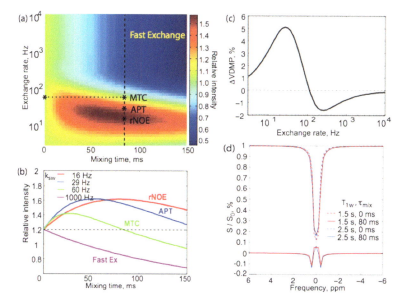

Figure 6.3 Bloch simulations: (a) effects of mixing time and exchange rate (log plot) on the normalized CEST signal intensity $S(\tau_{mix})/S(\tau_{mix}=0)$ (color scale) for the VDMP sequence; (b) projections of the VDMP-CEST signal ratio as a function of mixing time for four exchange rates corresponding approximately to typical rNOE (16 Hz), APT (29 Hz), MTC (60 Hz), and fast exchange (1000 Hz). Reprinted with permission from Ref. [56], Copyright 2015, John Wiley and Sons.

decay becomes independent of the exchange rate and decays due to the relaxation time constant T_1. Characteristic buildup and decay curves of VDMP-CEST signal for protons with different exchange rates are shown in Fig. 6.3, where a two-pool model was simulated with 16 saturation pulses at 3.5 ppm, each with duration of 20 ms and peak B_1 of 1.5 µT.

In order to separate exchange rates, the following procedure can be performed. Two images are acquired with two different mixing times: $\tau_{exch} = 0$ ($S_{1,\tau_{exch}=0}$) and τ_{null}, ($S_{2,\tau_{exch}=t}$), for which the faster exchange rate contribution to signal intensity is equal. The difference image of the two individual images with a new resultant CEST signal contrast is called the VDMP difference, with the signal intensity defined by [56].

$$\Delta VDMP(t2, t1) = (S_{2,\tau_{exch}=t2} - S_{1,\tau_{exch}=t1}) \quad (6.19)$$

6.2.7 Utilizing Labeling Flip Angle to Filter Contrast

Another method to manipulate saturation transfer is to modify the labeling pulse flip angle. Based on this realization, a pulsed-CEST method has been described based on collecting data using two separate flip angles ($\theta = \pi, 2\pi$) and termed chemical exchange rotation transfer (CERT) [38, 57]. The flip angles are modified by increasing the duration of the saturation pulse and not the average field strength, which will result in differences in the saturation efficiency for protons with different exchange rates. For protons that exchange slower than the duration of the labeling pulses, maximal saturation efficiency will occur at $\theta = \pi$, and a minima will occur at $\theta = 2\pi$. As exchange rate increases, or T_2 reduces, the impact of θ on saturation efficiency is reduced. This has been simulated by Zu et al. [38], with an alternative formula for MTR_{asym} proposed for analyzing this data:

$$MTR_{double} = \left. \frac{S_+(2\pi) - S_+(\pi)}{S_o} \right|_{B_{avg\ power}} \quad (6.20)$$

where $S_+(2\pi)$ is the water signal intensity with saturation pulse on resonance with the exchangeable protons and $\theta = 2\pi$. In practice, direct water saturation is also a consideration in adjusting the saturation field strength and duration as well.

6.2.8 OPARACHEE

While the pulsed methods described earlier are based on centering the labeling pulses on exchangeable solute protons, another option is to develop methods with the labeling pulses centered on bulk water. One example is on-resonance paramagnetic agent chemical exchange effect (OPARACHEE) [58], with a train of weak pulses with their phases cycled in a manner to produce a net 360° nutation for spins that remain within a band around the center of the spectrum (i.e., water protons that do not exchange). Decreases in Z magnetization are produced by the exchange of water protons to solute sites resonating outside this envelope, resulting in a <360° nutation for these protons. Low-power WALTZ-16 composite pulse trains are well suited for this purpose and compensate for B_0 inhomogeneities present in real imaging applications.

This method was first evaluated for detecting accumulation of the TmDOTA-4MC-paraCEST agent in the kidney of mice [59]. Contrast was detected based on monitoring the signal with waltz16* pulses versus signal without this preparation.

6.2.9 FLEX

A completely different method of measuring the CEST effect is by encoding the chemical shift evolution of exchanging solute molecules using a binomial pair of excitation pulses. As mentioned above, the FLEX signal is collected as a function of the time variables (t_{evol} or t_{mix} in Fig. 6.2). This signal can be either Fourier transformed to obtain a spectrum or quantified by fitting the time domain signal. FLEX offers some important advantages over saturation transfer. These advantages include the capability to simultaneously excite multiple groups of protons using the excitation pulses, and since the water frequency is also measured, the FLEX measurement contains an internal B_0 reference, thus avoiding the need to acquire additional B_0 maps. Further, the short excitation pulses can efficiently label rapidly exchanging protons [36] Lin et al. [37] showed in a study that even components decaying with a rate above 100,000 s^{-1} ($k_{sw} + R_2^*$) can be observed using the FLEX method. The exchange time is analogous to the mixing time in VDMP experiment and thus, together with the excitation pulse length, can be used to filter out exchanging components based on their exchange and transverse relaxation rates.

Since FLEX data are acquired as a function of either the t_{evol} and/or t_{mix}, powerful time domain analysis methods can be used to analyze and interpret FLEX data. For FLEX experiments acquired as a function of t_{evol}, the water signal intensity ($S(t_{evol})$) is described by

$$S(t_{evol}) = \sum_s PTR_s \cdot \exp[-(k_{sw} + R_{2s}^*)t_{evol}] \cdot \cos(\Delta\omega_{s,o1} \cdot t_{evol} + \phi_s)$$

(6.21)

where R_{2s}^* is the effective transverse relaxation rate of the solute pool (s), $\Delta\omega_{s,o1}$ is the frequency offset of the labeling pulse from the solute proton, ϕ_s is the phase, and the PTR$_s$ is the magnitude of each component given by

$$\text{PTR}_s x_s \cdot \lambda_s \cdot [1-\exp(-\kappa_{sw} \cdot t_{exch})] \cdot \sum_{i=1}^{n} \exp\{[-1+(i+1)/n] \cdot t_{prep}/T_{1w}\} \quad (6.22)$$

where λ_s is the excitation efficiency of the binomial pair of FLEX excitation pulses. From Eq. 6.22, we can see that the FLEX signal modulates as a superposition of all the labeled solute magnetization that has transferred to the water pool. Fitting Eq. 6.22 to experimental data allows the easy separation of components that may overlap in the frequency domain but have different decay rates ($k_{sw} + R_{2s}^*$).

6.2.10 Alternative Ways for CEST Acquisition

The typical pre-clinical CEST experiment includes a preparation/saturation period followed by an image acquisition readout period. In the preparation/saturation period, the desired contrast is generated by varying either a combination of the frequency, B_1 field strength, duration of the RF pulses, or inter-pulse delay. Most advances in CEST MRI techniques to date have in the optimization of these saturation/preparation pulse parameters for specific exchangeable species with different chemical shifts and/or different exchange rates. Recently, there have been several advances in the acquisition strategies of CEST spectra/images. Some of these acquisition strategies will be outlined in this section.

6.2.10.1 LOVARS

The Length and Offset VARied Saturation (LOVARS) method is a little different in that it can be employed either long CW pulses or shorter pulses for signal acquisition. The idea is that in order to discriminate CEST contrast from MTC, the saturation length (t_{sat}), number of pulses, or shape of saturation pulses can be varied in a systematic fashion to create saturation-transfer contrast patterns that depend on the chemical exchange properties of exchanging protons and then processing the resulting images using the fast Fourier transform (FFT), the general linear model (GLM), or another signal-conditioning method to discriminate CEST contrast agents from endogenous MTC or CEST contrast in tissue. At this stage, this approach has been used to highlight tumor tissue using either a RARE image acquisition [60] or a low flip-angle gradient-echo readout with flip back pulse [61].

6.2.10.2 Steady-state CEST

Steady-state CEST is referred to an approach that employs short repetitive segments consisting of a short saturation period and a quick imaging acquisition. The signal collected within the first several segmentations is discarded because saturation is being built up. Once a saturation steady state is reached, the loss of saturation due to T_1 relaxation during the time period of image acquisition can be effectively compensated by a short saturation period preceding the next image acquisition period.

Steady-state CEST provides an alternative way to obtain efficient saturation without the need to use a long RF pulse prior to every imaging acquisition step (each k-space line). This has a number of advantages. First, it makes CEST preparation applicable for a variety of fast imaging pulse sequences that employ short repetition time (TR). For example, Liu et al. [62] showed that the temporal resolution of CEST MRI can be markedly improved by filling the TR times in a FLASH sequence with short repetitive saturation pulses, which produced a more than 20 times faster acquisition times than the traditional gradient-echo-based method. This approach is particularly useful for monitoring the dynamic change of CEST at a single offset [62]. In addition, it can be easily implemented on a human MRI scanner where the implementation of a long, high B_1 RF pulse is difficult due to SAR restrictions. For example, Jones et al. have successfully used this approach to measure the 3D whole-brain APT signal in normal human subjects on a 7T human scanner [27]

Another advantage of steady-state CEST is the ability to reduce SAR since only short compensatory saturation pulses are needed once the steady state is reached. It was first demonstrated by Sun et al. using an acquisition method that utilizes uneven segmented

RF irradiation [63]. In this approach, a long RF pulse was first applied to generate the steady-state CEST contrast and after each slice acquisition, a short secondary RF pulse was applied to maintain CEST contrast at the previous saturation steady-state level. Since the secondary RF pulse can be quite brief, the repetition time for multi-slice acquisitions can be significantly reduced compared to the conventional CEST method. The steady-state approach can also be utilized to maintain constant saturation if a time interval has to be inserted between the long saturation pulse and the imaging acquisition. This can be exemplified by a recent CEST study of lung fibrosis, in which respiration gating has to be used to avoid respiration-induced motion artifacts [2]. The authors used a short RF pulse immediately before each imaging acquisition to effectively compensate the CEST signal loss during the waiting time for respiration stabilization.

A drawback of the steady-state approach is, however, that the efficiency of saturation may be relatively low. The efficiency of the steady-state saturation is determined by the water T_1 relaxation time and the time interval for pure water relaxation and exchange rate [43, 62]. Therefore, caution has to be taken in applying steady-state CEST.

6.2.10.3 SWIFT-CEST

The sweep imaging with Fourier transfer (SWIFT) technique was introduced to CEST acquisition when Soesbe et al. developed a class of Eu^{3+}-based paraCEST agents [64]. They observed a significant reduction in bulk water T_2 due to the T_2 exchange (T_{2ex}) mechanism when studying the exchange properties of these agents. When using conventional imaging sequences such as gradient echo or fast spin echo, the minimal echo time still results in a significant reduction in signal in regions of agent uptake. This reduction in signal makes the difference between the CEST ON image (saturation at ω_+) and the reference OFF image (saturation at $-\omega_-$). A sequence that utilizes ultrashort echo time (TE) is needed to prevent a loss of bulk water signal due to T_{2ex}.

The SWIFT technique uses frequency swept RF pulses interleaved with nearly simultaneous signal acquisition. Therefore,

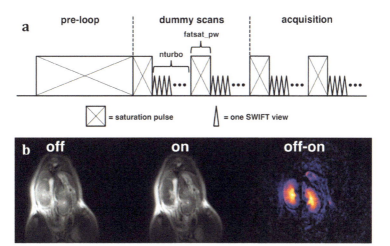

Figure 6.4 (a) An illustration of the SWIFT-CEST pulse sequence. The modified fat saturation pulses are included in both the dummy scans and the acquisition scans. An optional pre-loop saturation pulse (up to 5 s long) was added to ensure steady-state saturation before the imaging pulses. (b) SWIFT-CEST image after a 1.0 mmol/kg intravenous dose of EuDOTA-(gly)$_4^-$. Reprinted with permission from Ref. [64], Copyright 2011, John Wiley and Sons.

the TE can be extremely short (<10 μs). The technique was first published in 2006 by Garwood et al. [65] To incorporate the technique for CEST experiments, the fat saturation pulse in the original SWIFT technique is used as the frequency-selective saturation pulse (Fig. 6.4). After each saturation pulse, a few spokes in the spherical k-space spiral are acquired. The number of k-space spokes (SWIFT views) can be optimized and usually ranges from 16 to 32. The saturation and acquisition are repeated until the requested number of SWIFT views is acquired.

When studying the in vivo uptake of paraCEST agent EuDOTA-(gly)$_4^-$, a significant T_{2ex} effect was observed. However, using the optimized SWIFT-CEST technique, the uptake of agent by the kidney can be clearly seen in the CEST difference image, which is not readily observed using conventional multi-slice fast spin echo sequence. In addition to delineating T_{2ex} effect from CEST effect and improving SNR, SWIFT-CEST also reduces the motion and flow artifacts and is robust against B_0 inhomogeneities from changes in susceptibility.

6.2.10.4 Ultrafast gradient-encoded Z-spectroscopy

CEST experiments can be time consuming since a conventional Z-spectrum is generated by saturating protons at many discrete frequencies and recording the polarization of bulk water with and without RF saturation as a function of saturating frequency offset. Back in the early 1990s, Swanson [66] utilized a gradient method to obtain the MT Z-spectrum, originally referred to as a cross-relaxation spectrum, using just two acquisitions. However, due to the strong dipolar couplings of the semi-solid protons, the MT Z-spectra usually needs to cover a large range of frequencies. Therefore, the method is very demanding on gradient strength. Recently, Xu et al. [67] and Döpfert et al. [68] demonstrated that the gradient-encoding method is well suited to CEST studies due to the relatively narrow spectral width compared to MT. By spatially encoding a homogeneous sample using magnetic field gradients during the saturation pulse, a full Z-spectrum can be recorded with just two scans eliminating the process of mapping saturation signal at different frequencies. This method greatly expedites the acquisition of CEST Z-spectra, hence dubbed ultrafast Z-spectrum (UFZ) methodology.

Figure 6.5 shows the simplest implementation of the UFZ technique. When a gradient is switched on, the precession frequency of the spins becomes a function of their spatial location. For example, as illustrated in Fig. 6.5(b), a slice of the sample at a distance d from the center of the gradient experiences an additional external field (B_{sat}) amounting to

$$\Delta B_{sat}(d) = G_{sat} d, \tag{6.23}$$

where G_{sat} is the strength of the gradient during saturation. The resonance frequency of this slice will be shifted by

$$\Delta \omega_{sat}(d) = \gamma G_{sat} d, \tag{6.24}$$

where γ is the gyromagnetic ratio of the nucleus of interest. If the saturation RF irradiation is set on-resonance, only the central slice perpendicular to the gradient direction will experience direct saturation, while all other slices along the gradient direction experience off-resonant irradiation at a frequency of $\Delta \omega_{sat}(d)$. If the exchange process takes place at certain frequencies, the intensity

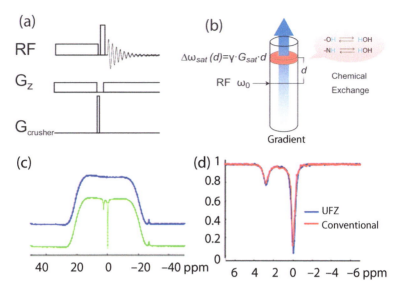

Figure 6.5 (a) Sequences for acquiring a UFZ-spectrum. (b) An illustration of the saturation frequency of a slice in the sample during the saturation period. (c) Reference spectrum without saturation (upper curve) and the raw UFZ-spectrum (lower curve). (d) A comparison of the conventional Z-spectrum and the UFZ-spectrum. Reprinted with permission from Ref. [67], Copyright 2015, John Wiley and Sons.

of the signal from the slices corresponding to those frequencies will be reduced. The profile containing the saturation and exchange information is then read out by applying a gradient G_{acq} during the acquisition period. Notice that this readout gradient does not need to be equal in strength to the saturation gradient, since

$$\Delta\omega_{acq}(d) = \gamma G_{acq} d = -(G_{acq}/G_{sat})\Delta\omega_{sat}(d) \qquad (6.25)$$

In both conventional and UFZ-spectra acquisition, a scan without the RF saturation is commonly acquired for normalization to remove the proton density. When normalizing the saturation spectrum to the reference spectrum, the proton density differences caused by sample shape irregularities, field inhomogeneity, and differences in coil sensitivities can be removed.

Depending on the hardware, if the gradient ramp up/down time is relatively long, it might be necessary to acquire an echo since the FID signal acquired during the gradient ramp up time maybe distorted. In principle, an echo acquisition also improves the SNR of the spectrum [68].

Due to its simplicity, the method can be readily incorporated into methods developed to study hyperpolarized systems and into existing imaging acquisition techniques [69, 70] On the other hand, each frequency of the UFZ spectrum corresponds to the location of labeled proton density instead of the whole sample volume. The SNR of the UFZ spectrum is inherently lower than the conventional Z-spectrum of the same sample. For phantom studies, this usually would not be an issue since sufficient SNR can always be achieved by adjusting the concentration or saturation parameter. However, the SNR issue will affect in vivo applications of the UFZ method, since a homogeneous region needs to be found and often these regions are too small to provide sufficient SNR for UFZ acquisition. Another factor to take into account is the gradient amplifier duty cycle since the gradient field needs to be switched on during the saturation period. Saturation is sometimes applied as a pulse train when the RF amplifier duty cycle reaches its limit. The same principle would apply to the gradient amplifier as well. Before using strong gradients over a long saturation duration, one must make sure that a protection mechanism is in place to avoid damaging the gradient amplifier.

6.3 Post-Processing

The post-processing of CEST MRI data plays a crucial role in the visualization and possible quantification of CEST contrast. Because CEST contrast is an offset-dependent behavior, the first thing one should do after acquiring a set of CEST data over a range of frequency offsets is to correct the shift of offsets spatially using a pre-acquired B_0 map or by fitting the frequency dependence of the image intensity in each voxel to a lineshape. Without this step, the acquired CEST contrast at a nominal offsets may present the contrast at other offset, and, even worse, the magnitude of the error varies spatially. The second step in the post-processing is to extract the CEST contrast from other effects that could interfere with the accuracy of CEST quantification. Finally, to differentiate the CEST contrast from interfering signals and to increase the CEST signal to noise, a number of model-free data-processing steps can be added. It should be noted that, due to the limited scope of this chapter, only post-processing procedures for the conventional CEST acquisition will be discussed.

6.3.1 B_0 Correction

More or less, B_0 inhomogeneity, the shift in the center frequency of water protons from their resonance frequency (often set to zero in MRI studies), exists in almost all in vitro and in vivo MRI acquisitions. This could cause a serious problem in the quantification of CEST contrast because, to a certain extent, the actual offsets of saturation pulses are different from the nominal values, and the signal differences vary spatially. Therefore, CEST MRI often requires correcting B_0 inhomogeneity in the post-data-processing steps to achieve an accurate quantification. Two common approaches are given here.

6.3.1.1 Using a pre-acquired B_0 map

The most commonly used B_0 correction strategy for CEST imaging is to retrospectively correct the offsets of each voxel using a pre-acquired B_0 map. Gradient-echo phase imaging[71] is the method most widely used to determine local B_0 shifts. The local B_0 shift in each voxel is determined by the measured phase Φ and echo time TE by a simple linear relation described by

$$\Phi = \lambda T_\mathrm{E} \Delta B = \lambda T_\mathrm{E} (B_\mathrm{loc} - B_0) \qquad (6.26)$$

However, using phase-mapping methods for the purpose of B_0 correction in a CEST MRI study may require an additional step of image registration when the CEST data collection uses a different pulse sequence. Another method WAter Saturation Shift Referencing (WASSR) has been developed specifically for B_0 correction, which employs the same pulse sequence for B_0 mapping and CEST data collection [72]. The basic principle is that the saturation of water signal is only dominated by water direct saturation (DS), not CEST or other mechanisms such as MTC, if a weak saturation field is applied, which minimizes the other contributions. The frequency-dependent DS can then be described by a Lorentzian lineshape (Eq. 6.28), in which maximal saturation occurs when the saturation offset equals the B_0 shift.

$$S_Z^{SS} = \frac{S_0^{SS}}{1 + \left(\dfrac{\omega_1}{\Delta\omega - (B_{\text{loc}} - B_0)}\right)^2 \dfrac{T_1}{T_2}} \quad (6.27)$$

where S_Z^{SS} is the steady-state longitudinal magnetization, and S_0^{SS} is the initial longitudinal magnetization, ω_1 is the strength of the applied RF pulse, which is resonating at $\Delta\omega$ (frequency with respect to the proton center frequency), and ΔB_0 is the B_0 shift.

In practice, the assumption of signal losses produced mostly through DS can be fulfilled using a CW pulse at ~0.5 µT, t_{sat} ~500 ms [73]. The ΔB_0 at each voxel can be determined using a non-model-based maximal symmetric fitting algorithm [72] or a Lorentz lineshape fitting algorithm [74].

Based on the measured B_0 map, we can then correct the offsets of the acquired CEST data set from their nominal values to the true ones, effectively moving the center of Z-spectrum for each pixel back to 0 Hz. In order to determine the signal at the offsets of interest, the data set will be interpolated into a new data set corresponding to the offset(s) of interest using data-processing tools such as MATLAB, IDL, or Python. A simple spline interpolation (in MATLAB) often works well, especially when the SNR is sufficiently high [74]. When the data are noisy, sophisticated interpolation methods may be needed. For example, high-order polynomial fitting and interpolation of Z-spectra have been shown to significantly improve spectrum resolution (up to 1 Hz) [75]. Another approach is to use a smoothing spline-based algorithm to process noisy Z-spectra [76]. While these model-free interpolation methods are time efficient and widely used in CEST studies, model-based fitting methods may also be of great interest, which will be introduced in the next section.

6.3.1.2 Fitting Z-spectral data

When the central peak in the Z-spectrum is sharp (a relatively weak saturation field strength used, i.e., <2 µT), or if the CEST agents exchangeable protons are sufficiently far from water so that small errors in B_0 correction will not cause problems (MTC and DS contrast are small) [76], one can directly fit the acquired Z-spectral data and move the center of the spectrum to zero, providing a much simpler way to correct B_0 inhomogeneity than that based on B_0 map. Such an approach has been shown robust in some particular applications. For example, one can fit the saturation profile using a Lorentzian lineshape accounting for the DS component [77]. The CEST effect can be extracted by subtracting the actual experimental Z-spectrum and the fitted DS spectrum. This is the case when MTC is negligible such as in aqueous phantom samples [46] or tissues irradiated with a low saturation power [78]. It is worth noting that the CEST proton pool can also be modeled by a Lorentzian function [46] and so does MT pool (under the weak-saturation-pulse approximation). As a result, one can linearly combine two or multiple models together to extract the CEST effect out of experiment data [79–81].

In addition to fitting the Z-spectrum, as it is acquired in the frequency domain, CEST data can be analyzed in the time domain. Yadav et al. [82] demonstrated that by modeling the Z-spectrum as a sum of Lorentzian or Gaussian lineshapes, large nuisance signal components could be fitted out in the time domain signal more consistently in the time domain. This approach called time domain removal of irrelevant magnetization (TRIM) was demonstrated on a wide range of data acquired from phantoms and in vivo. Due to the linearity between the frequency and time domains, fitting the data in either should be analogous. However, the more consistent results from TRIM could be due to the natural scaling of large signal components in the time domain, noting that the amplitude of each signal component in the time domain is the integral of that component in the frequency domain.

6.3.2 Asymmetry Analysis

Mathematically, the total saturation of water $S_{\text{sat}}(\Delta\omega)$ can be described by

$$\frac{S_{\text{sat}}(\Delta\omega)}{S_0} = CEST^{\text{exp}}(\Delta\omega) + CEST^{\text{endo}}(\Delta\omega) + MTC(\Delta\omega) + DS(\Delta\omega)$$

(6.28)

in which "exo" and "endo" refer to exogeneous and endogenous contributions. Given the fact that DS is symmetric with respect to the water resonance and MTC often assumed to be symmetric as well, one can use MTR_{asym} defined to separate the CEST effect from MT and DS:

$$MTR_{asym} = \frac{S_{sat}(-\Delta\omega) - S_{sat}(\Delta\omega)}{S_{0w}} = CEST^{exo}(\Delta\omega) + CEST^{endo}(\Delta\omega)$$
(6.29)

If the measured MTR_{asym} was caused purely by exchange, it would equal the PTR (Eq. 6.3). Alternatively, MTR_{asym} can be calculated by replacing the denominator S_0 with $S_{sat}(\Delta\omega)$:

$$MTR_{asym} = \frac{S_{sat}(-\Delta\omega) - S_{sat}(\Delta\omega)}{S_{sat}(-\Delta\omega)}$$
(6.30)

Using this definition, DS and MTC on the opposite side of water from CEST contrast are included in the denominator, resulting in an MTR_{asym}, which is "normalized" according to the signal losses produced by DS and MTC at $-\Delta\omega$. This definition produces higher MTR_{asym} values than Eq. 6.30. The advantage of using this definition is that, assuming the same macromolecular content, the impact of B_1 inhomogeneity can be reduced because, intuitively, an increase or decrease in B_1 will result in a raising or lowering in the denominator through an increase or decrease in MTC. While MTR_{asym} is the most widely used metric to quantify CEST effects, it has several intrinsic weaknesses. First, as many studies have shown, MTC is indeed not symmetric with respect to water resonance [83], making MTR_{asym} data contaminated by MTC in the frequency range close to the water resonance. This is problematic for most diaCEST and lipoCEST agents whose offsets are typically less than 5 ppm from water. In addition, MTR_{asym} is not a physically meaningful parameter and is not linearly correlated with the concentration or exchange rate. As such, a number of alternatives have been proposed to improve the quantification of CEST contrast.

6.3.3 Integration of CEST Effect over a Range of Offsets

Instead of comparing the saturated water signal at particular offsets, one can compare the signal over a range of offsets, presumably with a higher SNR [84, 85]. For example, based on a given Z-spectrum, one can determine the areas under the spectrum around the CEST peak (A_{INT_PEAK}) or its complimentary (above the spectrum, $1 - A_{INT\_PEAK}$), or determine the areas under the spectrum from the CEST peak to water resonance (A_{INT_BULK}) or its complimentary (above the spectrum, $1 - A_{INT\_BULK}$). Moreover, a so-called enhanced mode, MTR^{ENC} with $[S(-\Delta\omega) - S(+\Delta\omega)]/S(-\Delta\omega)$, can also be used. This is the inverse version of MTR_{asym} defined by Eq. 6.31 with a relation of $MTR^{ENC} = 1/MTR_{asym} - 1$. It is indicated that the enhanced mode is more suitable for high CEST contrast, i.e., CEST effect >30%. While all the approaches using an integrated area have been shown to improve the CNR, the approach using the area under spectrum (INT_PEAK) is believed more suitable for paraCEST agents with distant CEST peaks, and that using the areas above the spectrum (complimentary) may be ideal for diaCEST agents with small CEST effect and close to water resonance.

6.3.4 Non-MTR$_{asym}$ Metrics

While MTR_{asym} is the most widely used metric for CEST quantification, it has an inherent disadvantage: MTR_{asym} is not equal to PTE and is also not linearly dependent on either the solute concentration (i.e., the thermal equilibrium magnetization of CEST proton pool (M_{0b}), or $n[CA]$, where n is the number of exchangeable protons of each agent, or the fraction of exchangeable protons, $f = n[CA]/[H_2O]$) or exchange rate (k_{ex}). For example, in one of the earliest works carried out by Ward et al. [86], rearrangement has to be made from Eq. 6.32a to Eqs. 6.32b,c in order to directly compare the ratio of exchange rates of the two-type exchangeable protons on the same contrast agent.

$$\frac{M_S}{M_0} = \frac{1}{1 + n \cdot [CA] \cdot \kappa_{SW} T_{1w}} \quad (6.32a)$$

$$\left(\frac{M_0}{M_S} - 1\right) = [CA] \cdot n \cdot T_{1sat} \cdot \kappa_{SW} \quad (6.32b)$$

$$\frac{\left(\frac{M_0}{M_S} - 1\right)_{site1}}{\left(\frac{M_0}{M_S} - 1\right)_{site2}} = \frac{[CA] \cdot n_{site1} \cdot T_{1sat}}{[CA] \cdot n_{site2} \cdot T_{1sat}} \cdot \frac{\kappa_{SW}^{site1}}{\kappa_{SW}^{site2}} = \text{constant} \cdot \frac{\kappa_{SW}^{site1}}{\kappa_{SW}^{site2}}$$

$$(6.32c)$$

In a later study by Ali et al. [87], the CEST metrics were reformulated as analogs to the Lineweaver–Burke, Eadie–Hofstee, or Hanes equation forms of the Michaelis–Menten equation. This study showed that, regardless of the form of equation used, (M_0/M_S-1) has to be used as a whole in order to obtain a linear calibration of the CEST effect with respect to the concentration of CEST agent. Both the studies above suggested that M_0/M_S-1 should be used when the concentration will be extracted from the CEST measurements.

Very recently, Zaiss et al. have further developed the theoretical framework for CEST MRI [88] and proposed a new metric for CEST quantification, called MTR_{rex} (Eq. 6.33a) [89, 90]. Basically, the authors extend the spin-lock relaxation theory to CEST contrast, and with that, the Z-spectral data at each offset ($\Delta\omega$) now are rewritten in the form of several relaxation terms

$$z(\Delta\omega) = \frac{M_S}{M_0} = \cos^2\theta \frac{R_{1w}}{R_{1\rho}} \tag{6.32a}$$

where $R_{1\rho}$ is the longitudinal relaxation rate of the water pool in the rotating frame upon application of a saturation (or spin lock) RF pulse, defined by $R_{1\rho} = R_{eff} + R_{ex}$. R_{ex} and R_{eff} represent the exchange-dependent and -independent relaxations of the water pool in the rotating frame, respectively.

Assuming a small CEST population ($f_b \ll 1$) that has a sufficiently large chemical shift, the exchange-dependent relaxation term R_{ex} can be approximated as

$$R_{ex}^{max} \approx \alpha \cdot f_b \cdot k_{ex} = \frac{\omega_1^2}{\omega_1^2 + l_{ex}(k_{ex} + R_{2b})} \cdot f_b \cdot k_{ex} \tag{6.32b}$$

where f_b is the fraction of exchangeable protons ($f_b = M_{0b}/M_{0a}$), and α is the labeling factor and approaches 1 for strong B_1 and small R_{2b} and k_{ex}, making Eq. 6.32b

$$R_{ex}^{max} \approx f_b \cdot k_{ex} \tag{6.32c}$$

similar to the basic analytical solution [91, 92]. The non-exchange term R_{eff} is determined by the inherent properties of water pool (T_1, T_2, MT, $\Delta\omega$, and ω_1), and hence can be approximately cancelled by its counterpart at the opposite offset, i.e., $R_{eff}(\Delta\omega) - R_{eff}(-\Delta\omega) = 0$. Rearrangement of Eq. 6.33a at $+\Delta\omega$ and $-\Delta\omega$ leads to Eq. 6.33d

$$MTR_{rex} = 1/Z(\Delta\omega) - 1/Z(-\Delta\omega) = \frac{T_{1w}}{\cos^2\theta} \cdot f_b \cdot k_{ex} \tag{6.32d}$$

Equation 6.33d can be further simplified into the product of f_b, k_{ex}, and T_{1w}, when $\cos\theta$ approaches 1, which is valid when the chemical shift difference $\Delta\omega \gg \omega_1$. Based on that, a so-called apparent exchange-dependent relaxation (AREX, AREX = MTR_{rex}/T_{1w}) metric can be used if T_{1w} is experimentally measured. Thus, the physical meaningful parameters f_b and k_{ex} can now be directly quantified using AREX, simply by measuring the CEST data at two offsets, $\Delta\omega$ and $-\Delta\omega$ (or three offsets if $-\Delta\omega$ cannot be accurately measured due to the contaminations by other CEST agents) and T_{1w}. The new metrics, MTR_{rex} and AREX, thus may be more powerful than the conventional MTR_{asym} metric, especially for the applications of pH measurement, where k_{ex} is the function of pH. However, as indicated by the authors, statistical error of $1/Z$ is considerably higher than that of Z, which may limit a broad application of MTR_{rex} and AREX, especially for exchangeable protons with small chemical shifts such as the glucose –OH protons. Nevertheless, MTR_{rex} and AREX could be the merit of choice for those exchangeable species with $\Delta\omega \gg k_{ex}$ or ω_1.

6.3.5 Additional Image-Processing Steps

6.3.5.1 SNR/CNR filtering

Voxels with low SNR can compromise the accuracy of data interpolation, which has to be considered during data processing. There are several image-processing methods to address this problem. For instance, a priori determined SNR maps can be used to calculate CNR maps and filter out pixels with low CNR [93–95] based on the assumption that low CNR is due to either low SNR or low contrast. Using a CNR filter has the advantage that not only noisy pixels are filtered out but also those containing a low "background" CEST contrast, allowing the filtered maps to highlight pixels possessing more reliable CEST contrast. However, the CNR threshold needs to be determined manually, which may result in a bias.

6.3.5.2 Filters for image de-noising

Another method that has been employed is the so-called median filter method [93]. which is effective to remove "salt and pepper" noise and can reduce the variations raised from interpolation. Very recently, an R^2 (square of the correlation coefficient) filter was introduced [96]. In this approach, the R^2 for the interpolation curve with respect to the raw data is calculated pixel-wise. Pixels with low R^2, i.e., <0.99, are discarded, assuming noisy Z-spectra produce low R^2 values.

6.3.5.3 MTC-based image filtering

It is possible to use other MRI contrasts to improve the in vivo CEST quantification by discriminating the fake CEST signal induced by certain pathophysiological properties from those by CEST probes. For example, we have proposed a conventional MTC-based segmentation technique for selectively filtering CEST contrast maps based on using the signal at two resonance frequency offsets $\Delta\omega_1$ and $\Delta\omega_2$ to calculate the NOrmalized MAgnetization Ratio (NOMAR) [97], defined by

$$NOMAR(\Delta\omega_1/\Delta\omega_2) = (1 - MTR(\Delta\omega_1))/(1 - MTR(\Delta\omega_2)) \quad (6.33)$$

Simply by acquiring two additional MT-weighted images at saturation frequencies of -12.5 ppm and -50 ppm, the NOMAR values are calculated to differentiate voxels with low MTC (such as fat, CSF, edema, or blood) from voxels in the target tissue using a global threshold determined by histogram analysis. Segmentation techniques based on other types of MRI contrast can also be used to create tissue-selective CEST contrast maps, avoiding the potential complications caused by tissue saturation transfer properties.

6.3.5.4 B_1 correction

While B_1 inhomogeneity may seriously influence the CEST quantification, the correction of B_1 inhomogeneity in CEST contrast is challenging [71]. An empirical approach has been reported recently to improve measurement of CEST contrast in the presence of severe B_1 inhomogeneity [98]. In this approach, B_1 field maps were measured using a double angle method (30° and 60°), which are used to correct the MTR_{asym} maps at the frequency of glutamate using a calibration curve described by a second-order polynomial. It is indicated in this report, however, that the accurate determination of the calibration coefficients highly depends on the saturation and imaging parameters and the type of tissue.

6.4 Conclusion

CEST is a powerful tool for studying molecular systems and for generating MRI contrast. Here we have provided a brief theoretical and experimental framework for understanding CEST experiments. This framework was then applied to some examples of CEST pulse sequences. Also, since CEST data are composed of multiple signal sources, a brief guide for processing CEST experimental data was also given.

References

1. Xu J, Yadav NN, Bar-Shir A, et al. Variable delay multi-pulse train for fast chemical exchange saturation transfer and relayed-nuclear overhauser enhancement MRI. *Magnetic Resonance in Medicine*, 2014; 71(5): 1798–1812.
2. Jones CK, Huang A, Xu J, et al. Nuclear Overhauser enhancement (NOE) imaging in the human brain at 7 T. *Neuroimage*, 2013; 77(0): 114–124.
3. Zaiss M, Kunz P, Goerke S, Radbruch A, Bachert P. MR imaging of protein folding in vitro employing nuclear-Overhauser-mediated saturation transfer. *NMR in Biomedicine*, 2013; 26(12): 1815–1822.
4. Hoffman RA and Forsén S. Transient and steady-state Overhauser experiments in the investigation of relaxation processes. Analogies between chemical exchange and relaxation. *Journal of Chemical Physics*, 1966; 45(6): 2049–2060.
5. McConnell HM. Reaction rates by nuclear magnetic resonance. *Journal of Chemical Physics*, 1958; 28(3): 430–431.
6. Zhou J, Wilson DA, Sun PZ, Klaus JA, van Zijl PCM. Quantitative description of proton exchange processes between water and endogenous and exogenous agents for WEX, CEST, and APT experiments. *Magnetic Resonance in Medicine*, 2004; 51(5): 945–952.
7. Sun PZ, van Zijl PCM, and Zhou JY. Optimization of the irradiation power in chemical exchange dependent saturation transfer experiments. *Journal of Magnetic Resonance*, 2005; 175(2): 193–200.

8. Woessner DE, Zhang S, Merritt ME, and Sherry AD. Numerical solution of the Bloch equations provides insights into the optimum design of PARACEST agents for MRI. *Magnetic Resonance in Medicine*, 2005; 53(4): 790–799.
9. Li AX, Hudson RH, Barrett JW, Jones CK, Pasternak SH, and Bartha R. Four-pool modeling of proton exchange processes in biological systems in the presence of MRI-paramagnetic chemical exchange saturation transfer (PARACEST) agents. *Magnetic Resonance in Medicine*, 2008; 60(5): 1197–1206.
10. McMahon MT, Gilad AA, Zhou J, Sun PZ, Bulte JW, and van Zijl PC. Quantifying exchange rates in chemical exchange saturation transfer agents using the saturation time and saturation power dependencies of the magnetization transfer effect on the magnetic resonance imaging signal (QUEST and QUESP): Ph calibration for poly-L-lysine and a starburst dendrimer. *Magnetic Resonance in Medicine*, 2006; 55(4): 836–847.
11. van Zijl PCM, Jones CK, Ren J, Malloy CR, and Sherry AD. MRI detection of glycogen in vivo by using chemical exchange saturation transfer imaging (glycoCEST). *Proceedings of the National Academy of Sciences USA*, 2007; 104(11): 4359–4364.
12. van Zijl PCM and Yadav NN. Chemical exchange saturation transfer (CEST): What is in a name and what isn't? *Magnetic Resonance in Medicine*, 2011; 65(4): 927–948.
13. Ward KM, Aletras AH, and Balaban RS. A new class of contrast agents for MRI based on proton chemical exchange dependent saturation transfer (CEST). *Journal of Magnetic Resonance*, 2000; 143(1): 79–87.
14. Goffeney N, Bulte JWM, Duyn J, Bryant L.H, Jr., and van Zijl PCM. Sensitive NMR detection of cationic-polymer-based gene delivery systems using saturation transfer via proton exchange. *Journal of the American Chemical Society*, 2001; 123(35): 8628–8629.
15. Zhang S, Winter P, Wu K, and Sherry AD. A novel europium(III)-based MRI contrast agent. *Journal of the American Chemical Society*, 2001; 123(7): 1517–1518.
16. Aime S, Barge A, Delli Castelli D, et al. Paramagnetic lanthanide(III) complexes as pH-sensitive chemical exchange saturation transfer (CEST) contrast agents for MRI applications. *Magnetic Resonance in Medicine*, 2002; 47(4): 639–648.
17. Gilad AA, McMahon MT, Walczak P, et al. Artificial reporter gene providing MRI contrast based on proton exchange. *Nature Biotechnology*, 2007; 25(2): 217–219.

18. McMahon MT, Gilad AA, DeLiso MA, Berman SM, Bulte JW, and van Zijl PC. New "multicolor" polypeptide diamagnetic chemical exchange saturation transfer (DIACEST) contrast agents for MRI. *Magnetic Resonance Medicine*, 2008; 60(4): 803–812.
19. Snoussi K, Bulte JW, Guéron M, and van Zijl PC. Sensitive CEST agents based on nucleic acid imino proton exchange: Detection of poly(rU) and of a dendrimer-poly(rU) model for nucleic acid delivery and pharmacology. *Magnetic Resonance in Medicine*, 2003; 49(6): 998–1005.
20. Wu YK, Zhou YF, Ouari O, et al. Polymeric PARACEST agents for enhancing MRI contrast sensitivity. *Journal of the American Chemical Society*, 2008; 130(42): 13854–13855.
21. Aime S, Delli Castelli D, and Terreno E. Highly sensitive MRI chemical exchange saturation transfer agents using liposomes. *Angewandte Chemie*, 2005; 117(34): 5649–5651.
22. Zhao JM, Har-el YE, McMahon MT, et al. Size-induced enhancement of chemical exchange saturation transfer (CEST) contrast in liposomes. *Journal of the American Chemical Society*, 2008; 130(15): 5178–5184.
23. Liu G, Moake M, Har-el YE, et al. In vivo multicolor molecular MR imaging using diamagnetic chemical exchange saturation transfer liposomes. *Magnetic Resonance in Medicine*, 2012; 67(4): 1106–1113.
24. Trott O and Palmer III AG. R1ρ relaxation outside of the fast-exchange limit. *Journal of Magnetic Resonance*, 2002; 154(1): 157–160.
25. Jin T, Wang P, Zong X, and Kim S-G. Magnetic resonance imaging of the Amine-Proton EXchange (APEX) dependent contrast. *Neuroimage*, 2012; 59(2): 1218–1227.
26. Zaiss M and Bachert P. Exchange-dependent relaxation in the rotating frame for slow and intermediate exchange: Modeling off-resonant spin-lock and chemical exchange saturation transfer. *NMR in Biomedicine*, 2013; 26(5): 507–518.
27. Jones CK, Polders D, Hua J, et al. In vivo three-dimensional whole-brain pulsed steady-state chemical exchange saturation transfer at 7 T. *Magnetic Resonance in Medicine*, 2012; 67(6): 1579–1589.
28. Xu J, Yadav NN, Bar-Shir A, et al. Variable delay multi-pulse train for fast chemical exchange saturation transfer and relayed-nuclear overhauser enhancement MRI. *Magnetic Resonance in Medicine*, 2014; 71(5): 1798–1812.
29. Gang X, Phillip Zhe S, and Renhua W. Fast simulation and optimization of pulse-train chemical exchange saturation transfer (CEST) imaging. *Physics in Medicine and Biology*, 2015; 60(12): 4719.

30. Sun PZ, Benner T, Kumar A, and Sorensen AG. Investigation of optimizing and translating pH-sensitive pulsed-chemical exchange saturation transfer (CEST) imaging to a 3T clinical scanner. *Magnetic Resonance in Medicine*, 2008; 60(4): 834–841.
31. Singh A, Cai KJ, Haris M, Hariharan H, and Reddy R. On B_1 inhomogeneity correction of in vivo human brain glutamate chemical exchange saturation transfer contrast at 7T. *Magnetic Resonance in Medicine*, 2013; 69(3): 818–824.
32. Klomp DWJ, Dula AN, Arlinghaus LR, et al. Amide proton transfer imaging of the human breast at 7T: Development and reproducibility. *NMR in Biomedicine*, 2013; 26(10): 1271–1277.
33. Zhe Sun P, Benner T, Kumar A, and Sorensen AG. An investigation of optimizing and translating pH-sensitive pulsed-chemical exchange saturation transfer (CEST) imaging to a 3 T clinical scanner. *Magnetic Resonance in Medicine*, 2008; 60(4): 834–841.
34. Zu Z, Janve VA, Li K, Does MD, Gore JC, and Gochberg DF. Multi-angle ratiometric approach to measure chemical exchange in amide proton transfer imaging. *Magnetic Resonance in Medicine*, 2012; 68(3): 711–719.
35. Friedman JI, McMahon MT, Stivers JT, and van Zijl PCM. Indirect detection of labile solute proton spectra via the water signal using frequency-labeled exchange (FLEX) transfer. *Journal of the American Chemical Society*, 2010; 132(6): 1813–1815.
36. Yadav NN, Jones CK, Xu J, et al. Detection of rapidly exchanging compounds using on-resonance frequency-labeled exchange (FLEX) transfer. *Magnetic Resonance in Medicine*, 2012; 68: 1048–1055.
37. Lin C-Y, Yadav NN, Friedman JI, Ratnakar J, Sherry AD, and van Zijl PCM. Using frequency-labeled exchange transfer to separate out conventional magnetization transfer effects from exchange transfer effects when detecting ParaCEST agents. *Magnetic Resonance in Medicine*, 2012; 67: 906–911.
38. Zu Z, Janve VA, Li K, Does MD, Gore JC, and Gochberg DF. Multi-angle ratiometric approach to measure chemical exchange in amide proton transfer imaging. *Magnetic Resonance in Medicine*, 2012; 68(3): 711–719.
39. Zu Z, Li K, Janve VA, Does MD, and Gochberg DF. Optimizing pulsed-chemical exchange saturation transfer imaging sequences. *Magnetic Resonance in Medicine*, 2011; 66(4): 1100–1108.
40. Friedman JI, McMahon MT, Stivers JT, and van Zijl PC. Indirect detection of labile solute proton spectra via the water signal using frequency-

labeled exchange (FLEX) transfer. *Journal of the American Chemical Society*, 2010; 132(6): 1813–1815.
41. Yadav NN, Jones CK, Xu J, et al. Detection of rapidly exchanging compounds using on-resonance frequency labeled exchange (FLEX) transfer. *Magnetic Resonance Imaging in Medicine*, 2012; 68: 1048.
42. Dixon WT, Ren J, Lubag AJ, et al. A concentration-independent method to measure exchange rates in PARACEST agents. *Magnetic Resonance in Medicine*, 2010; 63(3): 625–632.
43. Dixon WT, Hancu I, Ratnakar SJ, Sherry AD, Lenkinski RE, and Alsop DC. A multislice gradient echo pulse sequence for CEST imaging. *Magnetic Resonance in Medicine*, 2010; 63(1): 253–256.
44. Aime S, Delli Castelli D, and Terreno E. Novel pH-reporter MRI contrast agents. *Angewandte Chemie*, 2002; 114(22): 4510–4512.
45. Zhang S, Merritt M, Woessner DE, Lenkinski RE, and Sherry AD. PARACEST agents: Modulating MRI contrast via water proton exchange. *Accounts of Chemical Research*, 2003; 36(10): 783–790.
46. Liu G, Li Y, and Pagel MD. Design and characterization of a new irreversible responsive PARACEST MRI contrast agent that detects nitric oxide. *Magnetic Resonance in Medicine*, 2007; 58(6): 1249–1256.
47. Schmitt B, Zaiss M, Zhou J, and Bachert P. Optimization of pulse train presaturation for CEST imaging in clinical scanners. *Magnetic Resonance in Medicine*, 2011; 65(6): 1620–1629.
48. Meldrum T, Bajaj VS, Wemmer DE, and Pines A. Band-selective chemical exchange saturation transfer imaging with hyperpolarized xenon-based molecular sensors. *Journal of Magnetic Resonance*, 2011; 213(1): 14–21.
49. Scheidegger R, Vinogradov E, and Alsop DC. Amide proton transfer imaging with improved robustness to magnetic field inhomogeneity and magnetization transfer asymmetry using saturation with frequency alternating RF irradiation. *Magnetic Resonance in Medicine*, 2011; 66(5): 1275–1285.
50. Sun PZ, Benner T, Kumar A, and Sorensen AG. Investigation of optimizing and translating pH-sensitive pulsed-chemical exchange saturation transfer (CEST) imaging to a 3T clinical scanner. *Magnetic Resonance in Medicine*, 2008; 60(4): 834–841.
51. Liu GS, Song XL, Chan KWY, and McMahon MT. Nuts and bolts of chemical exchange saturation transfer MRI. *NMR in Biomedicine*, 2013; 26(7): 810–828.
52. Scheidegger R, Vinogradov E, and Alsop DC. Amide proton transfer imaging with improved robustness to magnetic field inhomogeneity

and magnetization transfer asymmetry using saturation with frequency alternating RF irradiation. *Magnetic Resonance in Medicine*, 2011; 66(5): 1275–1285.

53. Lee JS, Regatte RR, and Jerschow A. Isolating chemical exchange saturation transfer contrast from magnetization transfer asymmetry under two-frequency RF irradiation. *Journal of Magnetic Resonance*, 2012; 215: 56–63.

54. Muller S. Multifrequency selective RF pulses for multislice MR imaging. *Magnetic Resonance in Medicine*, 1988; 6(3): 364–371.

55. McMahon MT, Gilad AA, Zhou J, Sun PZ, Bulte JWM, and van Zijl PCM. Quantifying exchange rates in chemical exchange saturation transfer agents using the saturation time and saturation power dependencies of the magnetization transfer effect on the magnetic resonance imaging signal (QUEST and QUESP): pH calibration for poly-L-lysine and a starburst dendrimer. *Magnetic Resonance in Medicine*, 2006; 55(4): 836–847.

56. Xu X, Yadav NN, Zeng H, et al. Magnetization transfer contrast-suppressed imaging of amide proton transfer and relayed nuclear Overhauser enhancement chemical exchange saturation transfer effects in the human brain at 7T. *Magnetic Resonance in Medicine*, 2016; 75(1): 88–96.

57. Zu Z, Janve VA, Xu J, Does MD, Gore JC, and Gochberg DF. A new method for detecting exchanging amide protons using chemical exchange rotation transfer. *Magnetic Resonance in Medicine*, 2013; 69(3): 637–647.

58. Vinogradov E, Zhang S, Lubag A, Balschi JA, Sherry AD, and Lenkinski RE. On-resonance low B_1 pulses for imaging of the effects of PARACEST agents. *Journal of Magnetic Resonance*, 2005; 176(1): 54–63.

59. Vinogradov E, He H, Lubag A, Balschi JA, Sherry AD, and Lenkinski RE. MRI detection of paramagnetic chemical exchange effects in mice kidneys in vivo. *Magnetic Resonance in Medicine*, 2007; 58(4): 650–655.

60. Song X, Gilad AA, Joel S, et al. CEST phase mapping using a length and offset varied saturation (LOVARS) scheme. *Magnetic Resonance in Medicine*, 2012; 68(4): 1074–1086.

61. Song XL, Xu JD, Xia SL, et al. Multi-Echo Length and Offset VARied Saturation (MeLOVARS) method for improved CEST imaging. *Magnetic Resonance in Medicine*, 2015; 73(2): 488–496.

62. Liu G, Ali MM, Yoo B, Griswold MA, Tkach JA, and Pagel MD. PARACEST MRI with improved temporal resolution. *Magnetic Resonance in Medicine*, 2009; 61(2): 399–408.

63. Sun PZ, Cheung JS, Wang E, Benner T, and Sorensen AG. Fast multislice pH-weighted chemical exchange saturation transfer (CEST) MRI with unevenly segmented RF irradiation. *Magnetic Resonance in Medicine*, 2011; 65(2): 588–594.
64. Soesbe TC, Togao O, Takahashi M, and Sherry AD. SWIFT-CEST: A new MRI method to overcome T2 shortening caused by PARACEST contrast agents. *Magnetic Resonance in Medicine*, 2012; 68(3): 816–821.
65. Idiyatullin D, Corum C, Park J-Y, and Garwood M. Fast and quiet MRI using a swept radiofrequency. *Journal of Magnetic Resonance*, 2006; 181(2): 342–349.
66. Swanson SD. Broadband excitation and detection of cross-relaxation NMR spectra. *Journal of Magnetic Resonance*, 1991; 95(3): 615–618.
67. Xu X, Lee J-S, and Jerschow A. Ultrafast scanning of exchangeable sites in NMR spectroscopy. *Angewandte Chemie*, 2013; 52(32): 8281–8284.
68. Döpfert J, Witte C, and Schröder L. Slice-selective gradient-encoded CEST spectroscopy for monitoring dynamic parameters and high-throughput sample characterization. *Journal of Magnetic Resonance*, 2013; 237(0): 34–39.
69. Boutin C, Léonce E, Brotin T, Jerschow A, and Berthault P. Ultrafast Z-spectroscopy for 129Xe NMR-based sensors. *Journal of Physical Chemistry Letters*, 2013; 4(23): 4172–4176.
70. Döpfert J, Witte C, and Schröder L. Fast gradient-encoded CEST spectroscopy of hyperpolarized xenon. *ChemPhysChem*, 2014; 15(2): 261–264.
71. Sun PZ, Farrar CT, and Sorensen AG. Correction for artifacts induced by B_0 and B_1 field inhomogeneities in pH-sensitive chemical exchange saturation transfer (CEST) imaging. *Magnetic Resonance in Medicine*, 2007; 58(6): 1207–1215.
72. Kim M, Gillen J, Landman BA, Zhou J, and van Zijl PC. Water saturation shift referencing (WASSR) for chemical exchange saturation transfer (CEST) experiments. *Magnetic Resonance in Medicine*, 2009; 61(6): 1441–1450.
73. Mulkern RV and Williams ML. The general solution to the Bloch equation with constant RF and relaxation terms: Application to saturation and slice selection. *Medical Physics*, 1993; 20(1): 5–13.
74. Liu G, Gilad AA, Bulte JW, van Zijl PC, and McMahon MT. High-throughput screening of chemical exchange saturation transfer MR contrast agents. *Contrast Media and Molecular Imaging*, 2010; 5(3): 162–170.

75. Zhou J, Lal B, Wilson DA, Laterra J, and van Zijl PCM. Amide proton transfer (APT) contrast for imaging of brain tumors. *Magnetic Resonance in Medicine*, 2003; 50(6): 1120–1126.
76. Stancanello J, Terreno E, Castelli DD, Cabella C, Uggeri F, and Aime S. Development and validation of a smoothing-splines-based correction method for improving the analysis of CEST-MR images. *Contrast Media and Molecular Imaging*, 2008; 3(4): 136–149.
77. Morrison C and Henkelman RM. A model for magnetization transfer in tissues. *Magnetic Resonance in Medicine*, 1995; 33(4): 475–482.
78. Jones CK, Polders D, Hua J, et al. In vivo three-dimensional whole-brain pulsed steady-state chemical exchange saturation transfer at 7 T. *Magnetic Resonance in Medicine*, 2012; 67(6): 1579–1589.
79. Zaiss M, Schmitt B, and Bachert P. Quantitative separation of CEST effect from magnetization transfer and spillover effects by Lorentzian-line-fit analysis of z-spectra. *Journal of Magnetic Resonance*, 2011; 211(2): 149–155.
80. Zaiss M, Schnurr M, and Bachert P. Analytical solution for the depolarization of hyperpolarized nuclei by chemical exchange saturation transfer between free and encapsulated xenon (HyperCEST). *Journal of Chemical Physics*, 2012; 136(14): 144106.
81. Chen LQ, Howison CM, Jeffery JJ, Robey IF, Kuo PH, and Pagel MD. Evaluations of extracellular pH within in vivo tumors using acidoCEST MRI. *Magnetic Resonance in Medicine*, 2014; 72(5): 1408–1417.
82. Yadav NN, Chan KW, Jones CK, McMahon MT, and van Zijl PC. Time domain removal of irrelevant magnetization in chemical exchange saturation transfer Z-spectra. *Magnetic Resonance in Medicine*, 2013; 70(2): 547–555.
83. Hua J, Jones CK, Blakeley J, Smith SA, van Zijl PC, and Zhou J. Quantitative description of the asymmetry in magnetization transfer effects around the water resonance in the human brain. *Magnetic Resonance in Medicine*, 2007; 58(4): 786–793.
84. Terreno E, Stancanello J, Longo D, et al. Methods for an improved detection of the MRI-CEST effect. *Contrast Media and Molecular Imaging*, 2009; 4(5): 237–247.
85. Shah T, Lu L, Dell KM, Pagel MD, Griswold MA, and Flask CA. CEST-FISP: A novel technique for rapid chemical exchange saturation transfer MRI at 7T. *Magnetic Resonance in Medicine*, 2011; 65(2): 432–437.

86. Ward KM and Balaban RS. Determination of pH using water protons and chemical exchange dependent saturation transfer (CEST). *Magnetic Resonance in Medicine*, 2000; 44(5): 799–802.
87. Ali MM, Liu G, Shah T, Flask CA, and Pagel MD. Using two chemical exchange saturation transfer magnetic resonance imaging contrast agents for molecular imaging studies. *Accounts of Chemical Research*, 2009; 42(7): 915–924.
88. Zaiss M and Bachert P. Chemical exchange saturation transfer (CEST) and MR Z-spectroscopy in vivo: A review of theoretical approaches and methods. *Physics in Medicine and Biology*, 2013; 58(22): R221–R269.
89. Zaiss M and Bachert P. Exchange-dependent relaxation in the rotating frame for slow and intermediate exchange: Modeling off-resonant spin-lock and chemical exchange saturation transfer. *NMR in Biomedicine*, 2013; 26(5): 507–518.
90. Zaiss M, Xu J, Goerke S, et al. Inverse Z-spectrum analysis for spillover-, MT-, and T1-corrected steady-state pulsed CEST-MRI–application to pH-weighted MRI of acute stroke. *NMR in Biomedicine*, 2014; 27(3): 240–252.
91. Goffeney N, Bulte JW, Duyn J, Bryant LH, Jr., and van Zijl PC. Sensitive NMR detection of cationic-polymer-based gene delivery systems using saturation transfer via proton exchange. *Journal of the American Chemical Society*, 2001; 123(35): 8628–8629.
92. Zhou J and Zijl PCM. Chemical exchange saturation transfer imaging and spectroscopy. *Progress in Nuclear Magnetic Resonance Spectroscopy*, 2006; 48(2–3): 109–136.
93. Liu G, Moake M, Har-el YE, et al. In vivo multicolor molecular MR imaging using diamagnetic chemical exchange saturation transfer liposomes. *Magnetic Resonance in Medicine*, 2012; 67(4): 1106–1113.
94. Liu G, Ali MM, Yoo B, Griswold MA, Tkach JA, and Pagel MD. PARACEST MRI with improved temporal resolution. *Magnetic Resonance in Medicine*, 2009; 61(2): 399–408.
95. Liu D, Zhou J, Xue R, Zuo Z, An J, and Wang DJJ. RF power dependence of human brain CEST, NOE and metabolite MT effects at 7T. Paper presented at: *International Society of Magnetic Resonance in Medicine 2012*; Melbourne, Australia.
96. Longo DL, Dastru W, Digilio G, et al. Iopamidol as a responsive MRI-chemical exchange saturation transfer contrast agent for pH mapping of

kidneys: In vivo studies in mice at 7T. *Magnetic Resonance in Medicine*, 2011; 65(1): 202–211.

97. Liu G, Chan KW, Song X, et al. NOrmalized MAgnetization Ratio (NOMAR) filtering for creation of tissue selective contrast maps. *Magnetic Resonance in Medicine*, 2013; 69(2): 516–523.

98. Singh A, Cai K, Haris M, Hariharan H, and Reddy R. On B_1 inhomogeneity correction of in vivo human brain glutamate chemical exchange saturation transfer contrast at 7T. *Magnetic Resonance in Medicine*, 2013; 69(3): 818–824.

Chapter 7

Uniform-MT Method to Separate CEST Contrast from MT Effects

Jae-Seung Lee,[a,b] Ravinder R. Regatte,[a] and Alexej Jerschow[b]

[a]*Department of Radiology, New York University Langone Medical Center, 550 First Avenue, New York, NY 10016, USA*
[b]*Department of Chemistry, New York University, 100 Washington Square E, New York, NY 10003, USA*
jaeseung.lee@nyumc.org, ravinder.regatte@nyumc.org, alexej.jerschow@nyu.edu

Magnetization transfer (MT) effects in tissues broadly classify saturation transfer effects arising from protons in macromolecules and semisolid components, the saturation of which is transferred to the water pool via spin exchange and/or cross relaxation [1, 2]. A key feature of these MT effects is that they are broadband with respect to the frequency offset for the saturation, which may be a result of the broad dipolar spectra that are characteristic of rigid components. In addition, it has been reported that MT effects are generally asymmetric around the water resonance, thereby complicating their elimination by a simple symmetrization or subtraction technique in evaluating chemical exchange saturation transfer (CEST) effects. Such interferences have been reported when in vivo CEST effects were measured [3–5], which can result in the false identification of

---

Chemical Exchange Saturation Transfer Imaging: Advances and Applications
Edited by Michael T. McMahon, Assaf A. Gilad, Jeff W. M. Bulte, and Peter C. M. van Zijl
Copyright © 2017 Pan Stanford Publishing Pte. Ltd.
ISBN 978-981-4745-70-3 (Hardcover), 978-1-315-36442-1 (eBook)
www.panstanford.com

CEST effects and incorrect interpretations regarding metabolic and pathologic processes.

When such MT effects originate from MT pools consisting of strongly coupled proton spins, the MT effects can be modified by irradiating the MT pools simultaneously at more than one frequency position. This procedure can make the MT effects uniform over a significant frequency range, and thus asymmetric MT effects can be mitigated [6]. Such uniform MT (uMT) phenomena are possible because the saturating radio frequency (RF) irradiation, when applied simultaneously at multiple frequency positions, can efficiently and completely saturate MT pools regardless of the frequency positions [7].

In this chapter, the saturation phenomena of spin-1/2 systems will be reviewed with special emphasis on the components needed for the discussion of the saturation of homogeneously broadened resonances, followed by a discussion of the saturation of strongly coupled proton spins under saturating RF irradiation with multiple frequency components. A saturation transfer scheme based on the uMT phenomena will be described, which can be used to separately measure MT and CEST effects. As an example, the technique's application to brain and knee magnetic resonance imaging (MRI) is discussed.

7.1 Saturation of a Spin-1/2 System

Saturation of a spin-1/2 system under RF irradiation has been extensively studied in the field of nuclear magnetic resonance (NMR) [8–10], before the pulsed NMR method became prevailing. Under conditions relevant to MT and CEST effects, the steady-state solution to the Bloch equations [11] and the kinetic equations from Provotorov's theory of partial saturation [9, 12] are applicable as presented in the following paragraphs.

In many NMR textbooks, the steady-state solution to the Bloch equations is given under the assumption of weak continuous RF irradiation. Here, the full steady-state solution [13, 14] is given by setting the direction of the continuous RF irradiation in the rotating frame to the y-axis:

$$M_x = M_0 \frac{\omega_1/T_2}{(\Delta\omega)^2 + (T_1/T_2)\omega_1^2 + 1/T_2^2}, \qquad (7.1)$$

$$M_y = M_0 \frac{\omega_1 \cdot \Delta\omega}{(\Delta\omega)^2 + (T_1/T_2)\omega_1^2 + 1/T_2^2}, \text{ and} \qquad (7.2)$$

$$M_z = M_0 \frac{(\Delta\omega)^2 + 1/T_2^2}{(\Delta\omega)^2 + (T_1/T_2)\omega_1^2 + 1/T_2^2}, \qquad (7.3)$$

where $\Delta\omega$ is the frequency offset of the RF irradiation from the RF of a spin system, ω_1 is the amplitude of the RF irradiation, M_0 is the spin polarization of the thermal equilibrium state, and T_1 and T_2 are, respectively, the longitudinal and transverse relaxation times. The assumption of weak RF irradiation allows one to ignore the term including ω_1, which leads to the familiar expressions of the absorption and dispersion Lorentzian line shapes measured in the continuous wave (CW) NMR method.

The dependence of M_z in the steady state on the frequency offset $\Delta\omega$ is important, which is the quantity recorded in a Z-spectrum [15]. From Eq. (7.3), the minimum of the Z-spectrum is $M_0/(1 + T_1 T_2 \omega_1^2)$ when $\Delta\omega = 0$ (in the absence of exchange), and the maximum is M_0 when $|\Delta\omega| \to \infty$. The half width at the half minimum [16] can be found by equating M_z to $M_0 - [M_0 - M_0/(1 + T_1 T_2 \omega_1^2)]/2$:

$$(\Delta\omega \text{ at half minimum}) = \pm \sqrt{\frac{1 + T_1 T_2 \omega_1^2}{T_2^2}}. \qquad (7.4)$$

This expression shows two contributions to the broadness of a Z-spectrum: the intrinsic line width depending on T_2 only and the off-resonance effect due to ω_1. When $T_1 T_2 \omega_1^2 \gg 1$, the contribution of the off-resonance effect represents the major component that determines the broadness of a Z-spectrum. When measuring CEST and MT effects, one may want to keep ω_1 relatively small to reduce this so-called "direct saturation" effect. Notice that $\Delta\omega$ at half minimum becomes $\pm T_2^{-1}$ when ω_1 is zero.

The aforementioned results may not be applicable to the cases when the internal interactions between the spins cannot be ignored, since the Bloch equations do not properly describe the behavior of such spin systems. One example relevant to MT effects would be the saturation of a strongly coupled spin system, to which a well-known thermodynamic interpretation can be applied [12, 17–20]. In this thermodynamic interpretation, one can regard the Zeeman and interaction Hamiltonians as separate subsystems and can treat a weak RF irradiation as a thermal contact between these subsystems. Traditionally, those subsystems are called "orders" since the nuclear spins are in a kind of ordered states [17, 19, 20]. In the Zeeman order, for example, the nuclear spins tend to align along the external magnetic field. The interactions between spins should be strong enough to efficiently propagate small perturbations throughout the entire spin system, so a quasi-equilibrium between subsystems can be established at a timescale much shorter than the spin-lattice relaxation. This quasi-equilibrium state is usually characterized by the so-called "spin temperature" [17], which is uniquely determined by the frequency offset of the weak RF irradiation.

The kinetic equations to describe the saturation of a strongly coupled system can be derived by solving the evolution of the density operator under the assumption that at all times the density operator has the quasi-equilibrium form

$$\rho(t) = \beta_Z(t)\omega_0 S_z - \beta_d(t)\mathcal{H}_d, \tag{7.5}$$

where ρ is the spin density operator, S_z is the z component of the spin angular momentum, $\mathcal{H}_d$ is the dipolar interaction Hamiltonian, $\omega_0 = \gamma B_0$ is the resonance frequency, β_Z and β_d are the Zeeman and dipolar spin temperatures, and $\mathcal{H}_d$ is the secular part of the dipolar interactions between spins. This assumption is valid when $\omega_1 \ll \omega_0$ and $\omega_1 \ll \omega_{loc}$, where $\omega_{loc}^2 \equiv \mathrm{tr}\{H_d^2\}/\mathrm{tr}\{S_z^2\}$ is the measure of the local dipolar field. Notice that this theoretical approach [9, 12] does not assume any specific forms for $\mathcal{H}_d$ except that it commutes with the Zeeman Hamiltonian. The term "dipolar" is used because the dipolar interaction would be the most common reason why a spin-1/2 system gets a homogeneously broadened spectrum. For the observables S_z and $\mathcal{H}_d$, the kinetic equations can be obtained and conveniently expressed in terms of polarizations [21]:

$$\frac{dP_S}{dt} = -W\left(P_S - \frac{\Delta\omega}{\omega_{loc}}P_d\right) - \frac{P_S - P_{S,0}}{T_{1,S}}, \tag{7.6}$$

$$\frac{dP_d}{dt} = W\frac{\Delta\omega}{\omega_{loc}}\left(P_S - \frac{\Delta\omega}{\omega_{loc}}P_d\right) - \frac{P_d}{T_{1,d}}, \tag{7.7}$$

where $P_S \equiv (2/N)\text{tr}\{S_z\rho\}$ and $P_d \equiv (2/N)(\text{tr}\{\mathcal{H}_d\rho\}/\omega_{\text{loc}})$ are the spin and dipolar polarizations for N spins, $W \equiv \pi\omega_1^2 g(\Delta\omega)$ with the normalized absorption line shape $g(\Delta\omega)$, and $T_{1,s}$ and $T_{1,d}$ are the longitudinal relaxation times of the Zeeman and dipolar orders, respectively. While P_S can be measured from the conventional spectrum recorded following a single RF pulse, P_d can be converted into the transverse magnetization through the adiabatic remagnetization in the rotating frame [21] or the Jeener–Broekaert sequence consisting of two RF pulses [22]. When $WT_{1,s} \gg 1$ and $WT_{1,d} \gg 1$, the quasi-equilibrium states established by the weak RF irradiation satisfy the relation $P_S = (\Delta\omega/\omega_{\text{loc}})P_d$. Note that the negative P_S or population inversion may not be reached from the initial positive P_S by saturation, but P_d changes its sign when $\Delta\omega$ crosses zero. The quasi-equilibrium state is reached at the characteristic rate of $\pi\omega_1^2 g(\Delta\omega)[1 + (\Delta\omega/\omega_{\text{loc}})^2]$, which is the non-zero eigenvalue of the kinetic equations given in Eqs. (7.6) and (7.7). In solids and liquid crystals, this characteristic rate can be very large, hence the quasi-equilibrium state can be established more or less instantaneously.

As an example, the linear-response spectra of a 5CB liquid crystal after weak saturating RF irradiation are shown in Fig. 7.1 [7]. Due to anisotropic molecular motion, a spin system in a liquid crystal usually exhibits residual dipolar interactions [23]. The ^{1}H NMR spectrum of 5CB spans about 40 kHz along the frequency offset, and $\omega_{\text{loc}}/2\pi$ was estimated to be about 6 kHz. The amplitude of RF irradiation $(\gamma B_1/2\pi)$ was 500 Hz, which is much smaller than the strength of the residual dipolar interactions. The quasi-equilibrium state established by weak saturating RF irradiation was monitored by a linear-response spectrum, in which the NMR signal is excited by a strong RF pulse with a flip angle satisfying the small-angle approximation (typically 1°–5°).

In a linear-response spectrum, the spectral intensity at a particular frequency offset is proportional to the differences between the populations of the energy eigenstates that are connected by single-quantum transitions at the particular frequency offset. As the flip angle increases, an RF pulse would mix the populations among more energy eigenstates, which are linked through several single-quantum transitions, so the spectrum will depend on P_S only

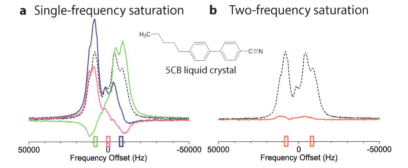

Figure 7.1 Single-frequency versus two-frequency saturation on a 5CB liquid crystal. The frequency positions of the saturating RF irradiation are indicated on the frequency axis by the boxes with the colors matched to the corresponding spectra. The linear-response spectrum of the thermal equilibrium state is shown as the black dashed line. (a) Linear-response ^{1}H NMR spectra acquired after a 5° pulse following single-frequency saturation. (b) Linear-response ^{1}H NMR spectrum following two-frequency saturation. The duration of the saturating RF irradiation was 50 ms.

while the detailed information about the population differences is smeared out. For example, P_S may be similar between the quasi-equilibrium states established by RF irradiations at the certain frequency offsets. If NMR spectra are recorded with a 90° pulse, they will look similar. On the other hand, the difference between the population distributions of the two states can be revealed in linear-response spectra. In Fig. 7.1(a), as an example, the linear-response spectra obtained following RF irradiations at -8 kHz and 8 kHz are presented as the blue and green lines, respectively. Their shapes are quite dissimilar, indicating two distinct internal equilibria. However, their integrated intensities are similar, so spectra recorded with a 90° pulse may not separate them clearly.

7.2 Uniform Saturation of a Dipolar-Coupled Spin-1/2 System

As seen in the aforementioned example, there is a way to efficiently saturate the system without having to use RF irradiation strong enough to cover the whole spectral range. The kinetic equations from Provotorov's theory, Eqs. (7.6) and (7.7), have in fact two steady-state solutions: the non-trivial solution that can be established by weak RF irradiation and the trivial one with $P_S = P_d = 0$, which is in fact reached when $\Delta\omega$ is exactly zero. Since the steady states with the distinct frequency offsets are distinguishable from each other, the system cannot find a common non-trivial steady state under weak RF irradiation at multiple frequency offsets. Instead, only the trivial solution with $P_S = P_d = 0$ will be allowed [24].

Some concrete insights into the effect of multi-frequency RF irradiation on the saturation of a strongly dipolar-coupled spin-1/2 system can be obtained by extending Provotorov's theory to the case of weak RF irradiation applied simultaneously at two frequency positions [7].

Incorporating one additional weak RF irradiation into Provotorov's theory is straightforward in the laboratory frame. There exists a cross-effect between two RF components at the frequencies ω and ω', which is proportional to the time integral $\iint dt'dt'' \cos\omega t' \cos\omega' t''$ and will be negligible unless the frequency difference $\omega - \omega'$ is comparable to the amplitude of the RF components. Otherwise, each weak RF irradiation contributes to the kinetics in the same way as in Eqs. (7.6) and (7.7):

$$\frac{dP_S}{dt} = -W\left(P_S - \frac{\Delta\omega}{\omega_{\text{loc}}}P_d\right) - W'\left(P_S - \frac{\Delta\omega'}{\omega_{\text{loc}}}P_d\right) - \frac{P_S - P_{S,0}}{T_{1,S}},$$
(7.8)

$$\frac{dP_d}{dt} = W\frac{\Delta\omega}{\omega_{\text{loc}}}\left(P_S - \frac{\Delta\omega}{\omega_{\text{loc}}}P_d\right) + W'\frac{\Delta\omega'}{\omega_{\text{loc}}}\left(P_S - \frac{\Delta\omega'}{\omega_{\text{loc}}}P_d\right) - \frac{P_d}{T_{1,d}}.$$
(7.9)

It is straightforward to show that the eigenvalues governing the dynamics described by Eqs. (7.8) and (7.9) are always real and negative [7], which means that P_S and P_d always decay to zero, regardless of their initial values and the frequency offsets $\Delta\omega$ and $\Delta\omega'$ if W and W' are much larger than $1/T_{1,S}$ and $1/T_{1,d}$. For this reason, it is called "uniform saturation." RF irradiation applied simultaneously at two frequency positions is sufficient to fully saturate a given system, because there are two orders connected by the RF irradiation. In general, the minimum number of frequency components for uniform saturation is same as the number of independent orders or spin temperatures in a given system.

The eigenvalues governing the dynamics described by Eqs. (7.8) and (7.9) determine how quickly the Zeeman and dipolar orders can be saturated. It can be shown that the most efficient saturation can happen when $W\Delta\omega + W'\Delta\omega' = 0$, which suggests that the two frequency offsets $\Delta\omega$ and $\Delta\omega'$ have opposite signs. In addition, larger W and W' are generally favorable for faster saturation, which can be implemented by increasing the RF amplitudes ω_1 and ω_1' under the condition $\omega_1, \omega_1' \ll \omega_{loc}$ and locating the frequency positions $\Delta\omega$ and $\Delta\omega'$ where the spectral intensities $g(\Delta\omega)$ and $g(\Delta\omega')$ are higher. In other words, faster saturation can happen when more transitions are directly irradiated.

As an example, the linear-response spectrum of 5CB following two weak RF irradiations applied simultaneously at ±8 kHz is shown in Fig. 7.1(b). It demonstrates that the system approaches the state with equal populations across all the energy levels. In contrast, the linear-response spectrum following single RF irradiation at 0 Hz reveals unequal populations across the energy levels, as seen in Fig. 7.1(a) (the magenta line), although the similarity of the integrated intensities of the two linear-response spectra indicates that P_S's of the two states are almost the same, which is close to zero. The RF irradiation simultaneously at ±8 kHz was conveniently implemented by a cosine-modulated pulse with a modulation frequency of 8 kHz and with a frequency offset at 0 Hz. This uniform saturation can be used to suppress unwanted signals in solid state and liquid-crystalline NMR [24] and to separate CEST contrast from MT effects as described in Section 7.3 [6, 25, 26].

7.3 Uniform-MT Methodology

Suppose that an MT pool consists of strongly dipolar-coupled proton spins. As seen in Section 7.1, the saturation of the MT pool depends on the frequency offset of the saturating RF irradiation, which is supposed to be weak in comparison with the spectral range of those proton spins; therefore, the amount of the MT effects would vary with the frequency offset of the saturating RF irradiation [2, 27, 28].

If the saturation of such an MT pool can be made independent of the frequency offset of saturating RF irradiation and if such saturation can be attained quickly before MT effects substantially develop, the amount of the MT effects may not depend much on the frequency offset of the saturating RF irradiation. As seen in Section 7.2, uniform saturation can efficiently and completely saturate strongly dipolar-coupled proton spins independently of the parameters of saturating RF irradiation. Hence, this approach has a potential to modify MT effects toward making them equal along the frequency offset of saturating RF irradiation, which can be beneficial to the clean measurement of CEST effects.

There are several ways in which this so-called uMT phenomenon can be incorporated into the CEST measurement, including those presented in Fig. 7.2. The simplest modification to the conventional method, in which the water signal is measured against the varying frequency offset of the saturating RF irradiation with a single frequency component (the solid downward arrow under the dashed left right arrow in Fig. 7.2(a)), would be the inclusion of one or more additional RF irradiations at fixed frequency offsets (the downward

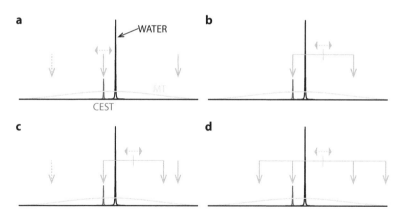

Figure 7.2 Examples of possible CEST measurement schemes exploiting uMT. (a) Conventional CEST measurement with one or more additional RF irradiations at fixed frequency offsets. (b) A scheme with two simultaneous RF irradiations at the fixed distance. (c) The scheme shown in (b) with one or more additional RF irradiations at fixed frequency offsets. (d) A scheme with four simultaneous RF irradiations at fixed distances.

solid and/or dashed arrows in Fig. 7.2(a)), which should be far off resonance so as not to affect CEST sites but to touch MT pools.

The second scheme, shown in Fig. 7.2(b), may look more complicated than the first scheme, but its implementation is often easier in practice. The frequency distance between two frequency components of the saturating RF irradiation is fixed, hence their frequency positions are decided by a single frequency offset, against which the water signal is recorded. In the scheme presented in Fig. 7.2(b), the frequency offset is defined as the average of two frequency positions. This scheme can provide information about not only CEST effects but also MT effects and B_0-field inhomogeneity. The details will be discussed as follows.

The second scheme may be extended with additional frequency components. Their frequency positions (Fig. 7.2(c)) or their frequency distances to the pre-existing frequency components (Fig. 7.2(d)) may be fixed. The scheme shown in Fig. 7.2(d) can appear as a non-ideal realization of the scheme shown in Fig. 7.2(b). For example, when the scheme shown in Fig. 7.2(b) is implemented by a shaped RF irradiation, due to digitization or truncation errors, the RF irradiation could have minor additional frequency components at the multiples of the modulation frequency [6].

On modern NMR spectrometers and MRI scanners, arbitrarily shaped RF pulses at desired frequency offsets and RF power levels can be easily generated. In this case, the scheme in Fig. 7.2(b) can be conveniently implemented with a single shaped RF pulse, while the scheme in Fig. 7.2(a) may require as many shaped RF pulses as the number of frequency offsets. Any shape function $S(t)$ can be split into two by multiplying a cosine function with the desired modulation frequency f_m:

$$S(t)\cos(2\pi f_m t) = \frac{1}{2}S(t)e^{i2\pi f_m t} + \frac{1}{2}S(t)e^{-i2\pi f_m t}. \qquad (7.10)$$

The frequency separation between two components is twice the modulation frequency f_m, and each frequency component carries half the total RF power.

Another simple and convenient way to generate RF irradiation with multiple frequency components is a square-shaped modulation, which consists of a sequence of two pulses with equal duration τ and opposite phases [7]. The sequence splits the RF power into the Fourier components at the frequencies $(2n + 1)/\tau$, where n is an integer and the frequency unit is Hz. Since the amplitude of the Fourier component decreases as $|n|$ increases, this sequence can be regarded as the non-ideal realization of the cosine-modulated pulse at the modulation frequency $f_m = 1/\tau$, as depicted in Fig. 7.2(d). A hardware-level implementation is, of course, also possible but appears to be less flexible.

Although similar types of sequences have been used in other methods, such as in the so-called Z-spectroscopy with alternating-phase irradiation (ZAPI) and Z-spectroscopy with alternating-phase irradiation and sine modulation (ZAPISM) [29], as well as in SAFARI [30], the underlying mechanisms of saturation within dipolar-broadened MT pools have not been investigated within the framework of Provotorov's theory. As a result, we believe important aspects regarding the isolation of CEST effects from asymmetric MT effects originating from dipolar-broadened MT pools have been missed in those works.

The modulation frequency f_m can be chosen in consideration of the spectral ranges of the water, CEST, and MT pools, as shown in Fig. 7.3(a). Since the goal is to induce CEST effects in the same way as the conventional method, while making MT effects uniform, any simultaneous irradiation of multiple proton species except for those in the MT pool should be avoided. This can be done if f_m is large enough such as not to touch any proton species within typical chemical shifts when the frequency offset of the cosine-modulated RF irradiation is zero (green arrows in Fig. 7.3(a)). Since the vast majority of the proton chemical shifts lies within 6 ppm from the water signal, it can be a good choice for f_m. The actual modulation frequency may need to be adjusted depending on the actual range of proton chemical shifts in a given tissue and the quality of uMT phenomena.

Whether MT effects are made uniform or not can be judged from the flatness of the Z-spectrum, which measures the water signal against the frequency offset of the cosine-modulated RF irradiation, around the zero frequency offset. When the Z-spectrum is flat around the zero frequency offset, the relative signal decrease may be used as MT contrast (Fig. 7.3(b)) since it is mostly due to MT effects,

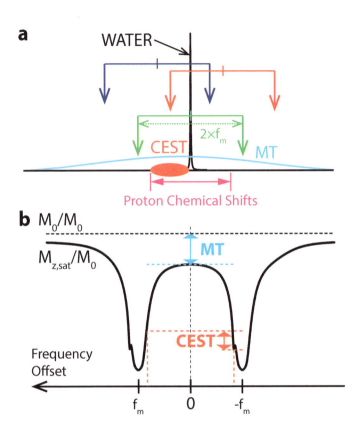

Figure 7.3 uMT CEST with a cosine-modulated RF irradiation at the modulation frequency f_m. (a) The spectral ranges for water protons (black), CEST (red), and MT (cyan) pools, and typical proton chemical shifts are presented together with the frequency positions of a cosine-modulated RF irradiation. (Green arrows) When the frequency offset is zero, only the MT pool will be irradiated, which will be uniformly saturated. (Red arrows) When the frequency offset is moved to the upfield position, CEST effects will be induced together with the direct saturation and uMT effects. (Blue arrows) For the corresponding frequency offset in the downfield position, the same amount of the direct saturation and uMT effects will be induced while there is no saturation of the CEST pool. (b) A typical Z-spectrum with a cosine-modulated RF irradiation, from which MT and CEST effects can be separately estimated. Two dips at the frequency offsets $\pm f_m$ are due to the direct saturation of the water protons. Reprinted from Ref. [26], Copyright 2014, with permission from Elsevier.

which can be called uniform MT ratio or u*MTR*:

$$\text{uMTR}(f_m) = 1 - M_{z,sat}(f_o = 0 \text{ Hz}, f_m)/M_0, \quad (7.11)$$

where M_0 and $M_{z,sat}$ are, respectively, the water signals without applying any saturating RF irradiation and after applying the cosine-modulated RF irradiation, and f_o is the frequency offset of the cosine-modulated RF irradiation.

When the frequency offset of the cosine-modulated RF irradiation is set to the upfield position (red arrows in Fig. 7.3(a)), CEST effects may be induced on top of uMT phenomena and the direct saturation of the water protons since one of two frequency components saturates the CEST pool. When the frequency offset is set to the downfield position (blue arrows in Fig. 7.3(a)), no CEST effects are expected but there are still uMT phenomena and direct saturation present. While the contribution due to uMT phenomena would be the same regardless of the frequency offset, the contribution from the direct saturation will be symmetric around the water signal. As often done in conventional CEST measurements, two signals measured at the frequency offsets symmetric around the water signal may be compared to remove the contribution due to direct saturation. Considering that the actual frequency positions of cosine-modulated RF irradiation at the frequency offset f_o is $f_o \pm f_m$, the following quantity called u*MTR* asymmetry will give the measure of isolated CEST effects, which is depicted as "CEST" in Fig. 7.3(b):

$$\text{uMTR}_{\text{asym}}(f_o, f_m)$$
$$= [M_{z,sat}(f_m - f_o, f_m) - M_{z,sat}(-f_m + f_o, f_m)]/M_0, \quad (7.12)$$

where $0 \leq f_o \leq f_m$. For $|f_m| > f_o$, it would be difficult to make sure that uMT phenomena are still intact, because one of the two frequency components could be out of the spectral range of the MT pools.

If B_0-field inhomogeneity exists, the Z-spectrum in Fig. 7.3(b) will be shifted along the axis of the frequency offset. The amount of this B_0 shift can be estimated by the average of the frequency positions of the two dips due to the direct saturation of the water protons, which should locate at the frequency offsets near $\pm f_m$ if the B_0-field inhomogeneity is not significant ($\Delta\omega_0 \ll f_m$).

7.4 Application to Brain MRI

As examples to demonstrate how the uMT technique works, the results from a brain MRI study will be discussed in this section [26] and those from a knee MRI study will be presented in the next section [25]. From the brain MRI study, a reference image without any saturating RF irradiation is shown in Fig. 7.4(a). The signal intensity of each pixel in the reference image plays the role of M_0 in Eq. (7.11) and Eq. (7.12). Then, a series of images are obtained, each with a different frequency offset of the cosine-modulated saturating RF irradiation. In this particular example, the modulation frequency f_m was 5 ppm or 1500 Hz at 7 T. As discussed in Section 7.3, f_m was chosen to avoid the simultaneous irradiation of any CEST and nuclear Overhauser effect (NOE) pools while ensuring the uMT phenomena. The frequency offsets were varied from -2500 Hz to $+2500$ Hz with a step size of 100 Hz. From this series of images, the Z-spectra for individual pixels can be constructed.

As discussed in Section 7.3, the Z-spectra measured with cosine-modulated saturating RF irradiation have two dips, and the averaged frequency positions of the two dips can be used to generate a B_0-field map, which is shown in Fig. 7.4(b). In order to find the locations of the two dips, the Z-spectra are interpolated first. In this example, a cubic spline interpolation was used. The B_0-field map is used to shift the interpolated Z-spectra, so that the locations of the two dips

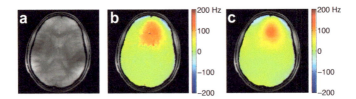

Figure 7.4 MRI images from the brain of a healthy volunteer. (a) A reference image acquired with a segmented GRE acquisition with centric phase encoding order. This image provides M_0 for the data processing. (b) A B_0-field map obtained from the averaged frequency position of two dips in the Z-spectrum acquired with cosine-modulated RF irradiation. (c) A B_0-field map obtained with the WASSR method. The B_0-field maps are color coded and overlaid on the reference image.

are $\pm f_m$ after the shift. For comparison, a conventional B_0-field map was obtained with the water saturation shift referencing (WASSR) method [31], which is shown in Fig. 7.4(c).

A map of uMTR defined in Eq. (7.11) provides a new type of MT contrast, as shown in Fig. 7.5(a). When using only one RF irradiation (conventional MT), the images in Fig. 7.5(b, c) were obtained, wherein either only the upfield or only the downfield RF irradiation was used. Since the offsets are much larger than typical chemical shift ranges, it is reasonable to assume that the contrast arises solely from MT phenomena. The difference between the two conventional MT contrast maps shown in Fig. 7.5(b, c) clearly indicates that the MT effects in brain are asymmetric around the water signal. Note that the MT effects were determined to be stronger on the left side in all the three contrast maps, which is probably due to the spatial variation of the B_1 field.

To quantify how much the uMT technique reduces the asymmetry of such MT effects, the Z-spectra and $uMTR_{asym}$ curves are compared with the Z-spectra and $MTR_{asym} (\equiv [M_{z,sat}(-f_o) - M_{z,sat}(+f_o)]/M_0)$ curves from the conventional method. Considering that the amount of the MT asymmetry may vary among different tissue environments, segmentations are made based on the $uMTR_{asym}$ values shown in Fig. 7.5(a). It is rather easy to make a segment with small MT effects (uMTR < 25%), which may correspond to ventricles filled with cerebrospinal fluid (CSF). On the other hand, it has been known that white matter produces higher MTR values than gray matter [32, 33], so the uMTR values are also expected to be higher with white matter than with gray matter. Although more studies should be performed to know how the uMTR values correlate with white matter and gray matter, the minimum uMTR value for the segment with higher uMTR values has been chosen such that the shape of the third segment with intermediate uMTR values produces the typical shape of the gray matter region. In the example shown in Fig. 7.6(a, b), the uMTR value to separate the segments with higher and intermediate uMTR values was chosen to be 40%.

The distributions of the quantities $M_{z,sat}/M_0$, MTR_{asym}, and $uMTR_{asym}$ against the frequency offset for the regions with higher and intermediate uMTR are presented as box plots in Fig. 7.6(c–f). The Z-spectra shown in Fig. 7.6(c, d) clearly depict the difference

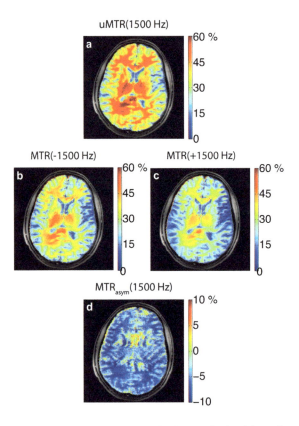

Figure 7.5 *MTR* contrast maps from the brain of a healthy volunteer. (a) A u*MTR* map acquired with saturating RF irradiation simultaneously at ± 1500 Hz. (b,c) An *MTR* map acquired with saturating RF irradiation at (b) −1500 Hz and (c) +1500 Hz. Figures (a–c) reprinted from Ref. [26], Copyright 2014, with permission from Elsevier. (d) The difference between two *MTR* maps. The B_0-field inhomogeneity is not applied. The contrast maps are color coded and overlaid on the reference image.

between the conventional (black boxes) and uMT (red boxes) techniques. Because they are plotted after correcting the B_0-field inhomogeneity, the Z-spectra from the conventional method have one dip centered at the zero frequency offset, and those from the uMT method have two dips located at the modulation frequency used for the cosine-modulated saturating RF irradiation. Notice that the Z-spectra from the conventional method show higher signal

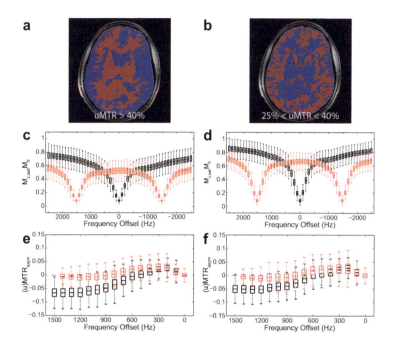

Figure 7.6 Segmentation based on u*MTR* and box plots of Z-spectra and *MTR*$_{asym}$ and u*MTR*$_{asym}$ curves from the brain of a healthy volunteer. (a,b) Segments wherein u*MTR* is (a) larger than 40% and (b) between 20% and 40%. (c) Box plots of Z-spectra for the segment shown in (a). The black and red boxes represent the data from the conventional and uMT methods, respectively. (d) Box plots of Z-spectra for the segment shown in (b). (e) Box plots of *MTR*$_{asym}$ (black) and u*MTR*$_{asym}$ (red) values from the segment shown in (a). (f) Box plots of *MTR*$_{asym}$ (black) and u*MTR*$_{asym}$ (red) values from the segment shown in (b). Figures (c–f) reprinted from Ref. [26], Copyright 2014, with permission from Elsevier.

intensities on the upfield side, which manifests the asymmetry of MT effects in those regions. On the other hand, the Z-spectra from the uMT method are almost flat around the zero frequency offset, which indicates that MT effects are made uniform.

The difference between the conventional and uMT methods is more noticeable by comparing the box plots of *MTR*$_{asym}$ and u*MTR*$_{asym}$ values, as shown in Fig. 7.6(e, f). For the frequency offsets from 1200 Hz to 1500 Hz or from 4 ppm to 5 ppm, where any saturation transfer effects other than MT effects might not be expected, the MTR_{asym} values from the conventional method are negative. In addition, the MTR_{asym} values from the regions with higher and intermediate uMTR values are different from each other, which indicates that the origins of MT effects may be different. For the same frequency offsets, on the other hand, the uMTR_{asym} values are close to zero. The discrepancy between the MTR_{asym} and uMTR_{asym} values exists even at a frequency offset as small as 400 Hz. It may indicate that the MTR_{asym} values for frequency offsets larger than 400 Hz contain the contribution from MT effects, which cannot be removed by evaluating MTR_{asym} because they are asymmetric around the water signal.

7.5 Application to Knee MRI

The human knee joint connects the femur, tibia, and patella and is a synovial joint consisting of synovial cavity, articular capsule, and articular cartilage. The extracellular matrix of articular cartilage is mainly composed of water (60–80%), type II collagen (15–20%), and negatively charged proteoglycans (PGs) (3–10%), which are complex macromolecules consisting of proteins and polysaccharides. The most common PG in articular cartilage is aggrecan, consisting of a bottle-brush-shaped protein core and a long extended domain to which many glycosaminoglycan (GAG) side chains are attached. GAGs are long unbranched polysaccharides consisting of a repeating disaccharide unit, which are made up of N-acetylgalactosamine and glucuronic acid or galactose and are functionally and structurally important constituents of articular cartilage and synovial fluid. The prevalent GAGs in articular cartilage are chondroitin sulfate (CS) and keratin sulfate (KS). They are covalently attached to a core protein to constitute PG aggregates, to which hyaluronic acid (HA), the only non-sulfated GAG, is non-covalently bound. HA is also found in synovial fluid.

The content of GAGs in vivo has been measured via contrast-enhanced MRI (dGEMRIC) [34], $T_{1\rho}$-weighted MRI [35], or ^{23}Na-MRI [36]. For clinical applications, dGEMRIC has the major drawback of a long temporal interval following the injection of the contrast agent (Gd-DTPA^{2-}), the requirement of joint exercise for efficient

Application to Knee MRI | 115

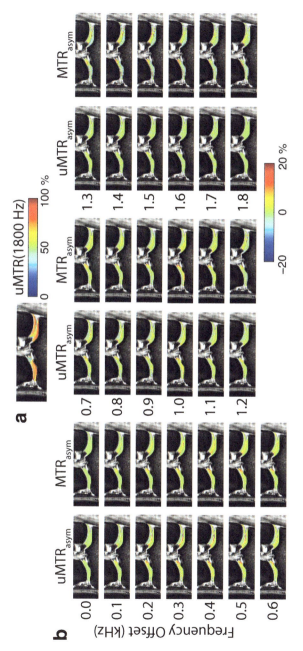

Figure 7.7 u*MTR* and gagCEST contrast maps from the knee joint of a healthy volunteer: (a) The u*MTR* contrast map measured with a simultaneous two-frequency (±1800 Hz) saturating RF irradiation. Reprinted with permission from Ref. [25], copyright 2013, Nature Publishing Group. (b) u*MTR*$_{asym}$ and *MTR*$_{asym}$ contrast maps after correcting B_0-field inhomogeneity. The contrast maps are color coded and overlaid on the reference image.

penetration of Gd into the tissue, and the knowledge of accurate measures of intratissue relaxivity of Gd. $T_{1\rho}$-weighted MRI mitigates most of these problems; however, it results in significant RF energy deposition and is sensitive to both GAG and collagen. Although ^{23}Na-MRI has high specificity toward PGs, its low gyromagnetic ratio and low tissue concentration yield low signal-to-noise ratio or low spatial resolution and demand hardware modifications and longer acquisition time.

CEST may be useful for measuring the content of GAGs in vivo since GAGs have exchangeable protons [37]. For example, each disaccharide unit of CS and KS has four exchangeable protons, one –NH and three –OH, and that of HA has one –NH and four –OH because HA is not sulfated. However, collagen has been known to produce MT effects [38], which can interfere with the measurement of CEST effects from GAGs (gagCEST). An experiment on an ex vivo bovine cartilage specimen [6] showed that the MT effects from cartilage tissue may be asymmetric around the water signal and can be made uniform by the uMT method, which makes it possible to isolate

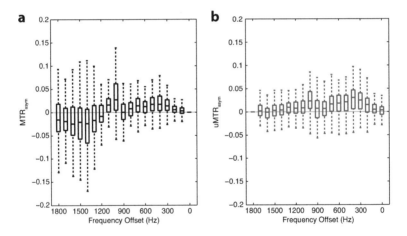

Figure 7.8 Box plots of MTR_{asym} and $uMTR_{asym}$ from the lateral femorotibial articular cartilage of a healthy volunteer. (a) The negative MTR_{asym} at frequency offsets larger than 1200 Hz indicates asymmetric MT phenomena in cartilage tissue. (b) $uMTR_{asym}$ may not be affected by such asymmetric MT effects, as evidenced by its close-to-zero values at the same frequency offsets.

the gagCEST effects. The same has been demonstrated in vivo [25], and an example of the comparison between the conventional and uMT methods on human knee joint is shown in Figs. 7.7 and 7.8. While the uMTR contrast clearly separates articular cartilage from the synovial cavity [Fig. 7.7(a)], the difference between the MTR_{asym} and uMTR_{asym} contrast maps is obvious at the frequency offsets from 1400 Hz to 1800 Hz, as seen in Fig. 7.7(b). The box plots of MTR_{asym} and uMTR_{asym} from the lateral femorotibial articular cartilage, where the uMTR values are high (>50%), confirm the asymmetric MT effects in articular cartilage and the separation of the CEST effects through the uMT phenomena (Fig. 7.8). The uMT approach could be very useful in assessing cartilage components by separating the MT and CEST pools better than previously possible.

7.6 Summary

We reviewed here the recently developed method of uMT CEST. The background of this methodology is described within the framework of saturation theory, whereby one can show that efficient saturation of dipolar-coupled pools can be achieved by simultaneous two-frequency (or more) irradiation. Such pools are thought to be present in most MT pools in tissues. When applied to CEST, this technique allows one to measure CEST without interferences from asymmetric MT, while the method can also be used to obtain MT contrast that is uniform over a large frequency range. Both ex vivo and in vivo evidences exist for the suppression of asymmetric MT in CEST measurements, and results from in vivo studies on brain and cartilage are shown in this chapter as well.

References

1. Wolff, S. D. and Balaban, R. S. (1989). Magnetization transfer contrast (MTC) and tissue water proton relaxation in vivo, *Magnetic Resonance in Medicine* **10**, 1, pp. 135–144.
2. Henkelman, R. M., Huang, X., Xiang, Q.-S., Stanisz, G. J., Swanson, S. D. and Bronskill, M. J. (1993). Quantitative interpretation of magnetization transfer, *Magnetic Resonance in Medicine* **29**, 6, pp. 759–766.

3. Zhou, J., Payen, J.-F., Wilson, D. A., Traystman, R. J. and van Zijl, P. C. M. (2003). Using the amide proton signals of intracellular proteins and peptides to detect pH effects in MRI, *Nature Medicine* **9**, 8, pp. 1085–1090.
4. Hua, J., Jones, C. K., Blakeley, J., Smith, S. A., van Zijl, P. C. M. and Zhou, J. (2007). Quantitative description of the asymmetry in magnetization transfer effects around the water resonance in the human brain, *Magnetic Resonance in Medicine* **58**, 4, pp. 786–793.
5. Ng, M.-C., Hua, J., Hu, Y., Luk, K. D. and Lam, E. Y. (2009). Magnetization transfer (MT) asymmetry around the water resonance in human cervical spinal cord, *Journal of Magnetic Resonance Imaging* **29**, 3, pp. 523–528.
6. Lee, J.-S., Regatte, R. R. and Jerschow, A. (2012). Isolating chemical exchange saturation transfer contrast from magnetization transfer asymmetry under two-frequency rf irradiation, *Journal of Magnetic Resonance* **215**, pp. 56–63.
7. Lee, J.-S., Khitrin, A. K., Regatte, R. R. and Jerschow, A. (2011). Uniform saturation of a strongly coupled spin system by two-frequency irradiation, *Journal of Chemical Physics* **134**, 23, p. 234504.
8. Abragam, A. (1961). *Principles of Nuclear Magnetism* (Oxford University Press, Oxford).
9. Goldman, M. (1970). *Spin Temperature and Nuclear Magnetic Resonance in Solids* (Clarendon, Oxford).
10. Slichter, C. P. (1996). *Principles of Magnetic Resonance, 3rd ed.* (Springer, Berlin).
11. Bloch, F. (1946). Nuclear induction, *Physical Review* **70**, 7-8, pp. 460–474.
12. Provotorov, B. N. (1962). Magnetic resonance saturation in crystals, *Journal of Experimental and Theoretical Physics* **41**, pp. 1582–1591.
13. Mulkern, R. V. and Williams, M. L. (1993). The general solution to the Bloch equation with constant rf and relaxation terms: Application to saturation and slice selection, *Medical Physics* **20**, 1, pp. 5–13.
14. Smith, S. A., Bulte, J. W. M. and van Zijl, P. C. M. (2009). Direct saturation MRI: Theory and application to imaging brain iron, *Magnetic Resonance in Medicine* **62**, 2, pp. 384–393.
15. Bryant, R. G. (1996). The dynamics of water-protein interactions, *Annual Review of Biophysics and Biomolecular Structure* **25**, 1, pp. 29–53.
16. Kim, M., Gillen, J., Landman, B. A., Zhou, J. and van Zijl, P. C. M. (2009). Water saturation shift referencing (WASSR) for chemical

exchange saturation transfer (CEST) experiments, *Magnetic Resonance in Medicine* **61**, 6, pp. 1441–1450.

17. Abragam, A. and Proctor, W. (1958). Spin Temperature, *Physical Review* **109**, 5, pp. 1441–1458.
18. Slichter, C. and Holton, W. (1961). Adiabatic demagnetization in a rotating reference system, *Physical Review* **122**, 6, pp. 1701–1708.
19. Anderson, A. and Hartmann, S. (1962). Nuclear magnetic resonance in the demagnetized state, *Physical Review* **128**, 5, pp. 2023–2041.
20. Jeener, J., Eisendrath, H. and Van Steenwinkel, R. (1964). Thermodynamics of spin systems in solids, *Physical Review* **133**, 2A, pp. A478–A490.
21. Lee, J.-S. and Khitrin, A. K. (2008). Thermodynamics of adiabatic cross polarization, *Journal of Chemical Physics* **128**, 11, p. 114504.
22. Jeener, J. and Broekaert, P. (1967). Nuclear magnetic resonance in solids: Thermodynamic effects of a pair of rf pulses, *Physical Review* **157**, 2, pp. 232–240.
23. Emsley, J. W. and Lindon, J. C. (1975). *NMR Spectroscopy using Liquid Crystal Solvents* (Pergamon, Oxford).
24. Lee, J.-S. and Khitrin, A. K. (2005). Pseudopure state of a twelve-spin system, *Journal of Chemical Physics* **122**, 4, p. 041101.
25. Lee, J.-S., Parasoglou, P., Xia, D., Jerschow, A. and Regatte, R. R. (2013). Uniform magnetization transfer in chemical exchange saturation transfer magnetic resonance imaging, *Scientific Reports* **3**, p. 1707.
26. Lee, J.-S., Xia, D., Ge, Y., Jerschow, A. and Regatte, R. R. (2014). Concurrent saturation transfer contrast in in vivo brain by a uniform magnetization transfer MRI, *NeuroImage* **95**, pp. 22–28.
27. Yeung, H. N., Adler, R. S. and Swanson, S. D. (1994). Transient decay of longitudinal magnetization in heterogeneous spin systems under selective saturation. IV. Reformulation of the spin-bath-model equations by the Redfield–Provotorov theory, *Journal of Magnetic Resonance, Series A* **106**, 1, pp. 37–45.
28. Morrison, C., Stanisz, G. and Henkelman, R. M. (1995). Modeling magnetization transfer for biological-like systems using a semi-solid pool with a super-Lorentzian lineshape and dipolar reservoir, *Journal of Magnetic Resonance, Series B* **108**, 2, pp. 103–113.
29. Närväinen, J., Hubbard, P. L., Kauppinen, R. A. and Morris, G. A. (2010). Z-spectroscopy with alternating-phase irradiation, *Journal of Magnetic Resonance* **207**, 2, pp. 242–250.
30. Scheidegger, R., Vinogradov, E. and Alsop, D. C. (2011). Amide proton transfer imaging with improved robustness to magnetic field inhomo-

geneity and magnetization transfer asymmetry using saturation with frequency alternating RF irradiation, *Magnetic Resonance in Medicine* **66**, 5, pp. 1275–1285.

31. Kim, M., Gillen, J., Landman, B. A., Zhou, J. and van Zijl, P. C. M. (2009). Water saturation shift referencing (WASSR) for chemical exchange saturation transfer (CEST) experiments, *Magnetic Resonance in Medicine* **61**, 6, pp. 1441–1450.

32. Mehta, R. C., Pike, G. B. and Enzmann, D. R. (1995). Magnetization transfer MR of the normal adult brain, *American Journal of Neuroradiology* **16**, 10, pp. 2085–2091.

33. Ge, Y., Grossman, R. I., Babb, J. S., Rabin, M. L., Mannon, L. J. and Kolson, D. L. (2002). Age-related total gray matter and white matter changes in normal adult brain. Part II: quantitative magnetization transfer ratio histogram analysis, *American Journal of Neuroradiology* **23**, 8, pp. 1334–1341.

34. Bashir, A., Gray, M. L., Boutin, R. D. and Burstein, D. (1997). Glycosaminoglycan in articular cartilage: In vivo assessment with delayed Gd(DTPA) (2-)-enhanced MR imaging, *Radiology* **205**, 2, pp. 551–558.

35. Regatte, R. R., Akella, S. V. S., Wheaton, A. J., Lech, G., Borthakur, A., Kneeland, J. B. and Reddy, R. (2004). 3D-T1ρ-relaxation mapping of articular cartilage, *Academic Radiology* **11**, 7, pp. 741–749.

36. Reddy, R., Insko, E. K., Noyszewski, E. A., Dandora, R., Kneeland, J. B. and Leigh, J. S. (1998). Sodium MRI of human articular cartilage in vivo, *Magnetic Resonance in Medicine* **39**, 5, pp. 697–701.

37. Ling, W., Regatte, R. R., Navon, G. and Jerschow, A. (2008). Assessment of glycosaminoglycan concentration in vivo by chemical exchange-dependent saturation transfer (gagCEST), *Proceedings of the National Academy of Sciences of the United States of America* **105**, 7, pp. 2266–2270.

38. Wolff, S. D., Chesnick, S., Frank, J. A., Lim, K. O. and Balaban, R. S. (1991). Magnetization transfer contrast: Mr imaging of the knee, *Radiology* **179**, 3, pp. 623–628.

Chapter 8

HyperCEST Imaging

Leif Schröder

ERC Project BiosensorImaging, Leibniz-Institut für Molekulare Pharmakologie (FMP), Campus Berlin-Buch, Robert-Rössle-Str. 10, 13125 Berlin, Germany
lschroeder@fmp-berlin.de

This chapter discusses applications of the chemical exchange saturation transfer (CEST) technique in the context of hyperpolarized (hp) exchanging xenon. For basic aspects of CEST with ^{1}H magnetic resonance imaging (MRI), the reader is referred to the previous chapters of this book. Herein we focus on systems with hp xenon that does not participate in covalent bonds (like in diaCEST) or ligand coordination (paraCEST). Instead, it is a somewhat special case of compartmental exchange as described in the proposed nomenclature by van Zijl et al. [1].

We will first give a short overview of the origins of hyperCEST and its links to the CEST development in general. This is followed by a section explaining the special conditions using hp nuclei for CEST, including the production, delivery, and data recording and evaluation for ^{129}Xe nuclear magnetic resonance (NMR). The design of hyperCEST agents is discussed next, and recent examples for applications in biomembrane studies and live cell NMR are given to illustrate the potential of this method.

Chemical Exchange Saturation Transfer Imaging: Advances and Applications
Edited by Michael T. McMahon, Assaf A. Gilad, Jeff W. M. Bulte, and Peter C. M. van Zijl
Copyright © 2017 Pan Stanford Publishing Pte. Ltd.
ISBN 978-981-4745-70-3 (Hardcover), 978-1-315-36442-1 (eBook)
www.panstanford.com

8.1 HyperCEST in the Historic Context of CEST Development

The development of hyperCEST imaging, pioneered by Pines and co-workers [2], since 2005 was largely driven by the desire to detect Xe biosensors at low concentrations. Such biosensors [3] were designed to be responsive contrast agents for targets of biological importance. They consist of a targeted host molecule (often a molecular cage such as cryptophane [4]) and Xe as the actual NMR-active probe that is temporarily bound to it. Caged Xe has a significantly different chemical shift (of the order of 100 ppm) from free Xe and is, therefore, easy to identify in the NMR spectrum (see Fig. 8.1a). The noble gas constantly undergoes reversible complex formation with the cage. This is the property that is used in hyperCEST to amplify the biosensor signal, but it was actually first used in a different context: Since the hyperpolarization is a non-renewable resource (see Section 8.2), its "consumption" should be well-considered, i.e., only the most necessary magnetization should be used for the encoding of each subset of information. This is called the "spin bank" concept where most of the precious hp magnetization is "deposited" in a condition or state (be it just a different chemical shift or really a different eigenstate of the system) until it is really needed for signal encoding and readout. The exchange properties of Xe were, therefore, used in the context of selective excitation [5] as follows: A selective detection pulse "consumes" only magnetization from the dilute pool of caged Xe, which is subsequently replaced by untouched hp free nuclei through chemical exchange with the abundant pool. The readout is repeated after a certain delay (tens of milliseconds) in a loop, and the correct timing for this loop is adjusted after identifying the exchange rate. This can be done by means of a selective inversion-recovery experiment, which illustrates that the Xe replacement happens on a timescale of tens of milliseconds [6] at room temperature in aqueous solution for a cage such as cryptophane-A (CrA) with a cavity of approximately 95 $Å^3$ van der Waals volume. This acquisition scheme was later used by Berthault et al. [7] who also applied it for image detection [8].

However, the concept of detecting the dilute pool after replenishment from the abundant pool becomes difficult to implement

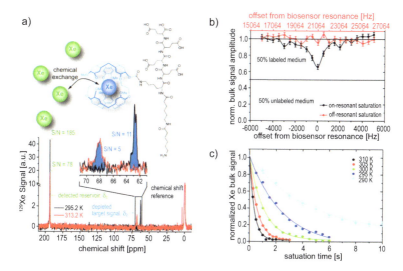

Figure 8.1 Basic hyperCEST system components and properties. (a) Hyperpolarized Xe atoms (green) reversibly bind to Xe hosts (here cryptophane-A cages, blue) that are coupled to a solubilizing peptide side chain (black). Free ^{129}Xe is observed at approximately 193 ppm in the NMR spectrum (relative to the gas phase signal). The caged Xe resonates at approximately 60–70 ppm and is very sensitive to temperature (reproduced from [9] with modifications) (b) First hyperCEST response of a cage-biotin conjugate bound to avidin-coated agarose beads that fill up 50% of a perfusion phantom. Sweeping of the resonance of caged Xe yields a signal decrease of free Xe (black data), whereas the off-resonant control (red) shows no response. (c) Depolarization curves of free Xe showing increased hyperCEST effect from caged Xe (20 μM cage concentration) at higher temperature. Reprinted with permission from Ref. [10], Copyright 2008, John Wiley and Sons.

when multiple signals should be detected separately or when a slice-selection gradient for imaging purposes is applied: In the first case, the concept of selective radiofrequency (RF) pulses with sufficient narrow band width makes them undesirably long in the time domain. The required pulse length might become longer than the residence time of the exchanging nuclei and, therefore, result in poor flip-angle accuracy. For the second case, the application of gradients smears out the resonance of caged Xe. This is problematic when multiple sensors are of interest at the same time ("multiplexing" approach). Such conditions require different sensors to have extremely large chemical shift separations. This is not realistic in final applications and was achieved only under rather artificial conditions [7].

It, therefore, became evident that it would be more beneficial to use the chemical exchange that connects the two pools in the inverse direction, i.e., detecting the solution pool of free Xe. Saturation transfer was a well-known technique since the days of Forsén's seminal paper [11], and both CEST [12] and paraCEST [13] techniques were already available for proton MRI. Saturation transfer with hp Xe was also known in terms of xenon transfer contrast (XTC) [14–16], albeit being developed for different conditions and not associated with the CEST concept; XTC is applied to gas phase detection at an interphase characterized by relative fast exchange. With regard to biosensor experiments in the solution phase, it was, therefore, tested if continuous wave (CW) saturation of caged Xe could induce a decrease in the signal from free Xe. Initial applications in the Pines lab were performed with avidin–agarose beads to which a biotinylized sensor had a high affinity. Such bead slurries were placed into a perfusion phantom that ensured delivery of hp Xe (which was bubbled into solution just prior to each measurement) [17]. Due to the nature of this setup, there were initial concerns whether the saturation effect of bound Xe would propagate fast enough into the solution pool within such a bead slurry. The feasibility of this kind of CEST experiment was first presented as a late breaking abstract at the ENC 2006 (poster abstract # 61 in [18]). Figure 8.1b shows one of the early Z-spectra obtained from that setup. The baseline noise illustrates that fluctuations in the delivery of hp gas for each data point were a non-negligible issue in those days and makes careful Xe delivery one of the important preparation steps for collecting high accuracy data (see also Section 8.2.3).

Another concern was how much information could be "stored" in the Xe magnetization, i.e., if the build-up of the CEST effect would happen fast enough for micromolar cage concentrations to be detected. Though being a transient, non-equilibrium condition, the hyperpolarization has a lifetime in solution that is often >100 s. Indeed, strong saturation transfer effects can be achieved under realistic conditions, as shown in simulations [19] and in experimental data [20, 21] such as Fig. 8.1c. The transient character of the hp magnetization is, therefore, not prohibitive, and build-up of the CEST effect is fast enough for significant CEST responses even at low agent concentrations with saturation times in test solutions of up to 20 s or more [9].

Following non-localized NMR acquisition, the implementation of imaging experiments was tackled. In order to differentiate this method from other CEST approaches, the quest for naming the technique became evident. It was common practice in the emerging field of CEST agents at that time to include the nature of the magnetism of the agent into the name to further distinguish various probes. To signalize the use of hp nuclei over para- or diamagnetic agents, the method was then coined hyperCEST. Implementation of the first imaging sequences differed in one particular aspect compared to the proton MRI work: The early imaging experiments for biosensors with dissolved Xe showed that the solution pool was not necessarily represented by a unique chemical shift [17]. Two solution signals occurred that represent different molecular environments (pure perfusion medium and bead slurry). Hence, not all the magnetization would necessarily be affected by saturation transfer. HyperCEST imaging was, therefore, first implemented with a chemical shift imaging sequence [2] that preserves the spectral separation of different solution peaks and allows for more selective evaluation of the CEST effect. Since this is a rather slow type of image encoding, it was later replaced by faster MRI sequences (see Section 8.2.3).

It should, nevertheless, be emphasized that it is one of the peculiarities of CEST experiments with Xe that the abundant spin pool may split up into more than one resonance and the exchange modeling has to be adapted accordingly. This can become an important aspect for experiments with phospholipid membranes where free Xe exists in both aqueous and lipid environments, but the pool of Xe that is specifically affected by RF saturation is formed by caged Xe in the lipid environment (see Section 8.4).

For the sake of completeness, it should also be mentioned that a signal transfer from the dilute pool of caged Xe to that of free Xe can also be achieved without saturation transfer but with exchange magnetization transfer. This method works with a pair of selective 90°

excitation pulses, separated by an evolution interval, τ_{evol}. During this interval, excited resonances evolve as transverse magnetization before being stored as longitudinal magnetization and detected after chemical exchange [5]. The spectral information is contained in the intensity modulation of the detected signal. The modulation is sampled by step-wise increasing τ_{evol} and applying a Fourier transformation with respect to this dimension. A comparison with CEST is briefly discussed in Ref. [22]. One important difference is that unlike for CEST (which allows identifying the presence of caged, exchangeable Xe in just two acquisitions), a whole series of data points must be acquired in order to perform the Fourier transformation. This principle has also been applied in the context of exchanging protons for ^{1}H NMR, where it has been termed frequency-labeled exchange (FLEX) [23, 24].

8.2 Hyperpolarized Xenon NMR

HyperCEST particularly benefits from the combination of two amplification strategies: hyperpolarization and saturation transfer. Both address the intrinsically poor sensitivity of NMR. Numerous techniques have been developed in this context. For the directly detected magnetization, the aim is to artificially increase the difference of the spin state populations, i.e., not relying on the thermal spin polarization that is governed by the Boltzmann distribution of the thermal equilibrium [25] (Fig. 8.2a,b). Such a system is then said to be hyperpolarized and provides a detectable magnetization that is increased by several orders of magnitude. In fact, such a preparation step is actually achieving quite the opposite effect of what the CEST saturation pulse is eventually doing. But this preparation comes with two benefits: (i) The starting magnetization is significantly increased; hence the ensemble can be detected at rather low spin density, and (ii) the intrinsic T_1 relaxation back to thermal equilibrium usually does not significantly counteract the saturation effect. Hence, the situation of saturation hp nuclei is conceptually different from proton CEST where the desired saturation competes with counteracting ^{1}H T_1 relaxation on the same timescale. The maximum achievable CEST effect for ^{1}H

Hyperpolarized Xenon NMR | 127

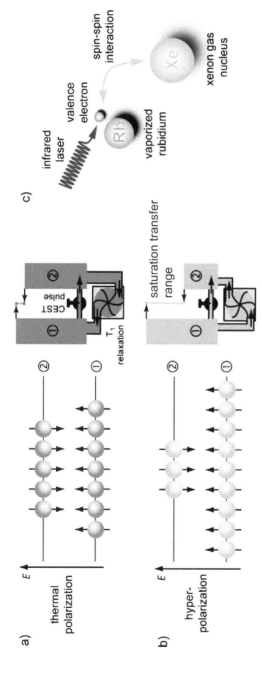

Figure 8.2 Properties of hyperpolarized nuclei compared to thermal polarization. (a) Population densities in spin-1/2 system for thermal polarization; the CEST pulse acts like a valve to equalize the population density of the ground state (①) and the excited state (②). Intrinsic T_1 relaxation at the same time acts as a pump to maintain the difference. (b) Hyperpolarized nuclei have a significantly increased population difference, and relaxation (which is also slower) does not counteract the saturation process. The dynamic range for saturation transfer is, therefore, larger than for thermally polarized nuclei. (c) Key elements of spin exchange optical pumping.

is usually limited to a fraction of the initial thermal magnetization and also reaches a steady state after a few seconds, which limits the dynamic range to encode information (Fig. 8.2a,b). Hyperpolarized Xe, therefore, provides excellent conditions for CEST after preparing a non-equilibrium magnetization.

Various techniques are available to generate hp spin systems [25]. The most commonly used ones are dynamic nuclear polarization (DNP), para-hydrogen induced polarization (PHIP), and spin exchange optical pumping (SEOP, Fig. 8.2c). They utilize either spin polarization or high spin order of a precursor (unpaired electron spins in organic free radicals in DNP, para-hydrogen in PHIP, or Rb valence electrons in SEOP) that is eventually converted into a hyperpolarization of the detected spin system. The achieved enhancement factors can be quite high ($>10^4$). DNP and PHIP produce hp tracers where the enhanced magnetization is not participating in chemical exchange (at least up to now). SEOP produces hp noble gases like He [26] but also Xe and Kr [27] of which Xe has been the candidate of choice for CEST applications.

8.2.1 Xenon NMR Conditions Compared to Protons

Xenon is ideal for saturation transfer NMR because, as a noble gas, it forms reversible complexes with certain (macro-)molecular structures. To this end, the design of the hosts (see Section 8.3) allows somewhat adjusting the exchange properties for tuning the CEST performance. RF saturation is also convenient in this context since the isotope ^{129}Xe (with a natural abundance of 26.4%) is a spin-1/2 nucleus and the two-level spin system is not involved in any scalar couplings. Hence, it can be easily saturated. Moreover, its large chemical shift range [28, 29] provides ideal conditions for separating the saturation and the detection pool (even at lower field strengths). Xe, therefore, unites several advantages that are sought after in ^{1}H CEST and that are only partially available through paraCEST agents [30]. In fact, the absolute chemical shift separations in xenon systems can be comparable to paraCEST agents despite a 3.6-fold lower gyromagnetic ratio compared to protons.

However, spin density and relative receptivity compared to proton experiments are rather low when working with thermal polarization. The mole fraction solubility for Xe in water is 7.89 × 10^{-5} at 25°C and 1 atm pressure (pages 5–149 of Ref. [31]). This corresponds to an Xe concentration of 4.38 mM if 1 atm of pure Xe is applied to water. Isotopically non-enriched Xe, therefore, provides only 1.16 mM of detectable ^{129}Xe under these conditions, which has to be compared to 110 M proton concentration in pure water. Experimental conditions can sometimes be chosen to provide a larger Xe pool, but even in non-polar solvents such as DMSO, the solubility is only approximately six-fold higher, which provides limited improvement for testing novel applications [21].

Another sensitivity aspect for any ^{129}Xe application is that the smaller gyromagnetic ratio reduces the detectable magnetization to 27.8% of the ^{1}H magnetization (at comparable spin ensemble size). It induces an RF signal at 3.6-fold reduced frequency, which further decreases the sensitivity. Both effects yield a relative receptivity of 2.16 × 10^{-2} compared to ^{1}H and make it impractical to perform xenon NMR with low-spin densities under conventional circumstances. Artificially increasing the population difference of the spin states by several orders of magnitude, through hyperpolarization, is the method that makes NMR of dissolved Xe possible and the aforementioned advantages accessible. The method to achieve this condition and the delivery of Xe to the site of interest will be described in the following sections.

8.2.2 Production of Hyperpolarized Xenon

The ultimate goal of manipulating the spin state populations is to achieve an enhanced macroscopic magnetization. In thermal equilibrium, the magnetization M of a spin-1/2 system is given by the total spin number N times the ensemble expectation value for the magnetic moment μ. This expectation value $<\mu>$ that is governed by the Boltzmann distribution yields the Curie law with $\mu = \gamma\hbar/2$

$$M = N<\mu> = N\mu \tanh(\gamma\hbar B_0/2k_B T) \qquad (8.1)$$

Rewriting this as $M = N\mu P$ yields the polarization

$$P = \tanh(\gamma\hbar B_0/2k_B T) = (N_\uparrow - N_\downarrow)/(N_\uparrow + N_\downarrow) \qquad (8.2)$$

which is eventually expressed as the population excess of one spin state (↑) over the other (↓), normalized by the sum of the populations. For thermal equilibrium at room temperature, this quantity is very small ($P_{th} \approx 10^{-8}\ldots10^{-6}$ at common field strengths of B_0 ~3–9 T). The hyperpolarization $P \gg P_{th}$ is achieved in a three-step polarization transfer process called SEOP (Fig. 8.2.c). A brief summary of the process will be given here, but for a detailed description of the physics behind the SEOP method, the reader is referred to Ref. [32].

The initial step is the production of circularly polarized light, i.e., the generation of photons with a well-defined angular momentum. This parameter is easy to control and can be prepared in a very pure state to subsequently pump a particular transition in an effective one-electron system. Herein, the photon angular momentum is used to influence the electron spins of an alkali metal vapor. This transfer represents the second step. Most polarizer setups work with rubidium vapor for which a laser emitting at 795 nm can drive the D1 transition to achieve high Rb electron spin polarization. A droplet of Rb is partially vaporized inside a glass cell and is exposed to a magnetic field of approximately 2–3 mT. The circularly polarized light traveling along the field axis selectively pumps the transition $5^2S_{1/2} \rightarrow 5^2P_{1/2}$ with a spin flip $m_s = +1/2 \rightarrow m_J = -1/2$. Within the $5^2P_{1/2}$ state, collisional mixing equalizes populations of the two sub-levels and is followed by radiation-free relaxation that populates both sub-levels of the $5^2S_{1/2}$ ground state. This relaxation is achieved through optical quenching with N_2 gas in the pumping cell. The result is a significant over-population of the spin-up level of the $5^2S_{1/2}$ state since the other one is continuously depleted by the laser. The third step is the transfer of this electron spin polarization onto the Xe nuclear spin system. To do so, the gas mixture containing the Xe must simply come in contact with the highly polarized Rb vapor. As the Xe flows through the vapor-containing pumping cell, dipolar interactions occur between Rb electron spins and Xe nuclear spins that come close enough to each other. They lead to flip-flop processes that convert many of the xenon spin-down states into the spin-up condition.

Recent progress in the production of hp Xe relies on high-power laser systems (>100 W CW output) that incorporate volume

Bragg gratings to achieve a line-narrowed laser emission [33]. Such gratings achieve a small bandwidth by reflecting a narrow section of the initial spectrum back onto the diode to force it to emit at this very frequency. This yields excellent pumping conditions for up to near unity polarizations of Xe under certain conditions [34].

However, careful design has to be applied when using such high photon densities, especially in the context of CEST experiments that require stable re-delivery of the hp gas (see below). Due to the required N_2 quench gas, most of the photon energy is eventually converted into heat. This can become a serious problem for compact systems where the Rb droplet resides inside the pumping cell that is illuminated by the laser. The laser-induced heating leads to temperatures that can significantly exceed that of the desired operating temperature [35]. Such overheating initiates further Rb vaporization with subsequent light absorption. This chain reaction is called "Rb runaway" and leads to inhomogeneous cell illumination with a decrease in Xe polarization and serious signal fluctuations. A dedicated temperature-monitoring system and heat dissipation device can prevent this deleterious side effect [36]. It allows operating an Xe hyperpolarizer under continuous flow conditions with excellent signal stability.

8.2.3 Delivery of hp Xe and Optimized Use of Magnetization

A limitation of hp nuclei is the fact that the valuable spin polarization does not recover by itself but rather requires a re-delivery of fresh hp Xe into the sample before the next measurement can be acquired. It is because of this different starting condition that hyperCEST faces different challenges than ^{1}H CEST: (a) Stable Xe delivery is crucial to eliminate shot-shot-noise that otherwise masks true CEST responses from false-positive effects and (b) the available magnetization should be used as efficiently as possible to reduce the repetitions of gas delivery to a minimum in the interests of reducing experimental time.

We will first focus on the delivery aspect since this is of higher importance for acquiring high-quality data, specifically because CEST measurements aim to detect a relative signal decrease from the starting signal. Immediate use of the gas from the polarizer as it leaves the pumping cell is possible and became the method of choice for many CEST applications. This method works only with the Xe partial pressure as it is in the pumping cell (typically around 100 mbar total Xe pressure, i.e., approximately 25 mbar of detectable ^{129}Xe) but has the advantage that it can be used in a continuous flow setup where the sample can remain inside the NMR magnet without further repositioning or manual addition of freshly produced hp Xe. Additionally, it comes with the convenience that the pressure from the polarizer can be used to bubble the gas into solution [37]. However, it requires keeping the conditions in the pumping cell rather stable. This includes maintaining a gas flow even when the bubbling is stopped for a moment during NMR signal acquisition. Based on triggered valves for controlled gas dispersion [37], a setup equipped with heat management of the pumping cell and with mass flow controllers and valves controlled through the NMR pulse sequence can achieve very precise Xe delivery [36, 38] with excellent resolution in Z-spectra (see Fig. 8.3a).

The second aforementioned aspect of optimum use of the magnetization is also somewhat applicable for para- and diaCEST agents due to the sometimes long repetition times that are needed for sufficient T_1 relaxation back to the initial thermal magnetization. Some of the solutions developed for Xe CEST experiments, therefore, also apply to CEST experiments in general. Nevertheless, the repetition cycles for each Xe re-delivery into sample solutions can be even longer than those usually applied in ^{1}H CEST. Optimized shared use of the Xe magnetization for multiple encodings can be used for certain applications to reduce this problem. Also, transverse magnetization should be refocused in echo trains as often as possible to make most efficient use of the signal for image encoding. Relatively long T_1 and T_2 relaxation times make it possible to apply only small excitation flip angles and/or acquire echo trains for encoding multiple lines of k-space. When single-shot acquisitions are used, an echo train with refocussing RF pulses after the initial 90° excitation can be used to encode multiple readout lines. Such schemes have been applied as RARE sequences [20]. Alternatively, gradient echoes can be applied if smaller excitation angles are used to preserve a certain longitudinal magnetization for the next

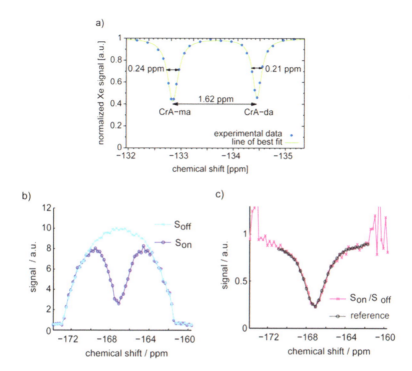

Figure 8.3 Different implementations of hyperCEST spectroscopy. (a) High-resolution spectrum with point-wise acquisition (reproduced with permission from *J. Chem. Phys.* **140**, 084203. Copyright 2014, AIP Publishing LLC. [36]). (b) Projection profiles for gradient-encoded CEST spectroscopy. The first profile (S_{off}) is the reference, and the second (S_{on}) carries the CEST information. (c) Reconstructed spectrum from (b) (pink) and conventional encoding (grey) for comparison. Reprinted with permission from Ref. [39], Copyright 2014, John Wiley and Sons.

acquisition [21, 40]. This scheme has also been used for acquiring Z-spectra as one-dimensional projection profiles [39, 41, 42] (which will be discussed in more detail in Section 8.2.4). The idea of using variable flip angles (VFAs) for hp gas was first implemented for lung imaging [43]. It generates equal amounts of transverse magnetization until the last encoding step can be done with a 90° excitation pulse. The minimalistic version of this idea can be used for CEST experiments that require two acquisitions (reference and saturation encoding): The so-called smashCEST approach (CEST with shared magnetization after single hyperpolarization, [21]) uses a 45°/90° excitation scheme. Careful calibration of the pulses allows acquisition of MRI scans with a single delivery of hp nuclei and encode the entire image in less than 1 s.

The acquisition of data for a whole series of saturation frequencies (with or without further spatial encoding) is more demanding. Conventional point-wise registration of such a hybrid of imaging and spectral information, therefore, benefits from sub-sampling and reconstruction techniques that are based on redundancies in the data sets. Two examples are principal component analysis and constrained reconstruction of undersampled CEST image data [40]. The correlations between data points in the spectral domain not only allow reduction in the acquisition time, but also provide each encoding step with more magnetization. A five-fold increase in SNR has been demonstrated with this technique.

8.2.4 Fast Spectral Encoding (Gradient-Encoded CEST)

Besides the desire for fast image encoding of the prepared Z-magnetization from hp nuclei, fast encoding of the spectral dimension of the participating CEST pools is also of interest. Conventional Z-spectroscopy applies a selective saturation pulse successively at variable carrier frequencies. This happens in the presence of a homogeneous magnetic field. The problem can be inverted by applying a well-defined field gradient for controlled inhomogeneous broadening in the presence of a single saturation frequency bandwidth. The result is a one-dimensional projection measurement (see Fig. 8.3b) that encodes the spectral information along the gradient's direction (subject to the condition that the available magnetization is sufficiently large to be spread over a certain frequency bandwidth). The use of multiple saturation frequencies becomes redundant, and a significant acceleration is achieved for Z-spectroscopy as the spectrum is retrieved after normalizing the on-resonant saturation data to the off-resonant control profile (see Fig. 8.3c). This approach was first introduced for magnetization transfer MRI in 1991 [41] and later for proton CEST [42, 44] and then extended to CEST with hp Xe [39, 45]. The conveniently slow T_2 relaxation of hp Xe was used in this context to increase the SNR by means of echo refocussing. Further acceleration was also achieved by using VFAs after a single Xe delivery. As such, the gradient encoding can yield an entire Z-spectrum with two excitation pulses [39] based on the smashCEST approach [21] or yield quantitative information about the exchange rate constant [45] based on the QUEST method [46].

8.3 Xenon Host Structures

The affinity of Xe for hydrophobic binding pockets and clefts of proteins has been used in many structural biology studies [47, 48]. Since the interaction time is often short, the NMR spectra in many cases just exhibit a shift of the Xe solution peak or a change in the line shape. These effects can be quantified by plotting their magnitude versus the protein concentration in solution. However, such conditions do not allow for CEST experiments. Various approaches have been implemented to design a distinct second NMR resonance that is in chemical exchange with the solution pool. The following discussion groups them into two categories, namely tailored hosts and compartmentalization of Xe.

8.3.1 Tailored Host Structures: Cryptophanes

The search for tailored, i.e., synthetic host structures for Xe atoms (van der Waals diameter of approximately 0.43 nm) with slow binding characteristics on the NMR timescale identified cryptophanes as potentially powerful tools for use in the development of xenon-based biosensors about 10 years ago [3]. Cryptophanes induce a change in chemical shift of dissolved Xe of more than 100 ppm and binding constants up to 42,000 M^{-1} have been reported depending on the solvent [49]. Since these cages come with different cavity sizes, optimization of this parameter was initially performed with the aim of increasing the binding constant [50]. In the context of CEST detection, however, this aspect must be seen in combination with the desire for a relative fast exchange to achieve efficient saturation transfer. To balance these two aspects, different versions of cages with intermediate cavity size are of particular interest [51].

Cryptophane-A (Cr-A, also coined Cr-222, see also Fig. 8.1a) has been used in many studies, in particular within the development of hyperCEST imaging [2, 20, 21]. The fractional occupancy for cryptophane-A conjugates in water is approximately 60% [37], but they induce nevertheless very efficient CEST effects. For a comparison with ^{1}H CEST performance, the reader is referred to Refs. [22, 30].

The high hydrophobicity of cryptophanes was first considered a problem for biological applications but can be addressed through chemical synthesis to modify cryptophanes [52–59]. Xenon in cryptophanes covers a convenient range of chemical shifts [4] that is of interest for facilitating the potential to multiplex several xenon hosts simultaneously, a concept which was already suggested in the original biosensor paper [3]. This concept is also known from proton CEST [60] but faces, at least for diaCEST, notable limitations due to significant overlap of CEST responses in ^{1}H spectra. The detection of multiple Xe CEST pools is much easier to implement, due to the larger chemical shift range, and has been successfully demonstrated under different conditions [7, 21, 36, 61]. CEST responses with frequency separations of 130–180 Hz could be resolved due to the fact that weak but long saturation pulses can be applied to ensure high spectral resolution [21, 36].

As for other processes of chemical exchange, the rate constant is increased with elevated temperature. The Xe@cryptophane system shows a convenient increase in this parameter in the physiological temperature range compared to room temperature and, therefore, yields improved saturation transfer for $T > 25°C$ [9, 10, 62]. This adjustable amplification has some similarities with a transistor and has led to the "transpletor" concept [10] in which the cages are used as controllable depolarization gates. HyperCEST has, therefore, been used for non-invasive NMR thermometry [62]. It should be mentioned that such Xe hosts come with adjustable CEST sensitivity ranges by using different concentrations: The large dynamic range allows to probe changes at higher temperatures with more dilute CEST pools [9]. The same behavior is also of great importance for applications with live cells. While conventional NMR detection of exchanging systems usually suffers from increased temperature due to line broadening, increased temperature has been used to detect fairly low total cage concentrations like 20 nM for imaging CD14 expression on macrophages [63]. Of course, increased temperature is also helpful to illustrate the potential of novel cages before they are studied in cellular environment. An example is hyperCEST detection of 1.4 pM of a triacetic acid cryptophane-A derivative (TAAC) at 320 K [52].

8.3.2 Compartmentalization of Xenon

An emerging class of Xe hosts is based on the concept that gas-binding nanocarriers can serve as compartments for larger numbers of gas atoms. It is well known that the lipophilicity of Xe causes easy uptake of the noble gas into phospholipid membranes. However, this type of compartmentalization is usually characterized by relative fast exchange [61], which has also been observed in experiments with different cell lines [64]. The result is a limited spectral resolution of the peak of xenon in lipid environment from the peak of free Xe in aqueous environment. Ideal carriers, therefore, come with a larger chemical shift separation that can be resolved despite the relative fast exchange. Compared to cryptophanes with their tailored cavities, these carriers can then induce very strong CEST effects at quite dilute concentrations. This was first demonstrated with PFOB nanodroplets [65] and an impressive sensitivity in the femtomolar range. A more recent application relies on genetically engineered gas vesicles [66] that even come with some parameter space to tune the frequency of the CEST response and options for functionalization. Another example for creation of a saturation pool of exchanging hp Xe without high-affinity xenon-binding sites such as cryptophanes is the recent investigation of bacterial spores [67]. Detection limits of 10^5 to 10^9 spores per mL were reported for this type of Xe carrier. Though no obvious way of functionalizing such spores for further targeting applications is on the horizon, it is noteworthy that hyperCEST NMR can help to elucidate spore composition and function that are otherwise difficult or impossible to access. In fact, proton CEST MRI turned out to be unsuccessful to detect spores [68].

A somewhat hybrid version of the compartmentalization approach is the use of cryptophanes as individual hosts in combination with nanocarriers. Partitioning of cryptophanes into lipid membranes was recognized early on [61, 69] and identified with additional chemical shifts (see Fig. 8.4.a), but it was originally considered an unwanted effect that causes unspecific binding [70]. However, it can also be used to form functionalized liposomal nanocarriers [71] (Fig. 8.4b) that provide very favorable conditions for saturation transfer (see Section 8.3.3).

8.3.3 Targeted Hosts

The large chemical shift range of Xe makes it an ideal candidate to design responsive CEST agents that detect the presence of other specific molecules or the change of parameters that influence the exchange rate or Xe-binding properties. For the sake of completeness, the reader is also referred to reviews focusing on biosensor synthesis [74–76].

In principle, all of the realized sensors are capable of performing hyperCEST experiments. However, only a certain subset has been tested with this detection method. These include biotinylated cages that bind to avidin [2] or cage conjugates that detect Pb and Cd ions [77]. Since some molecular targets like cell surface receptors occur at rather dilute concentrations (e.g., approximately 1.6×10^5 transferrin receptors on the plasma membrane per cell [70]), amplification strategies have been implemented that couple numerous sites for exchanging Xe to each targeting moiety (see also Table 8.1). A similar approach is known from paraCEST [78] and diaCEST [79] studies. The transferrin sensor study [70] is an example where this had already been demonstrated in a less extensive way by coupling a few cages to Lys residues albeit only measured via direct spectroscopy and not CEST. The recently reported modular design of a CD14-targeted sensor coupled multiple biotinylated CrA units to an avidin conjugated anti-CD14 antibody to achieve a higher payload of CrA per targeting unit [63]. Considering the natural abundance of ^{129}Xe and the typical fractional cage occupancy in water, the amount of NMR-active antibodies at any moment in time corresponds to 125 pM.

Higher cage payloads were demonstrated by grafting hundreds of cryptophane cages either onto viral capsids [80] or bacterial phages [73, 81]. This phage approach (see Fig. 8.4c) reached unprecedented

Xenon Host Structures | 139

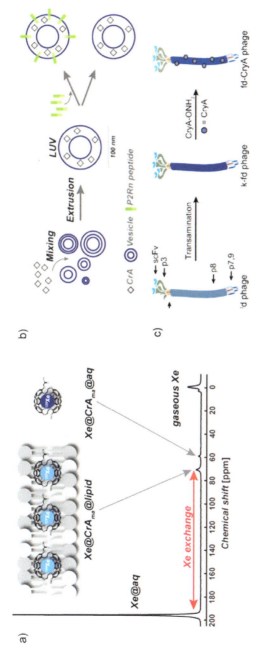

Figure 8.4 HyperCEST agents with high payloads of CEST sites. (a) Xe NMR spectrum with CrA cages in aqueous and lipid membrane environment (reproduced from [72]) shows two distinct cage-related resonances that can be used for CEST. (b) Mixing CrA cages (diamonds) with multilamellar vesicles and subsequent extrusion generates cage-loaded liposomes (large unilamellar vesicles, LUV) that can be functionalized with lipopeptides (green) for cell-specific uptake. Control experiments are done with unlabeled LUVs (reproduced from [71]) (c) The surface of fd bacteriophages, which provide an antibody-fragment against the EGF receptor, can be modified via transamination for coupling with CrA cages. Reprinted with permission from Ref. [73], Copyright 2013, John Wiley and Sons.

sensitivity of 230 fM detection threshold for the entire sensor construct in solution [81] and was subsequently demonstrated for targeted binding to EGF receptors [73]—a target also used for the first studies with the aforementioned gas vesicles [66]. Even higher cage payloads were achieved by incorporating unfunctionalized CrA into liposomes (approximately 3800 per liposome) that were decorated with lipopeptides for targeted uptake into microvascular brain endothelial cells [71]. Such peptide-modified nanoparticulate carriers show significantly reduced cell toxicity with respect to their high local CrA concentration and good internalization properties with 7.7×10^5 liposomes per cell corresponding to a total liposome concentration in the sample of approximately 1 nM. Such concentrations are still sufficient to achieve a CEST effect of approximately 80% with a 10 µT pulse for 8 s saturation. The performance of these constructs provides promising perspectives for future applications.

8.3.4 HyperCEST Modeling

Just as for ^{1}H CEST, there have been studies focusing on the quantification of the CEST effect with hp Xe. The Bloch–McConnell equations, which include terms for the chemical exchange between two pools during simultaneous application of a saturation pulse, give a general framework that can also be applied to hyperCEST. However, detailed studies in this field are still in their infancy. Exchange rate constants for cryptophanes have predominantly been determined with selective inversion/saturation recovery experiments [6], and binding constants have been calculated through the integration of the resonances from free and caged Xe [37] or alternatively by isothermal titration calorimetry measurements [82].

The selective excitation experiments are difficult for CEST systems with overlapping resonances, and pronounced spill-over effects occur, e.g., for PFOB nanodroplets that serve as Xe hosts (this system had been previously investigated by line width analysis [83]). The Xe exchange rate for PFOB nanodroplets was, therefore, quantified by modeling the Bloch–McConnell equations to experimental Z-spectra [65]. Simplifications for this system of coupled differential equations have been developed for ^{1}H CEST

[84] and were later also presented for hyperCEST in the context of Xe@cryptophanes [19]: An analytical solution is possible in this case because the magnetization is always larger than the stationary solution. One important result is that the shape of the Z-spectra along the chemical shift dimension δ can be approximated by the sum of exponential Lorentzians

$$\mathrm{ExpLor}(\delta) = \exp(-(2A/\pi)(\Gamma^2)/((\omega_{\mathrm{res}} - \delta)^2 + \Gamma^2)) \quad (8.3)$$

where the parameter Γ is the full width at half maximum of the Lorentzian (and, therefore, related to the exchange rate constant) and ω_{res} is the resonance frequency of the CEST pool. Fitting of experimental data to such line shapes found applications in various studies [20, 21, 36, 71] where it has not necessarily been used to quantify exchange but to facilitate decomposition of overlapping CEST responses from caged Xe in different molecular environments (see Section 8.5).

8.4 Phospholipid Membrane Studies/Delta Spectroscopy

As mentioned earlier, cryptophanes are well known to partition into membranes. This observation triggered experiments to investigate whether it would be possible to reveal differences in membrane composition [69]—an effect that might also contribute to effects observed for comparative cell studies [64]. However, the chemical shift differences for membrane-associated cages are rather small compared to the observed width of CEST responses, as also confirmed later in studies with different cell lines [85].

Despite this spectral overlap, the hyperCEST response of CrA cages in different model membranes differs in amplitude [69]. The effect was attributed to differences in membrane fluidity. POPC (1-palmitoyl-2-oleoyl-sn-glycero-3-phosphocholine) with its one unsaturated fatty acid showed significantly stronger effects compared to DPPC (1,2-dipalmitoyl-sn-glycero-3-phosphocholine; saturated fatty acids) under otherwise identical conditions. The effect was then studied in more detail, and a quantitative measure (previously used in saturation transfer context for amide protons [86]) was implemented as follows: The build-up of the CEST effect can be recorded in detail due to the large dynamic range of hyperCEST. The time domain of this build-up allows introduction of another dimension to resolve different hyperCEST pools even if they have the same chemical shift. The method is called DeLTA for Depolarization Laplace Transform Analysis of the CEST-driven depolarization of Xe atoms that have access to membrane-embedded CrA [72, 87]. The Laplace transformation with respect to this time domain yields the depolarization times that were then assigned to the different fluidity of the phospholipid membranes under investigation.

The fluidity itself is determined by the packing and ordering of the membrane's components. Permeation of Xe atoms in and out

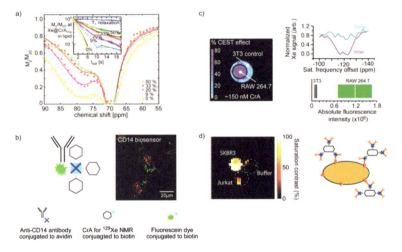

Figure 8.5 HyperCEST applications for membrane and cellular studies. (a) CEST responses and DeLTA analysis for different cholesterol levels in POPC liposomes (reproduced from [72]). (b) Modular setup of CD14-targeted biosensor and its cellular localization observed by fluorescence microscopy. (c) HyperCEST MRI of CD14 biosensor and FACS data for illustrating specificity (b,c reproduced from [63]) (d) Cancer cell labeling strategy with gas vesicles (GVs): saturation contrast image of three-compartment phantom that contained SKBR3 cells labeled with antibody-functionalized GVs, similarly labeled Jurkat cells and unlabeled SKBR3 cells. Anabaena flos-aquae GVs (black) are functionalized with anti-HER2 antibodies (orange) via biotin–avidin conjugation (grey, blue). The antibody recognizes the HER2 receptor (red) on SKBR3 cells. EGF-targeted gas vesicles. Reprinted with permission from Ref. [66], Copyright 2014, Nature Publishing Group.

of the membrane is hindered by low membrane fluidity as is the diffusion of CrA and Xe atoms within the membrane. Overall, such conditions induce a slow build-up of the hyperCEST effect with a long depolarization time. In POPC, the depolarization time was seven-fold faster than in DPPC at 303 K. The DeLTA method can be combined with imaging capabilities in a pixel-wise analysis to obtain an MRI contrast that illustrates differences in the fluidity of model membranes [87]. Different applications are currently under investigation, including the quantification of other membrane components such as cholesterol (see Fig. 8.5a) or the onset of domain fluctuations for various compositions of inhomogeneous membranes that undergo transitions between the liquid-disordered and solid-ordered phase [72]. Overall, DeLTA complements existing fluorescence-based methods for detecting membrane fluidity but is not limited by penetration depth for further applications and works with low concentrations of CrA as the NMR-sensitive probe (10^{-6} M).

8.5 Live Cell NMR of Exchanging Xenon

Xenon MRI applications in living organisms face special challenges compared to conventional ^{1}H MRI: Being an exogenous substance, Xe needs to be delivered to the point of interest and remains only detectable if the hyperpolarization life time is not dramatically reduced by the in vivo environment. Studies with live cells in the NMR magnet represent, therefore, an important intermediate step toward in vivo applications.

Early studies with hp gas on red blood cells [88] revealed transmembrane passage of Xe and determined the time constant for chemical exchange between plasma and erythrocytes to be 12 ms. This illustrates sufficient access for the NMR-active nuclei to the intracellular environment and triggered the development of Xe hosts for cellular uptake. Initial work focused on cryptophane cage delivery through cell penetrating peptides [89, 90]. These studies did not include Xe NMR data but illustrated already the successful delivery with fluorescent reporters attached to the construct. Detecting the cellular uptake of cryptophane cages with NMR is difficult to achieve without the CEST approach since most of the hp nuclei are not cage associated. However, the aforementioned stable re-delivery of hp Xe (see Section 8.2.3) is necessary to allow reliable identification of cell-associated host-guest complexes. The one-time delivery of a highly concentrated Xe atmosphere onto the cell suspension with subsequent shaking (as applied in Refs. [64, 70, 88]) is not suitable for high-resolution CEST. Repetitive bubbling of xenon into a cell suspension is one approach, but cells handle this process of direct gas bubbling very differently: Some studies report good survival for Jurkat cells [73], while other results indicate substantial cell damage due to shear forces induced by bursting gas bubbles for MDA-MB-231 cells [73], L929 fibroblasts, and RAW 264.7 macrophages [85].

Interpretation of such data, therefore, requires knowledge about potential degradation of the sample during the experiment. HyperCEST spectra from the supernatant can help to identify contributions from such cell debris because membrane-associated CrA has a distinct chemical shift that is also observed for Xe in hosts that are associated with cell fragments [20]. However, direct gas bubbling of Xe into cell suspensions represents a convenient high throughput method, which is suitable for quickly screening the performance of novel biosensor constructs. With respect to more detailed understanding, incorporation of a fluorescent probe to the xenon host serves as a useful tool through which characterization of the cellar uptake and intracellular distribution can be evaluated (via fluorescence microscopy and flow cytometry, FACS). Table 8.1 summarizes studies of different Xe biosensors that used hyperCEST detection and that were either already used in cellular environments or that have the potential to be used for cellular targeting/labeling. Some of these rely on the previously mentioned scaffolding approaches to increase the local Xe host concentration.

For untargeted CrA constructs, distribution is not limited to the outer cell membrane as one might expect from liposome studies [69]. A cryptophane–fluorescein conjugate, e.g., appears all over the cytoplasm [20]. Internalization occurs presumably in membranes of lysosomal vesicles following fluid phase pinocytosis, a mechanism already proposed for the case of the transferrin-based biosensor [70]. A flexible approach to combine NMR and fluorescence reporters is given by the modular setup that was

Table 8.1 HyperCEST studies with different Xe biosensors either already used in cellular environments or with the potential to be used for cellular targeting/labeling

Xe host	Scaffolding (# of Xe hosts)	Molecular target	In cellulo tests	Read-out Spectrosc.	Read-out MRI	Read-out Fluoresc.	Ref.
CrA	–	–		✓	✓	✓	[20]
CrA	MS2 viral capsid (125)	–	✓	✓			[80]
CrA	M13 bacteriophage (1050)	–		✓			[81]
CrA	Fd bacteriophage (330)	EGF receptor	✓	✓		✓	[73]
CrA	Avidin-labelled CD14 antibody (16*)	CD14	✓	✓	✓	✓	[63]
CrA	CrA-loaded liposomes (3800)	Glycocalyx of brain endothelial cells	✓	✓	✓	✓	[71]
PFOB nanodroplets	N/A	–		✓			[65]
Gas-binding protein nanostructures	N/A	EGF receptor	✓	✓	✓		[66]

*based on 4:1 avidin:antibody stoichiometry

Note: Some of these are based on scaffolding approaches to increase the local CEST site concentration and also include fluorescence reporters for independent uptake verification and quantification from cell lysates. Readout methods include CEST spectroscopy, CEST MRI, and fluorescence microscopy.

recently proposed for a CD14-targeted biosensor. It allows to "click" different reporter combinations onto the targeting unit as needed to perform both MRI and fluorescence studies (Fig. 8.5b).

The incorporation of fluorescence reporters is also valuable for other reasons: Quantification of the internalized hyperCEST agent is not possible by evaluation of the CEST response alone. Many studies, therefore, work with constructs where there is also a fluorescent reporter linked to the construct. By knowing the cage/dye ratio, the uptake of such bi-modal agents can be quantified by fluorescence measurements of the lysate [89, 90] or through FACS experiments [63, 85]. This is very helpful to carry out mockup imaging studies in which a concentration equivalent of "naked" cage is administered to cells for optimizing hyperCEST imaging parameters before the real study with a new agent is done.

Despite the lack of quantification through the CEST response, it contains qualitative information about cell-association of Xe hosts because this process can be confirmed by a change in chemical shift for both cryptophane-based agents and PFOB. This serves as an indicator for cellular uptake that is not available from a simple fluorescent reporter with unchanged emission wavelength. As mentioned earlier for cryptophanes, their hydrophobicity was early recognized as the driving force to embed into membranes that causes a approximately 10 ppm downfield shift [61, 69, 70]. PFOB studies, however, showed an upfield shift but also provide sensitivity to the internalization process [91]. It is, therefore, noteworthy that Xe NMR can, in principle, distinguish between free CEST agents and internalized ones despite the detection method does not provide spatial resolution on the cellular scale.

The chemical shifts detected upon cellular uptake when direct gas bubbling is applied to cell suspensions are consistent with those detected in a live cell bioreactor [20]. Such a reactor ensures high cell viability but is more complex to operate (see next paragraph). Promising CEST candidates, therefore, undergo initial tests in the direct bubbling experiments, and the results can subsequently be translated to a bioreactor setup at a later stage. The direct gas dispersion method has, therefore, been applied for demonstrating the abilities of various constructs such as cage-dye building blocks [85] and the aforementioned sensors for EGF receptors [73], the CD-

14 targeted modular sensor approach [63] (Fig. 8.5c), the test of targeted liposomal carriers [71], and the studies with targeted gas vesicles [66] (Fig. 8.5d).

Nevertheless, some studies require more sophisticated identification of cell-associated sensors, and compromising cell viability has to be avoided. This illustrates the need for more gentle gas delivery. This can be achieved with a type of perfusion setup [2, 17] that delivers Xe-saturated cell culture medium to encapsulated cells [20]. Such types of packed bead bioreactor allow important studies toward animal experiments for further optimization of the biosensor approach.

8.6 Conclusion

The original concept of individually caged Xe as an MRI biosensor approach has been expanded to exchanging Xe in a wider definition, which currently enables different variants of CEST detection with hp Xe. The nature of the noble gas facilitates design of agents that rely on reversible binding and supports moving the field of hyperCEST closer to in vivo applications. This has been made possible through progress in the production of hyperpolarized Xe, improved sensor design with enhanced CEST performance, and the implementation of faster imaging techniques that also take advantage of the spectral resolution. Applications of Xe MRI are, therefore, expanding beyond the classical imaging of the void spaces of the lung [14, 27, 92].

An important advantage for future animal applications of the hyperCEST approach is that the ability of generating image contrast is not limited by conventional detection thresholds as they apply, e.g., for metabolic (spectroscopic) imaging. As long as the distribution of free Xe in tissue can be imaged, there is the chance to induce CEST contrast with rather low Xe host concentrations. Xe MRI has been demonstrated under various conditions in vivo, in particular for the brain [93, 94]. Solubility in this tissue is high enough to detect signal changes of approximately 10% [95] that should also be achievable with hyperCEST agents.

Acknowledgments

This work has been supported by the European Research Council under the European Community's Seventh Framework Programme (FP7/2007-2013)/ERC grant agreement no. 242710, the Leibniz Association (WGL; grant SAW-2011-FMP-2), and the Human Frontier Science Program. The author thanks the whole biosensor team for great enthusiasm and creativity.

References

1. van Zijl, P. C. M. and Yadav, N. N. (2011). Chemical exchange saturation transfer (CEST): What is in a name and what isn't? *Magnetic Resonance in Medicine*, 65(4), 927–948. http://doi.org/10.1002/mrm.22761
2. Schröder, L., Lowery, T. J., Hilty, C., Wemmer, D. E., and Pines, A. (2006). Molecular imaging using a targeted magnetic resonance hyperpolarized biosensor. *Science*, 314(5798), 446–449. http://doi.org/10.1126/science.1131847
3. Spence, M. M., Rubin, S. M., Dimitrov, I. E., et al. (2001). Functionalized xenon as a biosensor. *Proceedings of the National Academy of Sciences of the United States of America*, 98(19), 10654–10657. http://doi.org/10.1073/pnas.191368398
4. Brotin, T. and Dutasta, J.-P. (2009). Cryptophanes and their complexes: Present and future. *Chemical Reviews*, 109(1), 88–130. http://doi.org/10.1021/cr0680437
5. Garcia, S., Chavez, L., Lowery, T. J., Han, S.-I., Wemmer, D. E., and Pines, A. (2007). Sensitivity enhancement by exchange mediated magnetization transfer of the xenon biosensor signal. *Journal of Magnetic Resonance*, 184(1), 72–77. http://doi.org/10.1016/j.jmr.2006.09.010
6. Spence, M. M., Ruiz, E. J., Rubin, S. M., et al. (2004). Development of a functionalized xenon biosensor. *Journal of the American Chemical Society*, 126(46), 15287–15294. http://doi.org/10.1021/ja0483037
7. Berthault, P., Bogaert-Buchmann, A., Desvaux, H., Huber, G., and Boulard, Y. (2008). Sensitivity and multiplexing capabilities of MRI based on polarized ^{129}Xe biosensors. *Journal of the American Chemical Society*, 130(49), 16456–16457.

8. Kotera, N., Tassali, N., Léonce, E., et al. (2012). A sensitive zinc-activated ^{129}Xe MRI probe. *Angewandte Chemie International Edition*, 51(17), 4100–4103. http://doi.org/10.1002/anie.201109194
9. Schröder, L., Meldrum, T., Smith, M., Lowery, T., Wemmer, D., and Pines, A. (2008). Temperature response of Xe129 depolarization transfer and its application for ultrasensitive NMR detection. *Physical Review Letters*, 100(25). http://doi.org/10.1103/PhysRevLett.100.257603
10. Schröder, L., Chavez, L., Meldrum, T., et al. (2008). Temperature-controlled molecular depolarization gates in nuclear magnetic resonance. *Angewandte Chemie International Edition*, 47(23), 4316–4320. http://doi.org/10.1002/anie.200800382
11. Forsén, S. and Hoffman, R. A. (1963). Study of moderately rapid chemical exchange reactions by means of nuclear magnetic double resonance. *The Journal of Chemical Physics*, 39(11), 2892. http://doi.org/10.1063/1.1734121
12. Ward, K. M., Aletras, A. H., and Balaban, R. S. (2000). A new class of contrast agents for MRI based on proton chemical exchange dependent saturation transfer (CEST). *Journal of Magnetic Resonance*, 143(1), 79–87. http://doi.org/10.1006/jmre.1999.1956
13. Zhang, S., Winter, P., Wu, K., and Sherry, A. D. (2001). A novel europium(III)-based MRI contrast agent. *Journal of the American Chemical Society*, 123(7), 1517–1518.
14. Dregely, I., Ruset, I. C., Mata, J. F., et al. (2012). Multiple-exchange-time xenon polarization transfer contrast (MXTC) MRI: Initial results in animals and healthy volunteers. *Magnetic Resonance in Medicine*, 67(4), 943–953. http://doi.org/10.1002/mrm.23066
15. Ruppert, K., Mata, J. F., Wang, H.-T. J., et al. (2007). XTC MRI: Sensitivity improvement through parameter optimization. *Magnetic Resonance in Medicine*, 57(6), 1099–1109. http://doi.org/10.1002/mrm.21241
16. Ruppert, K., Brookeman, J. R., Hagspiel, K. D., and Mugler, J. P. (2000). Probing lung physiology with xenon polarization transfer contrast (XTC). *Magnetic Resonance in Medicine*, 44(3), 349–357. http://doi.org/10.1002/1522-2594(200009)44:3<349::AID-MRM2>3.0.CO;2-J
17. Hilty, C., Lowery, T. J., Wemmer, D. E., and Pines, A. (2006). Spectrally resolved magnetic resonance imaging of a xenon biosensor. *Angewandte Chemie International Edition*, 45(1), 70–73. http://doi.org/10.1002/anie.200502693
18. 47th ENC: Experimental Nuclear Magnetic Resonance Conference: 47th ENC conference, April 23–28, 2006, Asilomar Conference Center, Pacific Grove, CA: final program. (2006). Santa Fe, N.M.: [ENC].

19. Zaiss, M., Schnurr, M., and Bachert, P. (2012). Analytical solution for the depolarization of hyperpolarized nuclei by chemical exchange saturation transfer between free and encapsulated xenon (Hyper-CEST). *The Journal of Chemical Physics*, 136(14), 144106–144110. http://doi.org/doi:10.1063/1.3701178

20. Klippel, S., Döpfert, J., Jayapaul, J., et al. (2014). Cell tracking with caged xenon: Using cryptophanes as MRI reporters upon cellular internalization. *Angewandte Chemie International Edition*, 53(2), 493–496. http://doi.org/10.1002/anie.201307290

21. Kunth, M., Döpfert, J., Witte, C., Rossella, F., and Schröder, L. (2012). Optimized use of reversible binding for fast and selective NMR localization of caged xenon. *Angewandte Chemie International Edition*, 51(33), 8217–8220. http://doi.org/10.1002/anie.201202481

22. Schröder, L. (2013). Xenon for NMR biosensing: Inert but alert. *Physica Medica*, 29(1), 3–16. http://doi.org/10.1016/j.ejmp.2011.11.001

23. Friedman, J. I., McMahon, M. T., Stivers, J. T., and van Zijl, P. C. M. (2010). Indirect detection of labile solute proton spectra via the water signal using frequency-labeled exchange (FLEX) transfer. *Journal of the American Chemical Society*, 132(6), 1813–1815. http://doi.org/10.1021/ja909001q

24. Yadav, N. N., Jones, C. K., Xu, J., et al. (2012). Detection of rapidly exchanging compounds using on-resonance frequency-labeled exchange (FLEX) transfer. *Magnetic Resonance in Medicine*, 68(4), 1048–1055. http://doi.org/10.1002/mrm.24420

25. Witte, C. and Schröder, L. (2013). NMR of hyperpolarised probes. *NMR in Biomedicine*, 26(7), 788–802. http://doi.org/10.1002/nbm.2873

26. Chen, W. C., Gentile, T. R., Ye, Q., Walker, T. G., and Babcock, E. (2014). On the limits of spin-exchange optical pumping of ^{3}He. *Journal of Applied Physics*, 116(1), 014903. http://doi.org/10.1063/1.4886583

27. Lilburn, D. M. L., Pavlovskaya, G. E., and Meersmann, T. (2013). Perspectives of hyperpolarized noble gas MRI beyond ^{3}He. *Journal of Magnetic Resonance*, 229, 173–186. http://doi.org/10.1016/j.jmr.2012.11.014

28. Goodson, B. M. (2002). Nuclear magnetic resonance of laser-polarized noble gases in molecules, materials, and organisms. *Journal of Magnetic Resonance*, 155(2), 157–216. http://doi.org/10.1006/jmre.2001.2341

29. Oros, A.-M., and Shah, N. J. (2004). Hyperpolarized xenon in NMR and MRI. *Physics in Medicine and Biology*, 49(20), R105–153.

30. Harel, E., Schröder, L., and Xu, S. (2008). Novel detection schemes of nuclear magnetic resonance and magnetic resonance imaging:

Applications from analytical chemistry to molecular sensors. *Annual Review of Analytical Chemistry*, 1(1), 133–163. http://doi.org/10.1146/annurev.anchem.1.031207.113018

31. Haynes, W. M. (2014). *CRC Handbook of Chemistry and Physics*, 95th Edition, CRC Press.

32. Walker, T. G. and Happer, W. (1997). Spin-exchange optical pumping of noble-gas nuclei. *Reviews of Modern Physics*, 69(2), 629.

33. Nikolaou, P., Whiting, N., Eschmann, N. A., Chaffee, K. E., Goodson, B. M., and Barlow, M. J. (2009). Generation of laser-polarized xenon using fiber-coupled laser-diode arrays narrowed with integrated volume holographic gratings. *Journal of Magnetic Resonance*, 197(2), 249–254. http://doi.org/10.1016/j.jmr.2008.12.015

34. Nikolaou, P., Coffey, A. M., Walkup, L. L., et al. (2013). Near-unity nuclear polarization with an open-source [129]Xe hyperpolarizer for NMR and MRI. *Proceedings of the National Academy of Sciences*, 110(35), 14150–14155. http://doi.org/10.1073/pnas.1306586110

35. Walter, D., Griffith, W., and Happer, W. (2001). Energy transport in high-density spin-exchange optical pumping cells. *Physical Review Letters*, 86(15), 3264–3267. http://doi.org/10.1103/PhysRevLett.86.3264

36. Witte, C., Kunth, M., Rossella, F., and Schröder, L. (2014). Observing and preventing rubidium runaway in a direct-infusion xenon-spin hyperpolarizer optimized for high-resolution hyper-CEST (chemical exchange saturation transfer using hyperpolarized nuclei) NMR. *The Journal of Chemical Physics*, 140(8), 084203. http://doi.org/10.1063/1.4865944

37. Han, S.-I., Garcia, S., Lowery, T. J., et al. (2005). NMR-based biosensing with optimized delivery of polarized [129]Xe to solutions. *Analytical Chemistry*, 77(13), 4008–4012. http://doi.org/10.1021/ac0500479

38. Witte, C., Kunth, M., Döpfert, J., Rossella, F., and Schröder, L. (2012). Hyperpolarized xenon for NMR and MRI applications. *Journal of Visualized Experiments*, (67). http://doi.org/10.3791/4268

39. Döpfert, J., Witte, C., and Schröder, L. (2014). Fast gradient-encoded CEST spectroscopy of hyperpolarized xenon. *ChemPhysChem*, 15(2), 261–264. http://doi.org/10.1002/cphc.201300888

40. Döpfert, J., Witte, C., Kunth, M., and Schröder, L. (2014). Sensitivity enhancement of (hyper-)CEST image series by exploiting redundancies in the spectral domain. *Contrast Media and Molecular Imaging*, 9(1), 100–107. http://doi.org/10.1002/cmmi.1543

41. Swanson, S. D. (1991). Broadband excitation and detection of cross-relaxation NMR spectra. *Journal of Magnetic Resonance (1969)*, 95(3), 615–618.

42. Xu, X., Lee, J.-S., and Jerschow, A. (2013). Ultrafast scanning of exchangeable sites by NMR spectroscopy. *Angewandte Chemie (International Edition in English)*, 52(32), 8281–8284. http://doi.org/10.1002/anie.201303255

43. Zhao, L., Mulkern, R., Tseng, C. H., et al. (1996). Gradient-echo imaging considerations for hyperpolarized ^{129}Xe MR. *Journal of Magnetic Resonance Series B*, 113, 179–183.

44. Döpfert, J., Witte, C., and Schröder, L. (2013). Slice-selective gradient-encoded CEST spectroscopy for monitoring dynamic parameters and high-throughput sample characterization. *Journal of Magnetic Resonance*, 237, 34–39. http://doi.org/10.1016/j.jmr.2013.09.007

45. Boutin, C., Léonce, E., Brotin, T., Jerschow, A., and Berthault, P. (2013). Ultrafast Z-spectroscopy for ^{129}Xe NMR-based sensors. *The Journal of Physical Chemistry Letters*, 4(23), 4172–4176. http://doi.org/10.1021/jz402261h

46. McMahon, M. T., Gilad, A. A., Zhou, J., Sun, P. Z., Bulte, J. W. M., and van Zijl, P. C. M. (2006). Quantifying exchange rates in chemical exchange saturation transfer agents using the saturation time and saturation power dependencies of the magnetization transfer effect on the magnetic resonance imaging signal (QUEST and QUESP): Ph calibration for poly-L-lysine and a starburst dendrimer. *Magnetic Resonance in Medicine*, 55(4), 836–847. http://doi.org/10.1002/mrm.20818

47. Locci, E., Dehouck, Y., Casu, M., et al. (2001). Probing proteins in solution by $^{(129)}$Xe NMR spectroscopy. *Journal of Magnetic Resonance*, 150(2), 167–174. http://doi.org/10.1006/jmre.2001.2325

48. Rubin, S. M., Lee, S.-Y., Ruiz, E. J., Pines, A., and Wemmer, D. E. (2002). Detection and characterization of xenon-binding sites in proteins by ^{129}Xe NMR spectroscopy. *Journal of Molecular Biology*, 322(2), 425–440.

49. Jacobson, D. R., Khan, N. S., Collé, R., et al. (2011). Measurement of radon and xenon binding to a cryptophane molecular host. *Proceedings of the National Academy of Sciences*, 108(27), 10969–10973. http://doi.org/10.1073/pnas.1105227108

50. Fogarty, H. A., Berthault, P., Brotin, T., Huber, G., Desvaux, H., and Dutasta, J.-P. (2007). A cryptophane core optimized for xenon encapsulation. *Journal of the American Chemical Society*, 129(34), 10332–10333. http://doi.org/10.1021/ja073771c

51. Kotera, N., Delacour, L., Traoré, T., et al. (2011). Design and synthesis of new cryptophanes with intermediate cavity sizes. *Organic Letters*, 13(9), 2153–2155. http://doi.org/10.1021/ol2005215

52. Bai, Y., Hill, P. A., and Dmochowski, I. J. (2012). Utilizing a water-soluble cryptophane with fast xenon exchange rates for picomolar sensitivity NMR measurements. *Analytical Chemistry*, 84(22), 9935–9941. http://doi.org/10.1021/ac302347y
53. Canceill, J., Lacombe, L., and Collet, A. (1987). Water-soluble cryptophane binding lipophilic guests in aqueous solution. *Journal of the Chemical Society, Chemical Communications*, 3, 219–221.
54. Delacour, L., Kotera, N., Traoré, T., et al. (2013). "Clickable" hydrosoluble PEGylated cryptophane as a universal platform for ^{129}Xe magnetic resonance imaging biosensors. *Chemistry (Weinheim an Der Bergstrasse, Germany)*, 19(19), 6089–6093. http://doi.org/10.1002/chem.201204218
55. Dubost, E., Kotera, N., Garcia-Argote, S., et al. (2013). Synthesis of a functionalizable water-soluble cryptophane-111. *Organic Letters*, 15(11), 2866–2868. http://doi.org/10.1021/ol4012019
56. Fairchild, R. M., Joseph, A. I., Holman, K. T., et al. (2010). A water-soluble Xe@cryptophane-111 complex exhibits very high thermodynamic stability and a peculiar $^{(129)}$Xe NMR chemical shift. *Journal of the American Chemical Society*, 132(44), 15505–15507. http://doi.org/10.1021/ja1071515
57. Hill, P. A., Wei, Q., Troxler, T., and Dmochowski, I. J. (2009). Substituent effects on xenon binding affinity and solution behavior of water-soluble cryptophanes. *Journal of the American Chemical Society*, 131(8), 3069–3077. http://doi.org/10.1021/ja8100566
58. Huber, G., Brotin, T., Dubois, L., Desvaux, H., Dutasta, J.-P., and Berthault, P. (2006). Water soluble cryptophanes showing unprecedented affinity for xenon: Candidates as NMR-based biosensors. *Journal of the American Chemical Society*, 128(18), 6239–6246. http://doi.org/10.1021/ja060266r
59. Traoré, T., Clavé, G., Delacour, L., et al. (2011). The first metal-free water-soluble cryptophane-111. *Chemical Communications*, 47(34), 9702–9704. http://doi.org/10.1039/c1cc13378k
60. Liu, G., Gilad, A. A., Bulte, J. W. M., van Zijl, P. C. M., and McMahon, M. T. (2010). High-throughput screening of chemical exchange saturation transfer MR contrast agents. *Contrast Media and Molecular Imaging*, 5(3), 162–170. http://doi.org/10.1002/cmmi.383
61. Meldrum, T., Schröder, L., Denger, P., Wemmer, D. E., and Pines, A. (2010). Xenon-based molecular sensors in lipid suspensions. *Journal of Magnetic Resonance*, 205(2), 242–246. http://doi.org/10.1016/j.jmr.2010.05.005

62. Schilling, F., Schröder, L., Palaniappan, K. K., Zapf, S., Wemmer, D. E., and Pines, A. (2010). MRI thermometry based on encapsulated hyperpolarized xenon. *ChemPhysChem*, 11(16), 3529–3533. http://doi.org/10.1002/cphc.201000507
63. Rose, H. M., Witte, C., Rossella, F., Klippel, S., Freund, C., and Schröder, L. (2014). Development of an antibody-based, modular biosensor for ^{129}Xe NMR molecular imaging of cells at nanomolar concentrations. *Proceedings of the National Academy of Sciences*, 111(32), 11697–11702. http://doi.org/10.1073/pnas.1406797111
64. Boutin, C., Desvaux, H., Carrière, M., et al. (2011). Hyperpolarized ^{129}Xe NMR signature of living biological cells. *NMR in Biomedicine*, 24(10), 1264–1269. http://doi.org/10.1002/nbm.1686
65. Stevens, T. K., Ramirez, R. M., and Pines, A. (2013). Nanoemulsion contrast agents with sub-picomolar sensitivity for xenon NMR. *Journal of the American Chemical Society*, 135(26), 9576–9579. http://doi.org/10.1021/ja402885q
66. Shapiro, M. G., Ramirez, R. M., Sperling, L. J., et al. (2014). Genetically encoded reporters for hyperpolarized xenon magnetic resonance imaging. *Nature Chemistry*, 6(7), 629–634. http://doi.org/10.1038/nchem.1934
67. Bai, Y., Wang, Y., Goulian, M., Driks, A., and Dmochowski, I. J. (2014). Bacterial spore detection and analysis using hyperpolarized ^{129}Xe chemical exchange saturation transfer (Hyper-CEST) NMR. *Chemical Science*, 5(8), 3197–3203. http://doi.org/10.1039/C4SC01190B
68. Liu, G., Bettegowda, C., Qiao, Y., et al. (2013). Noninvasive imaging of infection after treatment with tumor-homing bacteria using chemical exchange saturation transfer (CEST) MRI: CEST MRI of bacteria. *Magnetic Resonance in Medicine*, 70(6), 1690–1698. http://doi.org/10.1002/mrm.24955
69. Sloniec, J., Schnurr, M., Witte, C., Resch-Genger, U., Schröder, L., and Hennig, A. (2013). Biomembrane interactions of functionalized cryptophane-A: Combined fluorescence and ^{129}Xe NMR studies of a bimodal contrast agent. *Chemistry: A European Journal*, 19(9), 3110–3118. http://doi.org/10.1002/chem.201203773
70. Boutin, C., Stopin, A., Lenda, F., et al. (2011). Cell uptake of a biosensor detected by hyperpolarized ^{129}Xe NMR: The transferrin case. *Bioorganic and Medicinal Chemistry*, 19(13), 4135–4143. http://doi.org/10.1016/j.bmc.2011.05.002
71. Schnurr, M., Sydow, K., Rose, H. M., Dathe, M., and Schröder, L. (2015). Brain endothelial cell targeting via a peptide-functionalized liposomal

carrier for xenon hyper-CEST MRI. *Advanced Healthcare Materials*, 4(1), 40–45. http://doi.org/10.1002/adhm.201400224

72. Schnurr, M., Witte, C., and Schröder, L. (2014). Depolarization Laplace transform analysis of exchangeable hyperpolarized ^{129}Xe for detecting ordering phases and cholesterol content of biomembrane models. *Biophysical Journal*, 106(6), 1301–1308. http://doi.org/10.1016/j.bpj.2014.01.041

73. Palaniappan, K. K., Ramirez, R. M., Bajaj, V. S., Wemmer, D. E., Pines, A., and Francis, M. B. (2013). Molecular imaging of cancer cells using a bacteriophage-based ^{129}Xe NMR biosensor. *Angewandte Chemie (International Edition in English)*, 52(18), 4849–4853. http://doi.org/10.1002/anie.201300170

74. Berthault, P., Huber, G., and Desvaux, H. (2009). Biosensing using laser-polarized xenon NMR/MRI. *Progress in Nuclear Magnetic Resonance Spectroscopy*, 55(1), 35–60. http://doi.org/10.1016/j.pnmrs.2008.11.003

75. Palaniappan, K. K., Francis, M. B., Pines, A., and Wemmer, D. E. (2014). Molecular sensing using hyperpolarized xenon NMR spectroscopy. *Israel Journal of Chemistry*, 54(1–2), 104–112. http://doi.org/10.1002/ijch.201300128

76. Taratula, O. and Dmochowski, I. J. (2010). Functionalized ^{129}Xe contrast agents for magnetic resonance imaging. *Current Opinion in Chemical Biology*, 14(1), 97–104. http://doi.org/10.1016/j.cbpa.2009.10.009

77. Tassali, N., Kotera, N., Boutin, C., et al. (2014). Smart detection of toxic metal ions, Pb^{2+} and Cd^{2+}, using a ^{129}Xe NMR-based sensor. *Analytical Chemistry*, 86(3), 1783–1788. http://doi.org/10.1021/ac403669p

78. Wu, Y., Zhou, Y., Ouari, O., et al. (2008). Polymeric PARACEST agents for enhancing MRI contrast sensitivity. *Journal of the American Chemical Society*, 130(42), 13854–13855. http://doi.org/10.1021/ja805775u

79. Snoussi, K., Bulte, J. W. M., Guéron, M., and van Zijl, P. C. M. (2003). Sensitive CEST agents based on nucleic acid imino proton exchange: Detection of poly(rU) and of a dendrimer-poly(rU) model for nucleic acid delivery and pharmacology. *Magnetic Resonance in Medicine*, 49(6), 998–1005. http://doi.org/10.1002/mrm.10463

80. Meldrum, T., Seim, K. L., Bajaj, V. S., et al. (2010). A xenon-based molecular sensor assembled on an MS2 viral capsid scaffold. *Journal of the American Chemical Society*, 132(17), 5936–5937. http://doi.org/10.1021/ja100319f

81. Stevens, T. K., Palaniappan, K. K., Ramirez, R. M., Francis, M. B., Wemmer, D. E., and Pines, A. (2013). HyperCEST detection of a ^{129}Xe-based contrast agent composed of cryptophane-A molecular cages on a bacteriophage scaffold. *Magnetic Resonance in Medicine*, 69(5), 1245–1252. http://doi.org/10.1002/mrm.24371

82. Hill, P. A., Wei, Q., Eckenhoff, R. G., and Dmochowski, I. J. (2007). Thermodynamics of xenon binding to cryptophane in water and human plasma. *Journal of the American Chemical Society*, 129(30), 9262–9263. http://doi.org/10.1021/ja072965p

83. Wolber, J., Rowland, I. J., Leach, M. O., and Bifone, A. (1999). Perfluorocarbon emulsions as intravenous delivery media for hyperpolarized xenon. *Magnetic Resonance in Medicine*, 41(3), 442–449. http://doi.org/10.1002/(SICI)1522-2594(199903)41:3<442::AID-MRM3>3.0.CO;2-7

84. Zhou, J., Wilson, D. A., Sun, P. Z., Klaus, J. A., and van Zijl, P. C. M. (2004). Quantitative description of proton exchange processes between water and endogenous and exogenous agents for WEX, CEST, and APT experiments. *Magnetic Resonance in Medicine*, 51(5), 945–952. http://doi.org/10.1002/mrm.20048

85. Rossella, F., Rose, H. M., Witte, C., Jayapaul, J., and Schröder, L. (2014). Design and characterization of two bifunctional cryptophane? A-based host molecules for xenon magnetic resonance imaging applications. *ChemPlusChem*, 79(10), 1463–1471. http://org/10.1002/cplu.201402179

86. Koskela, H., Heikkinen, O., Kilpeläinen, I., and Heikkinen, S. (2007). Rapid and accurate processing method for amide proton exchange rate measurement in proteins. *Journal of Biomolecular NMR*, 37(4), 313–320. http://doi.org/10.1007/s10858-007-9145-y

87. Schnurr, M., Witte, C., and Schröder, L. (2013). Functionalized ^{129}Xe as a potential biosensor for membrane fluidity. *Physical Chemistry Chemical Physics*, 15(34), 14178. http://doi.org/10.1039/c3cp51227d

88. Bifone, A., Song, Y.-Q., Seydoux, R., et al. (1996). NMR of laser-polarized xenon in human blood. *Proceedings of the National Academy of Sciences*, 93(23), 12932–12936.

89. Seward, G. K., Bai, Y., Khan, N. S., and Dmochowski, I. J. (2011). Cell-compatible, integrin-targeted cryptophane-^{129}Xe NMR biosensors. Chemical Science, 2(6), 1103–1110. http://doi.org/10.1039/C1SC00041A

90. Seward, G. K., Wei, Q., and Dmochowski, I. J. (2008). Peptide-mediated cellular uptake of cryptophane. *Bioconjugate Chemistry*, 19(11), 2129-2135. http://doi.org/10.1021/bc8002265
91. Klippel, S., Freund, C., and Schröder, L. (2014). Multichannel MRI labeling of mammalian cells by switchable nanocarriers for hyperpolarized xenon. *Nano Letters*, 14(10), 5721-5726. http://doi.org/10.1021/nl502498w
92. Mugler, J. P. and Altes, T. A. (2013). Hyperpolarized ^{129}Xe MRI of the human lung. *Journal of Magnetic Resonance Imaging*, 37(2), 313-331. http://doi.org/10.1002/jmri.23844
93. Duhamel, G., Choquet, P., Grillon, E., et al. (2001). Xenon-129 MR imaging and spectroscopy of rat brain using arterial delivery of hyperpolarized xenon in a lipid emulsion. *Magnetic Resonance in Medicine*, 46(2), 208-212.
94. Zhou, X. (2012). Hyperpolarized xenon brain MRI. In *Advances in Brain Imaging*, Chaudhary, V. (ed.). USA: InTech Press. Retrieved from http://cdn.intechopen.com/pdfs/27258.pdf
95. Mazzanti, M. L., Walvick, R. P., Zhou, X., et al. (2011). Distribution of hyperpolarized xenon in the brain following sensory stimulation: Preliminary MRI findings. *PLoS ONE*, 6(7), e21607. http://doi.org/10.1371/journal.pone.0021607

Section III

DiaCEST/paraCEST/lipoCEST Contrast Probes

Chapter 9

Current Landscape of diaCEST Imaging Agents

Amnon Bar-Shir,[a] Xing Yang,[b] Xiaolei Song,[b] Martin Pomper,[b] and Michael T. McMahon[b,c]

[a] *Department of Organic Chemistry, Weizmann Institute of Science, Rehovot 76100, Israel*
[b] *Johns Hopkins University School of Medicine, 707 N Broadway, Baltimore, MD 21205, USA*
[c] *Kennedy Krieger Institute, 707 N Broadway, Baltimore, MD 21205, USA*
mcmahon@mri.jhu.edu

9.1 Introduction

One of the most attractive features of chemical exchange saturation transfer (CEST) imaging is its ability to detect naturally occurring compounds that are part of important metabolic pathways using labile protons already present on these compounds, allowing direct molecular imaging studies. This is a unique feature for magnetic resonance imaging (MRI) contrast agents, as paramagnetic and superparamagnetic MRI contrast agents instead require conjugation or association of metals to these compounds to impart signal contrast, which is less desirable. Furthermore, while conventional

---

Chemical Exchange Saturation Transfer Imaging: Advances and Applications
Edited by Michael T. McMahon, Assaf A. Gilad, Jeff W. M. Bulte, and Peter C. M. van Zijl
Copyright © 2017 Pan Stanford Publishing Pte. Ltd.
ISBN 978-981-4745-70-3 (Hardcover), 978-1-315-36442-1 (eBook)
www.panstanford.com

proton magnetic resonance spectroscopy (^{1}H MRS) and magnetic resonance spectroscopic imaging (MRSI) also allow the detection and quantification of metabolites, these techniques have always been limited by their relatively low sensitivity, and as such images produced using these techniques suffer from low spatial resolution in vivo. In addition, hyperpolarization has allowed the detection of organic compounds such as isotopically labeled pyruvate and lactate. However, this technique currently has challenges in imaging for more than several minutes after the administration of hyperpolarized substrate in live subjects due to T_1 relaxation of hyperpolarized spins. This combination of factors has led to a great enthusiasm for developing CEST imaging technologies for the amplified detection of a variety of organic compounds.

As mentioned in Chapter 1, the very first experiments that observed the transfer of saturation through chemical exchange were performed in 1963 by Sture Forsén and Ragnar Hoffman, with the transfer occurring for OH spins between two compounds—tert-butylalcohol and 2-hydroxyacetophenone—an intermolecular exchange in CS_2, leading them to put forward a theory that explained how the magnetization transferred in this intermolecular two-site exchange related to T_1 relaxation rate and exchange rate [1, 2]. In a series of papers from 1963 to 1980, Forsén et al. then proceeded to identify a number of organic compounds with suitable exchange properties, including alcohols, phenols, peptides, and proteins and further studied more complex types of spectra, including those that contained intramolecular exchange, three or more exchange sites, and interconversion between multiple conformations [3–8]. This pioneering work inspired a number of nuclear magnetic resonance (NMR) researchers to apply saturation transfer to study a wide range of dynamic processes, including probing the binding of ligands to proteins [9] and the structural dynamics of proteins and nucleic acids [10–18]. However, as described in detail in Chapter 2, it was not until Robert Balaban rediscovered magnetization transfer (MT) to water in the 1990s that MRI researchers became aware of the potential of saturation transfer for detecting organic molecules in tissue.

An enormous variety of organic and inorganic molecules possess labile protons, including strong acids such as hydrochloric acid and

Introduction

strong bases such as sodium hydroxide. Unfortunately, however, CEST contrast is not observable for all of these compounds; instead, as discussed in detail in Chapter 6, the transfer of signal loss from one spin to another via chemical exchange requires that the exchange rate is on the order of the chemical shift difference between the two spins ($\Delta\omega$) with the amount of contrast or proton transfer ratio (PTR) further dependent on the field strength of the saturation pulses and relaxation rates. When a particular compound is considered, it is convenient to use the proton transfer enhancement (PTE) to describe the contrast produced by this contrast agent [19]:

$$\text{PTE} = \frac{N_{CA} \cdot M_{CA}}{\chi_{CA}} \cdot \text{PTR} \qquad (9.1)$$

where N_{CA} is the number of labile protons per molecule, M_{CA} is the molecular weight of the agent, and χ_{CA} is the fractional concentration of labile protons of the contrast agent. As this expression suggests, N_{CA} and M_{CA} are important factors to consider for CEST agents in general [19]. Because of the relationship among exchange rate (k_{ex}), $\Delta\omega$, saturation field strength (ω_1), and PTE (as mentioned in detail in Chapter 6), the detection sensitivity varies widely across the chemical shift spectrum of diaCEST agents. To highlight the relationship among these parameters, we have performed calculations, which include continuous wave saturation conditions that might be appropriate at magnetic field strengths of 3 T and 7 T (Fig. 9.1). These calculations were performed by numerically solving the Bloch equations as described in detail elsewhere [20–22]. As is shown by these simulations, the maximum contrast attainable using $\omega_1 = 2$ µT is significantly lower than if $\omega_1 = 4$ µT is chosen. For example, at 3 T for agents with $\Delta\omega = 9.25$ ppm, $k_{ex} = 1000$ s^{-1}, we find that the MTR_{asym} increases from 3.6% to 5.3%, which corresponds to amplification factors for the 10 mM proton signal of 399 and 579, respectively. In addition, there is a significant advantage apparent in these simulations of increasing the magnetic field strength to 7 T for detecting compounds with $\Delta\omega < 6$ ppm. As $\Delta\omega < 6$ ppm for labile protons on a number of important metabolites, this is an important consideration. It should be noted, however, that these simulations set the labile proton concentration to 10 mM; however, for agents such as creatine and

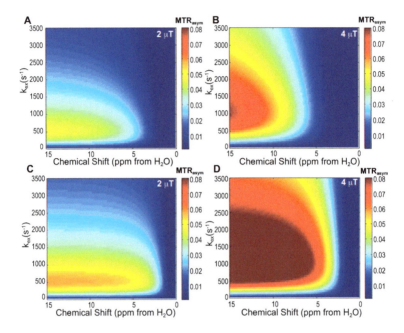

Figure 9.1 Simulated CEST contrast as a function of labile proton $\Delta\omega$ and k_{ex} for the following conditions: (A) $B_0 = 3$ T, $\omega_1 = 2$ μT; (B) $B_0 = 3$ T, $\omega_1 = 4$ μT; (C) $B_0 = 7$ T, $\omega_1 = 2$ μT; (D) $B_0 = 7$ T, $\omega_1 = 4$ μT. Simulation parameters: $\chi_{CA} = 10$ mM, $t_{sat} = 4$ s, and for 3 T: $T_{1w} = T_{1s} = 3$ s, $T_{2w} = T_{2s} = 0.1$ s; for 7 T: $T_{1w} = T_{1s} = 4$ s, $T_{2w} = T_{2s} = 0.1$ s.

myo-inositol, $N_{CA} = 4$ and 5, respectively, which then increases the effective labile proton concentrations by a factor of N_{CA} over their molar concentrations, as shown in Eq. 9.1, improving their detectability. While these simulations are suitable for analyzing saturation transfer through chemical exchange, one alternative to consider is that some studies employ alternative methods to detect diaCEST agents such as $T_{1\rho}$, T_2, FLEX, or by monitoring a single frequency point to mix CEST and T_2 effects together [23–26]. The contrast will be maximal at different $\Delta\omega$ and k_{ex} values than shown in Fig. 9.1 when using these alternative methods [27, 28].

The range of $\Delta\omega$ on known diaCEST agents at the time of this writing is depicted in Fig. 9.2, using water as the reference frequency as is the convention for CEST studies [29] instead of tetramethylsilane (TMS), which is the IUPAC standard used in the NMR literature.

9.2 Molecules with Alkyl Amines and Amides

As mentioned in Chapter 2, the first compounds of interest for medical imaging with exchangeable protons, which were shown to be detectable through saturation transfer imaging, were urea and ammonia [30, 31]. Urea possesses four labile protons resonating $\sim$1.6 ppm from water, and ammonia possesses four labile protons per molecule resonating $\sim$2.4 ppm from water, which has allowed an association of the CEST contrast detected at these frequencies in the kidney to these compounds [30, 32]. However, because these overlap with a number of other high concentration metabolites, it is difficult to assign all of the saturation transfer contrast at these $\Delta\omega$'s to these molecules. Another feature of urea and ammonia is that on many scanners they can act as T_{2ex} agents (producing broadening in the water line) because of their relatively rapid exchange rates and small $\Delta\omega$ [33]. Efforts are ongoing to improve the detection of ammonia on high field scanners [34].

Proton NMR spectroscopy has allowed the detection of a number of low molecular weight metabolites, which are present at high concentrations in biological tissue, with the NMR chemical shifts and coupling constants for 35 detectable metabolites presented in detail by Maudsley et al. [35]. This list includes a number of metabolites with labile amine and amide protons that, it would be reasonable to expect, could also be detected through CEST imaging. Creatine is one of the metabolites that occur at high concentration in brain, muscle, and blood and also possess suitable amine protons. Because of the importance of creatine, its CEST imaging is extensively covered in Chapter 18 In addition to creatine, L-amino acids, di-peptides, and tri-peptides have been detected at fairly high concentrations in biological tissue using proton NMR spectroscopy and possess amine or amide protons that might be amenable to CEST studies. Two such compounds with relatively slow exchanging amines are L-glutamate and *N*-acetylaspartylglutamate, present in concentrations of 3 mM or higher in tissue [35]. L-glutamate is a major neurotransmitter in the brain and spinal cord, which has a concentration that is altered in a range of neurologic disorders and has been found to be particularly well detected through CEST imaging via its two labile protons resonating at ∼3 ppm from water [36–40]. L-glutamate's amine protons have a k_{ex} ∼6700 s^{-1} at neutral pH. L-alanine is another amino acid that is present at moderate concentrations in the human brain (0.5 mmol/kg) [35], but it has amine protons with larger k_{ex} than glutamate at neutral pH, making this metabolite harder to detect endogenously. Besides these amino acids that are present at high concentrations in tissue, L-arginine and N-acetyl-L-arginine have been inserted into nanocarriers and hydrogels to allow their detection through CEST imaging [41–43]. Both compounds possess four amine protons resonating at 1.8 ppm, and a key feature of these labile protons is the strong dependence of their exchange with pH in the physiologically relevant range from 6 to 7.4 This feature has been exploited to allow CEST MRI sensing of changes in pH [41].

9.3 Molecules with Alkyl Hydroxyls

A wide range of small non-aromatic hydroxylcontaining compounds can generate CEST contrast, including myo-inositol [44], scyllo-inositol, glucose [45–47], threonine, serine, glycerophosphocholine [48] and lactate [49], which are all found at high enough concentrations to be detectable through proton NMR spectroscopy [35]. In water, non-aromatic OH protons resonate between 0.5 and 2.5 ppm from water depending on their environment, and at neutral pH, the hydroxyls in these compounds have k_{ex} ∼600–2000 s^{-1}. As the simulations in Fig. 9.2 show, 7 T clinical scanners are appropriate for detecting these compounds in patients. This is rapidly becoming the most widely studied group of compounds for CEST imaging, and Chapter 17 has been devoted to discuss the administration of one important member of this family of compounds, glucose, to animals and patients. While the number of compounds that produce contrast with $\Delta\omega = 1$–2 ppm probably limits the specificity of CEST measurements for endogenous compounds (without administration) falling within this range, CEST measurements at 1–2 ppm have been suggested as a potential marker of glial cell proliferation

Molecules with Alkyl Hydroxyls | **167**

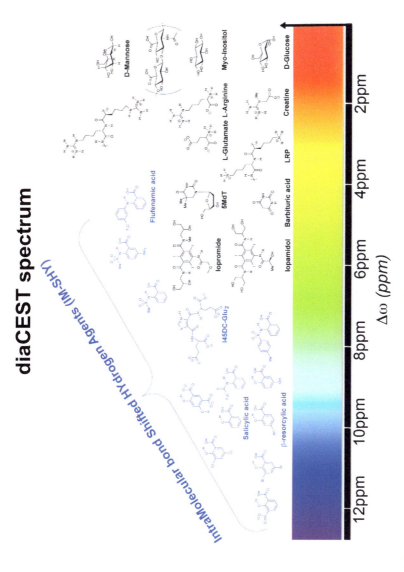

Figure 9.2 Spectral range for currently available diaCEST agents, including well-known sugars, non-aromatic amines, heterocycles, and intramolecular bondshifted hydrogen agents in blue.

and tumor progression. Alternatively, infusion of these compounds and monitoring the dynamic contrast in tissue allow more clear assignment of the contrast and is now being evaluated in patients [25].

9.4 N-H Containing Heterocyclic Compounds

As mentioned earlier, one main drawback of diaCEST agents is the small $\Delta\omega$ between their exchangeable protons (i.e., $\Delta\omega < 4$ ppm for the aforementioned amide, amine, and hydroxyl protons). This may lead to deleterious effects from direct saturation of the water protons and overlap with background signals from a number of endogenous exchangeable protons [22, 59–61], making it difficult to identify which compounds were the source of the contrast. One alternative is to design chelated paramagnetic-metal complexes as paramagnetic CEST (paraCEST) agents, which induce highly shifted exchangeable protons (or water molecules); indeed these agents are the subject of Chapters 11–14. Unfortunately, however, paraCEST agents are not based on natural compounds (or their analogues), are not biodegradable, and are potentially toxic when not rapidly cleared from the body. As a result, it is of interest to identify organic compounds that might possess labile protons with $\Delta\omega > 4$ ppm.

In fact, Balaban et al. anticipated this issue, and in their seminal paper [29], they demonstrated that nitrogen-containing heterocyclic compounds, with their exchangeable protons resonating at $\Delta\omega =$ 4.5–6.5 ppm, may be used as diaCEST-based agents. These proposed heterocyclic CEST agents (Fig. 9.3) are based on –NH groups in (hetero)cyclic compounds and include imino acids, indoles, nucleosides and their pyrimidine and purine bases, as well as derivatives of barbituric acid (BA) and imidazole.

Although a relatively large $\Delta\omega$ (4.5–6.5 ppm) is observed for –NH protons on these selected heterocyclic compounds (Fig. 9.3), the k_{ex} of these exchangeable protons with water may be too fast for CEST applications on 3 T to 11.7 T scanners and thus should be modified in order to make these agents suitable. In one particular example, it was shown that by changing the acid

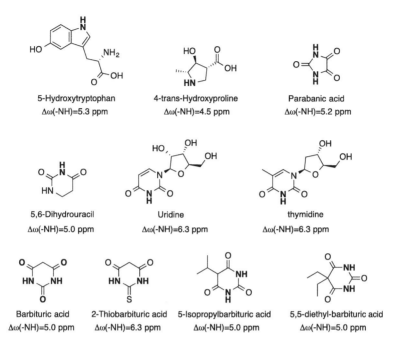

Figure 9.3 Chemical structures of heterocyclic compounds as potential diaCEST agents. The chemical shift offset ($\Delta\omega$) of the –NH exchangeable proton (shown in **bold**) of the compound is indicated.

dissociation constant (pK_a) value of the imino proton of thymidine (dT) by rational chemical modification, one could modulate its k_{ex} and thus its CEST characteristics [62, 63]. This set of modifications is summarized in Fig. 9.4. An appropriate chemical modification of dT increases its –NH exchangeable proton pK_a value, which leads to the reduction in its k_{ex} and renders the labile proton compliant with the slow exchange condition on the NMR timescale, improving its CEST MRI properties. Upon hydrogenation of the 5,6-double bond of the pyrimidine ring, the pK_a value of the exchangeable imino proton increases from about 9.8 for dT to 11.6 for 5,6-dihydrothymidine (compound 1, Fig. 9.4) due to the loss of the pyrimidine ring aromaticity. The substitution of the electron-donating methyl group of dT with an electron-withdrawing group Cl, to generate 5-chloro-2-deoxyuridine (compound 4, Fig. 9.4), reversed the inductive properties of the 5-substituent and considerably reduced the pK_a

value of the imino proton to about 7.9. Using an 11.7 T MRI scanner, the CEST contrast generated by the deoxynucleoside analogues in PBS (pH = 7.4, 37°C) was evaluated and compared with that of dT (Fig. 9.4b–j). The solid lines in (Fig. 9.4c–f) represent the CEST spectra, while the dashed lines represent MTR_{asym} plots. The imino protons of compounds 1–3 showed a local maximum contrast 5 ppm downfield of the water proton frequency, but only the dihydrothymidines (compounds 1 and 2) showed a well-defined sharp peak at that frequency (Fig. 9.4c–f). It is clearly seen that the highest the pK_a value calculated for the NH protons of thymidine analogue, the highest the CEST contrast obtained. This relationship between pK_a and CEST contrast could be further explained by the relationship between pK_a and k_{ex} and between k_{ex} and the observed CEST contrast.

Several methodologies have been proposed for measuring the k_{ex} of exchangeable protons for potential CEST agents [21, 64, 65]. These k_{ex} values in CEST agents are readily quantified using either saturation time or saturation power dependencies of the MT effect on the NMR signal (QUEST and QUESP) [21]. The plots in (Fig. 9.4g–j) demonstrate the use of the QUESP approach for measuring k_{ex} for the NH protons in dT and its analogues. It is clearly observed that for slow exchanging protons (such as those of compounds 1 and 2), the QUESP plots (Fig. 9.4g,h) level off upon increasing the saturation power beyond an optimal B_1. For protons with larger k_{ex}, the CEST effect continues to increase with B_1. This phenomenon is in good agreement with simulations performed by Woessner et al. [66] showing that the maximal CEST contrast (slow exchange regime) is achievable when an optimal saturation pulse (B_1) is applied. This optimal saturation pulse could be estimated for every exchangeable proton of potential CEST agent using the following equation: $B_1^{(optimal)} = k_{ex}/2\pi$. Therefore, one should know the k_{ex} value of the exchangeable proton on the probe of interest. The relationships between the pK_a values and k_{ex} and thus with the obtained CEST contrast are reflected in Fig. 9.4 and also summarized in Table 9.1. These observations and the ability to quantify pK_a values in silico may help one to design diaCEST agents with improved CEST capabilities.

N-H Containing Heterocyclic Compounds | 171

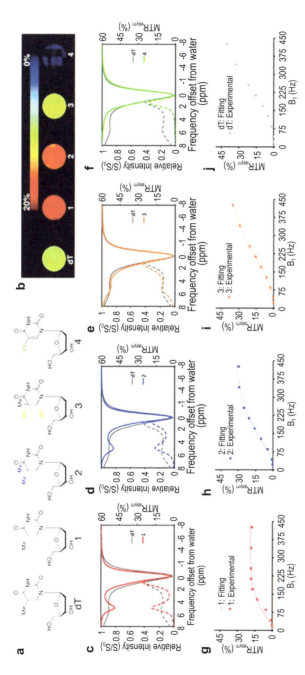

Figure 9.4 Evaluation of thymidine analogues as potential CEST agents. (a) Chemical structure of dT and compounds 1–4. (b) MTR_{asym} maps of dT and compounds 1–4 obtained at a 5 ppm frequency offset from the water resonance. (c–f) CEST spectra (solid lines) and MTR_{asym} plots (dashed lines). A $B_1 = 200$ Hz was used for b–f. (g–j) MTR_{asym} plotted as a function of saturation power (B_1, Hz), experimental (dots) and QUESP fitting (lines). In c–j, dT is plotted in gray, compound 1 in red, compound 2 in blue, compound 3 in orange, and compound 4 in green. Data were acquired at 11.7 T, pH = 7.4, and 37 °C for 20 mM CEST agent solutions. For compounds 1–3, MTR_{asym} values were calculated at 5 ppm (g–i), and for dT at 6 ppm (j). Reprinted with permission from Ref. [11], Copyright 2013, American Chemical Society.

Table 9.1 Properties of dT and its analogues

	dT	Compound 1	Compound 2	Compound 3	Compound 4
$pK_a^{(i)}$	9.96	11.60	11.57	11.0	7.97
$k_{ex} \ [s^{-1}]^{(ii)}$	5100	800	1700	3800	N.D.
$MTR_{asym}^{(iii)}$	10%	18%	16%	11%	N.D.

(i) pK_a values of the imino proton as calculated using MarvinSketch
(ii) Quantified exchange rates of imino and water protons calculated from QUESP
(iii) MTR_{asym} values (%) at 5 ppm for 20 mM agent with $B_1 = 170$ Hz/4 s
N.D. = not determined

These findings led to the identification of 5-methyl-5,6-dihydrothymidine (compound 2, Fig. 9.4) as an optimal probe for in vivo monitoring of the HSV1-tk reporter gene expression using CEST MRI [62, 63]. In light of this demonstration, the approach of rationally designed synthetically modified nucleosides as improved diaCEST agents can be generalized for designing other probes for the in vivo imaging of a variety biological targets.

Another promising example is BA, a heterocyclic pyrimidinetrione (Fig. 9.3). Because of electronic factors, BA possesses labile –NH, which not only resonate at $\Delta\omega = 5.0$ ppm [67, 68] but also exchange at an appropriate rate in buffered saline to produce excellent contrast. This has been exploited in several in vivo studies where BA was loaded into liposomal nanocarriers that have been shown to allow detection at very low concentrations. The developed BA-loaded liposomes could be detected using CEST imaging at a 5 ppm offset from water, far enough to be readily differentiated from the 1–4 ppm CEST contrast present in endogenous metabolites. The lead BA–liposome formulation exhibited 30% CEST contrast and 25% BA content in vitro, both of which were well maintained with ∼80% of the initial intensity/dose retained over 8 h of dialysis against PBS. In one particular example, the BA-diaCEST liposomes were used to monitor the response to tumor necrosis factor-alpha (TNF-α), an agent in clinical trials that increases vascular permeability and uptake of nanocarriers into tumors [67]. In a second approach, it was shown that liposome-based mucus-penetrating particles (MPP) loaded with BA (as a diaCEST MRI agent) with optimized PEG coating enable drug delivery and imaging at mucosal surfaces [68].

9.5 Salicylic Acid and Anthranilic Acid Analogues

Despite the success of the heterocyclic agents, as Fig. 9.1 suggests for 3 T CEST imaging, it would be nice to increase the $\Delta\omega > 6$ ppm. Salicylic acid is a natural compound that functions as a plant hormone. It is best known as the important active metabolite of its prodrug, aspirin. Structure-related salicylic acid analogues and derivatives have been widely used as non-steroidal anti-inflammatory drugs (NSAIDs). It has also been employed as a key ingredient in topical anti-acne products [69, 70]. Its chemical structure consists of a phenol proton hydrogen bonded to a carboxylate, which has been known with large chemical shift [71, 72]. These properties make salicylic acid a suitable contrast agent for MRI applications using CEST. Upon applying a saturation pulse, salicylic acid gave significant contrast at $\Delta\omega = 9.3$ ppm with $k_{ex} = 410$ s^{-1} at pH 7.4 [20, 73]. As shown in detail in Fig. 9.1 and discussed elsewhere [74–76], it is important to achieve a balance between chemical shift and k_{ex} to maximize CEST contrast. The figure evaluated application of long $\omega_1 = 2$ μT and 4 μT saturation pulses by numerically solving the Bloch–McConnell equations [21] for a two-pool system. It shows the great advantage of increasing the chemical shift beyond 6 ppm from water, with diminishing gains occurring by ~12 ppm when a 4 μT saturation pulse is utilized on 3 T clinical scanners. The optimum k_{ex} is about 1050 s^{-1} for that saturation field strength.

The excellent CEST properties of salicylic acid come from a combination of the hydrogen-bonding strength, phenolic proton pK_a, aromatic de-shielding, and water solvation of the labile protons. A systematic investigation of phenol-based intramolecular bond-shifted hydrogens (IM-SHYs) has been performed to illustrate the key factors for providing significant CEST contrast (Table 9.2) [20, 77]. The carboxylate anion was proven to be critical for buffering the proton k_{ex} of the ortho-phenol. Switching it to other hydrogen-bonding acceptors, such as ester, amide, aldehyde, nitro, quinoline N-oxide, tetrazole, and others, failed to result in protons that produce strong contrast. The R-2-OH group could be switched to several N-H groups while maintaining a suitable and tunable labile

Table 9.2 Stereo-electronic effect on the CEST properties of IM-SHYs

#	R-2	R-3	R-4	R-5	R-6	Signal (ppm)	k_{ex}
1	OH	H	H	H	H	9.3	410
2	NHPh	H	H	H	H	4.8	700
3	NHSO$_2$Me	H	H	NH$_2$	H	6.3	540
4	NHSO$_2$Me	H	H	H	H	7.3	470
5	NHTs	H	H	H	H	7.8	420
6	NHCOCF$_3$	H	H	H	H	9.3	180
7	OH	t-Bu	H	t-Bu	H	9.3	21
8	OH	OH	H	H	H	9.0	2,200
9	OH	Br	H	Br	H	10.5	550
10	OH	NO$_2$	H	H	H	12.0	1,400
11	OH	H	OH	H	H	9.5	450
12	OH	H	H	NO$_2$	H	10.3	6,000
13	OH	H	H	NH$_2$	H	8.5	370
14	OH	H	H	H	OMe	9.0	4,900

Source: Schema reprinted with permission from Ref. [20], Copyright 2014, John Wiley and Sons.

proton, including *N*-aryl-anthranilic acids (4.8 ppm), *N*-sulfonyl-anthranilic acids (6.3–7.8 ppm), and *N*-trifluoroacetyl-anthranilic acid (9.3 ppm) [77]. Bulky groups can be attached at R-3, which reduces the k_{ex}, and also fast exchangeable proton substitutions can be introduced to increase k_{ex} (e.g., entries 7 versus 8). Extra de-shielding can be introduced through the modification of the R-3 position with bromo, iodo, aryl, carbonyl, and nitro groups (e.g., entries 9 and 10). The farthest chemical shift could reach 12.0 ppm for 3-nitrosalicylic acid. Substitutions at the 4- and 5-

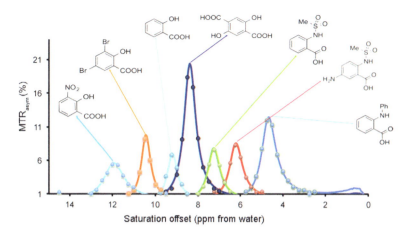

Figure 9.5 Representative examples of IM-SHYs with tunable exchangeable protons. Reprinted with permission from Ref. [20], Copyright 2014, John Wiley and Sons.

positions on 2-hydroxybenzoic acid slightly affect the chemical shift through electronic effects. Electron-withdrawing groups increase the chemical shift by 0.2–1.0 ppm, while electron-donating groups decrease the chemical shift 0.3–0.8 ppm (e.g., entries 12 versus 13). Normally, 4-/5- modifications on 2-hydroxybenzoic acid do not change the k_{ex} significantly with the exception of the strong electron-withdrawing nitro group. They present the best positions to conjugate targeting agents and other functions while maintaining CEST contrast. Keeping a proton at R-6 is required to provide the correct k_{ex} for low field CEST imaging. Substitutions will result in dramatic k_{ex} changes (e.g., entry 14). As general scaffolds, the salicylic acid/anthranilic acid IM-SHYs can produce strong contrast from 4.8 ppm to 12 ppm (Fig. 9.5). These probes were designed with multi-frequency CEST imaging in mind, with frequencies spanning a range that should allow discrimination between multiple agents within a single image [78]. The reliable CEST signal of IM-SHYs also enables their conjugation to nanoparticles [79] or enzyme substrates [80–82] for imaging applications. In addition, the concept of IM-SHY for shifting the chemical shift of water exchangeable protons could also be applied to other scaffolds, such as imidazole N-H [83].

9.6 Macromolecules with Labile Protons

Designing macromolecular diaCEST agents presents certain challenges in addition to the enormous potential benefits in molar sensitivity. One of the challenges for macromolecular agents is that their structure will have an enormous impact on k_{ex}, making the prediction of CEST contrast based on the knowledge of the moieties present alone challenging. Indeed, this dependence has enabled measurements on the degree of folding of proteins, which has been utilized extensively by numerous NMR research groups as mentioned earlier. A very detailed description of the relationship can be found in the work of Walter Englander et al. [10, 11].

As part of our ongoing efforts developing diaCEST reporter genes, we have focused on identifying peptides, which possess sufficient numbers of suitable labile protons to allow their detection upon cell expression. We focused on serine-, threonine-, arginine-, and lysine-rich peptides [50]. This study started by using the exchange prediction equations developed by Englander et al. [10] to identify a set of 12 residue peptides that would be expected to have fast k_{ex} values suitable for producing CEST contrast, with peptides rich in these four amino acids investigated because of their labile protons. We prepared 33 peptide sequences, and grouped these as NH (lysine rich), gNH$_2$ (arginine rich), and OH (serine or threonine rich) peptides. Many of our peptides displayed excellent contrast, with the members of the gNH$_2$ group displaying the highest sensitivity compared to others at 11.7 T. The NH group of which lysine-rich protein is a member had the second highest sensitivity. All 33 peptides displayed strong CEST contrast, which is what is required for in vivo applications.

In addition to determining which peptides produced the most contrast, we also investigated how well we could predict the CEST contrast produced by peptides based on protein secondary structure considerations. The backbone NH exchange rates were predicted based on the neighbors to the left (l) and right (r) using the combined base-catalyzed and acid-catalyzed equation [10]:

$$k_{ex} = k_{A,\,ref}(A_l \times A_r)[H^+] + k_{B,\,ref}(B_l \times B_r)[OH^-] + k_{W,\,ref}(B_l \times B_r) \tag{9.2}$$

where $k_{A,\,ref} = 1.83$ M/s, $k_{B,\,ref} = 3.82 \times 10^8$ M/s, and $k_{W,\,ref} = 5.27 \times 10^{-4}$ M/s are the acid, base, and water rate constants for the appropriate reference [10], and A and B are the acid and base rate factors, respectively. The general idea behind this equation is that smaller or basic side chains speed up exchange, while larger or acidic side chains reduce this rate. The predictive model proved to reasonably estimate the NH PTEs for the lysinerich peptides (R = 0.71) where the backbone amide protons dominate the CEST contrast. Unfortunately, the agreement for sidechain protons was poor.

Macromolecular CEST agents are expected to play an important role moving forward. As has been described, peptides and proteins can behave as excellent CEST MRI agents. A significant consideration for macromolecular agents is that their folding can significantly impact proton exchange. Indeed for human protamine 1, we have shown that upon binding of negatively charged nucleotides, there is a strong reduction in contrast [51]. Glycosaminoglycans have also been shown to display excellent contrast characteristics [52–55]. The overall glycosylation status of macromolecules will strongly impact contrast, with alterations shown to be important in cancer development [56]. Besides peptides and proteins, a number of carbohydrates such as glycogen can produce excellent contrast and are currently under consideration as agents [57]. In addition, dendrimers and poly nucleic acids can be constructed in a way to produce excellent contrast [19, 21, 58]. Indeed, because of the importance of macromolecular CEST agents, we have dedicated two additional chapters to discussing two specific classes further: proteins suitable for reporting on gene activity (Chapter 10) and glycosaminoglycans (Chapter 16). We believe macromolecular CEST agents will have a significant impact on medical imaging.

9.7 Fluorine and Chemical Exchange Saturation Transfer

Although the highly shifted ^{1}H diaCEST agents mentioned earlier successfully overcome some of the main obstacles of diaCEST probes by reducing both (i) the direct saturation of the water protons and (ii) the contributions from endogenous CEST contrast, such effects, in addition to the endogenous MT effect from macromolecules, were not totally eliminated. Another potential complication of such ^{1}H diaCEST agents is the large concentration of water protons for the target tissue. Such large water proton pool sizes in biological tissues although make MRI a robust imaging methodology with high signal-to-noise ratio (SNR), they also restrict the ability to sense low concentrations of compounds through interactions with the probe. This is particularly the case for diaCEST contrast agents, where hundreds of micromolar to a few millimolar concentrations of the contrast agents are required for a reliable, detectable CEST contrast. Some of these obstacles can be overcome by using non-^{1}H (heteronuclear) CEST agents instead.

Saturation transfer experiments have been performed using NMR spectroscopy and heteronuclear spins for a long time to observe and quantity the exchange between multiple spins [84–86], long before Balaban et al. proposed the diaCEST approach for molecular MRI [29]. Indeed, Balaban was initially inspired by ^{31}P saturation transfer studies of the creatine kinase reaction to perform ^{1}H-CEST experiments [87–89]. The combination of heteronuclear NMR and CEST opens a new avenue for the design of MRI sensors since it exploits the benefits of both methodologies, i.e., (i) the amplification intrinsic in the CEST mechanism; (ii) the large chemical shifts (several hundreds of ppm for heteronuclear spins); (iii) the high sensitivity of the $\Delta\omega$ to the local environment; and (iv) the lack of background signal.

In two specific examples, it was shown that the CEST experiment could be performed using ^{19}F-NMR spectroscopy to detect low concentrations of metal ions with biological relevance [90, 91]. In the early 1980s, Tsein proposed a Ca^{2+} indicator based on the metal ion chelate 1,2-bis(o-aminophenoxy)ethane-N,N,N',N'-tetraacetic acid (BAPTA) [92]. The ability to monitor real-time changes in the intracellular levels of Ca^{2+} revolutionized our understanding of Ca^{2+} physiology and initiated the development of a variety of synthetic fluorescent dyes, not only for Ca^{2+}, but also for other metal ions [93, 94]. Interestingly, synthetic fluorinated derivatives of BAPTA (nF-BAPTA, Fig. 9.6) show a unique $\Delta\omega$ on the ^{19}F-NMR spectrum upon binding of divalent metal ions (M^{2+}) [95]. This exclusive selectivity and the resulting ^{19}F-NMR confirmation of

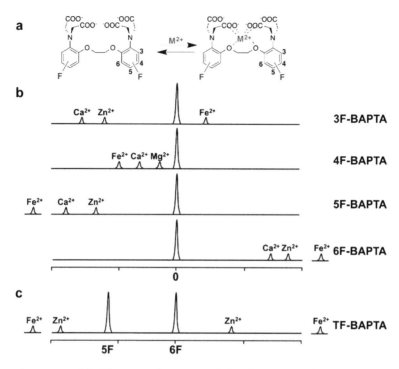

Figure 9.6 (a) Schematic depiction of the dynamic exchange process between free nF-BAPTA and bound [M^{2+}-nF-BAPTA]. (b) Illustration of ^{19}F chemical shifts for complexes of nF-BAPTA with M^{2+} ions relative to free nF-BAPTA. (c) Particular case of 5,5,6,6-tetrafluoro-BAPTA (TF-BAPTA), where ^{19}F chemical shifts for complexes of TF-FBAPTA with M^{2+} ions shown in two sets, downfield for the 5F position and upfield for the 6F position. Reprinted with permission from Ref. [90], Copyright 2013, American Chemical Society.

the binding of nF-BAPTA to metal ions, together with the spectral resolution of ^{19}F-NMR spectra, have been exploited for intracellular M^{2+} detection both in vitro and in vivo using MRS [95–98]. However, one main limitation of MRS techniques is that direct detection of the low concentration ^{19}F resonance of the M^{2+}-nF-BAPTA complex is required (schematically shown in Fig. 9.6). Such peak detections in ^{19}F-NMR spectra may suffer from limited SNR (especially in vivo) and, if indeed detectable, are also accompanied by a very low spatial resolution due to sensitivity constraints. To overcome these limitations, the use of a MT technique to indirectly detect the ion binding was proposed [86]. By radiofrequency labeling at the ^{19}F frequency of Ca^{2+}-5F-BAPTA and monitoring the label transfer to the ^{19}F frequency of 5F-BAPTA, the signal of low concentrations of Ca^{2+} could be indirectly detected with improved sensitivity. By combining ^{19}F- and CEST MRI, in an approach termed ion CEST (iCEST) [90], spatial monitoring and mapping of Ca^{2+} ions with high specificity are possible by capitalizing on the dynamic exchange between the ion-bound and free ^{19}F-chelate and the shift in the $\Delta\omega$ value in ^{19}F-NMR (Fig. 9.7). Figure 9.7a illustrates the dynamic exchange process between free 5F-BAPTA and its complex with M^{2+} [M^{2+}−5F-BAPTA]. The observed change in the ^{19}F $\Delta\omega$ for 5F-BAPTA upon M^{2+} binding is depicted in Fig. 9.7b. When the exchange between the M^{2+}-[5F-BAPTA] complex and free 5F-BAPTA is slow on the NMR timescale, as in the case of M^{2+} =Ca^{2+} or Zn^{2+}, a well-defined peak is observed for the [M^{2+}−5F-BAPTA] resonance. However, for fast exchange or loose binding of the M^{2+} by 5F-BAPTA, no peak can be resolved, as shown in Fig. 9.7b for Mg^{2+}. The ^{19}F-iCEST properties of 5F-BAPTA in the presence of Ca^{2+} (slow-to-intermediate exchange), Zn^{2+} (very slow exchange), and Mg^{2+} (fast exchange) were determined on a 16.4 T MRI scanner at pH 7.2 (Fig. 9.7c–e). A pronounced saturation transfer contrast was detected in the Ca^{2+}-containing solutions, but not in the solutions containing Zn^{2+} or Mg^{2+}. From the Bloch simulations (solid line in Fig. 9.7c), the k_{ex} between free and Ca^{2+}-bound 5F-BAPTA was found to be ∼200 s^{-1}. A molar ratio of 2000:1 between 5F-BAPTA and Ca^{2+} and signal amplification of a 1:20 change in the ^{19}F signal correspond to an amplification factor of ∼100. Therefore, Ca^{2+} ion distributions could be monitored and mapped at very low concentrations that do not allow direct ^{19}F-MRI of bound 5F-BAPTA (Fig. 9.7f).

Chemical modification of 5F-BAPTA to obtain 5,5′,6,6′-tetrafluoro-BAPTA (TF-BAPTA) dramatically changes the k_{ex} (from ∼200 s^{-1} to ∼10,000 s^{-1}) between free and Ca^{2+}-bound chelate [99–101]. The $\Delta\omega$ of the 6F position of TF-BAPTA is insensitive to different Ca^{2+} levels, while that of the 5F position responds to changes in Ca^{2+} binding with fast exchange kinetics. Thus, the shift difference between the 5F and 6F resonances can be used to measure intracellular Ca^{2+} [99, 100]. Figure 9.8b shows the ^{19}F-NMR spectrum of TF-BAPTA in the presence of either Zn^{2+} or

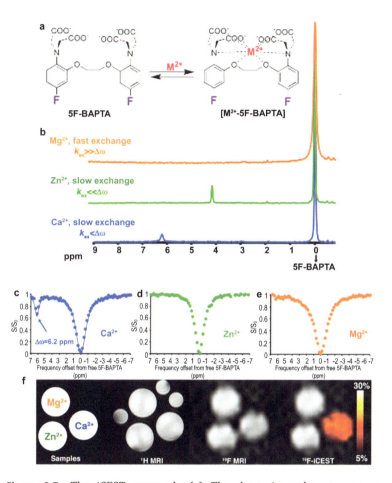

Figure 9.7 The iCEST approach. (a) The dynamic exchange process between free 5F-BAPTA and M2+-bound [M^{2+}−5F-BAPTA]. (b) ^{19}F-NMR spectra of 5F-BAPTA in the presence of Mg^{2+}, Zn^{2+}, or Ca^{2+}. (c–e) ^{19}F-iCEST Z-spectra of solutions containing 10 mM 5F-BAPTA and 50 μM M^{2+}; lines represent Bloch simulations. (f) ^{1}H-MRI, ^{19}F-MRI, and iCEST ($\Delta\omega = 6.2$) of M^{2+} solutions. Reprinted with permission from Ref. [90], Copyright 2013, American Chemical Society.

Fe^{2+}. The $\Delta\omega$ values of the ^{19}F atoms at the 5- (purple) and 6-positions (green) are shifted downfield and upfield, respectively, in the presence of either of the ions, with a larger effect for the paramagnetic Fe^{2+}. Interestingly, the different binding kinetics of

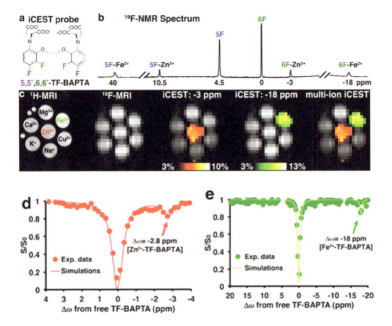

Figure 9.8 (a) The chemical structure of 5,5′,6,6′-tetrafluoro-BAPTA (TF-BAPTA). (b) The ^{19}F-NMR spectrum (470 MHz) of 10 mM TF-BAPTA in the presence of 1 mM Zn^{2+} or Fe^{2+}. (c) ^{1}H-MRI, ^{19}F-MRI, and iCEST ($\Delta\omega = -3$ ppm) overlaid on ^{19}F-MRI; iCEST ($\Delta\omega = -18$ ppm) overlaid on ^{19}F-MRI; and both iCEST results ($\Delta\omega = -3$ ppm, $\Delta\omega = -18$ ppm) overlaid on ^{19}F-MRI. ^{19}F-iCEST spectra for samples containing 10 mM TF-BAPTA and 200 μM Zn^{2+} (d) and Fe^{2+} (e). Circles represent experimental signal; solid lines represent Bloch simulations (two-pool model). Reprinted with permission from Ref. [91], Copyright 2015, American Chemical Society.

TF-BAPTA with selected metal ions, which lead to different $\Delta\omega$ and k_{ex}, allows the use of the iCEST approach to monitor both Zn^{2+} and Fe^{2+} (Fig. 9.8c). Figure 9.8c shows the ^{1}H and ^{19}F MR images of seven tubes containing 10 mM TF-BAPTA and 200 μM added ion, without any observable changes in ^{1}H or ^{19}F MR contrast. However, the ^{19}F-iCEST images show a clear differential MR contrast between the samples containing Zn^{2+} and Fe^{2+}, for a saturation pulse applied at $\Delta\omega = -2.8$ and -18 ppm, respectively. These $\Delta\omega$ values were chosen from the ^{19}F-NMR spectra, using the offset values of TF-BAPTA upon the addition of Zn^{2+} or Fe^{2+}, respectively. The corresponding ^{19}F-iCEST spectra for samples containing either

Zn^{2+} or Fe^{2+} are shown in (Fig. 9.8d,e), where Bloch simulations were used to estimate the k_{ex} between free and bound TF-BAPTA. The 20 S^{-1} found that k_{ex} is rather low, and much higher CEST contrast may be obtained for ^{19}F chelates with higher k_{ex} values.

In summary, the increased interest in ^{19}F-probes for molecular MRI [102, 103], in combination with the recent advances in CEST MRI, creates a unique MRI platform for studying previously undetectable targets and may open a window for the design of novel ^{19}F-CEST probes to investigate unrevealed biological processes with MRI.

References

1. Forsen S and Hoffman RA. Study of moderately rapid chemical exchange reactions by means of nuclear magnetic double resonance. *J. Chem. Phys.*, 1963; 39(11): 2892.
2. Forsen S and Hoffman RA. A new method for study of moderately rapid chemical exchange rates employing nuclear magnetic double resonance. *Acta Chem. Scand.*, 1963; 17(6): 1787.
3. Kordel J, Drakenberg T, Forsen S, and Thulin E. Peptidyl-prolyl cis-trans isomerase does not affect the Pro-43 cis-trans isomerization rate in folded calbindin D9k. *FEBS Lett.*, 1990; 263(1): 27–30.
4. Johansson MU, Nilsson H, Evenas J, et al. Differences in backbone dynamics of two homologous bacterial albumin-binding modules: Implications for binding specificity and bacterial adaptation. *J. Mol. Biol.*, 2002; 316(5): 1083–1099.
5. Forsen S and Hoffman RA. Exchange rates by nuclear magnetic multiple resonance 3. Exchange reactions in systems with several nonequivalent sites. *J. Chem. Phys.*, 1964; 40(5): 1189.
6. Forsen S and Hoffman RA. On use of partly resolved structure for analysis of NMR spectra. Application to determination of signs of hydroxyl proton-ring proton nuclear spin coupling constants in substituted phenols. *J. Mol. Spectrosc.*, 1966; 20(2): 168.
7. Dahlqvis K and Forsen S. Intramolecular exchange rates from complex NMR spectra. *Acta Chem. Scand.*, 1970; 24(2): 651.
8. Malmendal A, Evenas J, Forsen S, and Akke M. Structural dynamics in the C-terminal domain of calmodulin at low calcium levels. *J. Mol. Biol.*, 1999; 293(4): 883–899.

9. Mayer M and Meyer B. Characterization of ligand binding by saturation transfer difference NMR spectroscopy. *Angew. Chem.-Int. Edit.*, 1999; 38(12): 1784–1788.

10. Bai Y, Milne JS, Mayne L, and Englander SW. Primary structure effects on peptide group hydrogen exchange. *Proteins*, 1993; 17: 75–86.

11. Connolly GP, Bai Y, Jeng M-F, and Englander SW. Isotope effects in peptide group hydrogen exchange. *Proteins*, 1993; 17: 87–92.

12. McConnell BM and von Hippell PH. Hydrogen exchange as a probe of the dynamic structure of DNA: I. General acid-base catalysis. *J. Mol. Biol.*, 1970; 50: 297–316.

13. Phillips WD, Glickson JD, and Rupley JA. Proton magnetic resonance study of the indole NH resonances of lysozyme. Assignment, deuterium exchange kinetics, and inhibitor binding. *J. Am. Chem. Soc.*, 1971; 93(16): 4031–4038.

14. Johnston PD and Redfield AG. An NMR study of the exchange rates for protons involved in the secondary and tertiary structure of yeast tRNA Phe. *Nucleic Acids Res.*, 1977; 4(10): 3599–3615.

15. Englander SW and Kallenbach NR. Hydrogen exchange and structural dynamics of proteins and nucleic acids. *Quart. Rev. Biophys.*, 1984; 16(4): 521–655.

16. Bryant R. The dynamics of water-protein interactions. *Annu. Rev. Biophys. Biomol. Struct.*, 1996; 25(1): 29–53.

17. Roder H, Wagner G, and Wuthrich K. Amide proton exchange in proteins by EX1 kinetics: Studies of the basic pancreatic trypsin inhibitor at variable p2H and temperature. *Biochemistry*, 1985; 24(25): 7396–7407.

18. Roder H, Wagner G, and Wuthrich K. Individual amide proton exchange rates in thermally unfolded basic pancreatic trypsin inhibitor. *Biochemistry*, 1985; 24(25): 7407–7411.

19. Snoussi K, Bulte JW, Gueron M, and van Zijl PC. Sensitive CEST agents based on nucleic acid imino proton exchange: Detection of poly(rU) and of a dendrimer-poly(rU) model for nucleic acid delivery and pharmacology. *Magn. Reson. Med.*, 2003; 49(6): 998–1005.

20. Yang X, Yadav NN, Song X, et al. Tuning phenols with intramolecular bond shifted hydrogens (IM-SHY) as diaCEST MRI contrast agents. *Chemistry*, 2014; 20(48): 15824–15832.

21. McMahon MT, Gilad AA, Zhou J, Sun PZ, Bulte JW, and van Zijl PC. Quantifying exchange rates in chemical exchange saturation transfer agents using the saturation time and saturation power dependencies

of the magnetization transfer effect on the magnetic resonance imaging signal (QUEST and QUESP): Ph calibration for poly-L-lysine and a starburst dendrimer. *Magn. Reson. Med.*, 2006; 55(4): 836–847.
22. Liu G, Song X, Chan KW, and McMahon MT. Nuts and bolts of chemical exchange saturation transfer MRI. *NMR Biomed.*, 2013; 26(7): 810–828.
23. Jin T, Autio J, Obata T, and Kim S-G. Spin-locking vs. chemical exchange saturation transfer MRI for investigating chemical exchange process between water and labile metabolite protons. *Magn. Reson. Med.*, 2011; 65(5): 1448–1460.
24. Yadav NN, Jones CK, Xu J, et al. Detection of rapidly exchanging compounds using on-resonance frequency labeled exchange (FLEX) transfer. *Magn. Reson. Med.*, 2012; 68(4): 1048–1055.
25. Xu X, Yadav NN, Knutsson L, et al. Dynamic glucose-enhanced (DGE) MRI: Translation to human scanning and first results in glioma patients. *Tomography*, 2015; 1(2): 105–114.
26. Yadav NN, Xu J, Bar-Shir A, et al. Natural D-glucose as a biodegradable MRI relaxation agent. *Magn. Reson. Med.*, 2014; 72(3): 823–828.
27. Soesbe TC, Ratnakar SJ, Milne M, et al. Maximizing T(2)-exchange in Dy(3+)DOTA-(amide)(X) chelates: Fine-tuning the water molecule exchange rate for enhanced T(2) contrast in MRI. *Magn. Reson. Med.*, 2014; 71(3): 1179–1185.
28. Soesbe TC, Merritt ME, Green KN, Rojas-Quijano FA, and Sherry AD. T(2) exchange agents: A new class of paramagnetic MRI contrast agent that shortens water T(2) by chemical exchange rather than relaxation. *Magn. Reson. Med.* 2011; 66(6): 1697–1703.
29. Ward KM, Aletras AH, and Balaban RS. A new class of contrast agents for MRI based on proton chemical exchange dependent saturation transfer (CEST). *J. Magn. Reson.*, 2000; 143(1): 79–87.
30. Wolff SD and Balaban RS. NMR imaging of labile proton exchange. *J. Magn. Reson.*, 1990; 86(1): 164–169.
31. Dagher AP, Aletras A, Choyke P, and Balaban RS. Imaging of urea using chemical exchange-dependent saturation transfer at 1.5 T. *J. Magn. Reson. Imaging.*, 2000; 12(5): 745–748.
32. Guivel-Scharen V, Sinnwell T, Wolff SD, and Balaban RS. Detection of proton chemical exchange between metabolites and water in biological tissues. *J. Magn. Reson.*, 1998; 133(1): 36–45.
33. Daryei I and Pagel MD. Double agents and secret agents: The emerging fields of exogenous chemical exchange saturation transfer and T2-

exchange magnetic resonance imaging contrast agents for molecular imaging. *Res. Rep. Nucl. Med.*, 2015; 5: 19–32.
34. Desmond KL and Stanisz GJ. Understanding quantitative pulsed CEST in the presence of MT. *Magn. Reson. Med.*, 2012; 67(4): 979–990.
35. Govindaraju V, Young K, and Maudsley AA. Proton NMR chemical shifts and coupling constants for brain metabolites. *NMR Biomed.*, 2000; 13(3): 129–153.
36. Cai K, Haris M, Singh A, et al. Magnetic resonance imaging of glutamate. *Nat. Med.*, 2012; 18(2): 302–306.
37. Kogan F, Singh A, Debrosse C, et al. Imaging of glutamate in the spinal cord using GluCEST. *NeuroImage*, 2013; 77: 262–267.
38. Cai K, Singh A, Roalf DR, et al. Mapping glutamate in subcortical brain structures using high-resolution GluCEST MRI. *NMR Biomed.*, 2013; 26(10): 1278–1284.
39. Davis KA, Nanga RPR, Das S, et al. Glutamate imaging (GluCEST) lateralizes epileptic foci in nonlesional temporal lobe epilepsy. *Sci. Transl. Med.*, 2015; 7(309): 309ra161.
40. Haris M, Nath K, Cai K, et al. Imaging of glutamate neurotransmitter alterations in Alzheimer's disease. *NMR Biomed.*, 2013; 26(4): 386–391.
41. Chan KW, Liu G, Song X, et al. MRI-detectable pH nanosensors incorporated into hydrogels for in vivo sensing of transplanted cell viability. *Nat. Mater.*, 2013; 12(3): 268–275.
42. Liu G, Moake M, Har-el Y-e, et al. In vivo multicolor molecular MR imaging using diamagnetic chemical exchange saturation transfer liposomes. *Magn. Reson. Med.*, 2012; 67(4): 1106–1113.
43. Liu G, Chan KWY, Song X, et al. NOrmalized MAgnetization Ratio (NOMAR) filtering for creation of tissue selective contrast maps. *Magn. Reson. Med.*, 2013; 69(2): 516–523.
44. Haris M, Cai K, Singh A, Hariharan H, and Reddy R. In vivo mapping of brain myo-inositol. *NeuroImage*, 2011; 54(3): 2079–2085.
45. Walker-Samuel S, Ramasawmy R, Torrealdea F, et al. In vivo imaging of glucose uptake and metabolism in tumors. *Nat. Med.*, 2013; 19(8): 1067.
46. Chan KW, McMahon MT, Kato Y, et al. Natural D-glucose as a biodegradable MRI contrast agent for detecting cancer. *Magn. Reson. Med.*, 2012; 68(6): 1764–1773.
47. Rivlin M, Tsarfaty I, and Navon G. Functional molecular imaging of tumors by chemical exchange saturation transfer MRI of 3-O-methyl-D-glucose. *Magn. Reson. Med.*, 2014; 72(5): 1375–1380.

48. Chan KWY, Jiang L, Cheng M, et al. CEST-MRI detects metabolite levels altered by breast cancer cell aggressiveness and chemotherapy response. *NMR Biomed.*, 2016; 29(6): 806–816.
49. DeBrosse C, Nanga RPR, Bagga P, et al. Lactate chemical exchange saturation transfer (LATEST) imaging in vivo a biomarker for LDH activity. *Sci. Rep.*, 2016; 6: 19517.
50. McMahon MT, Gilad AA, DeLiso MA, Cromer Berman SM, Bulte JWM, and van Zijl PCM. New "multicolor" polypeptide diamagnetic chemical exchange saturation transfer (DIACEST) contrast agents for MRI. *Magn. Reson. Med.*, 2008; 60(4): 803–812.
51. Oskolkov N, Bar-Shir A, Chan KWY, et al. Biophysical characterization of human protamine-1 as a responsive CEST MR contrast agent. *ACS Macro Lett.*, 2015; 4(1): 34–38.
52. Ling W, Regatte RR, Navon G, and Jerschow A. Assessment of glycosaminoglycan concentration in vivo by chemical exchange-dependent saturation transfer (gagCEST). *Proc. Natl. Acad. Sci. U.S.A.*, 2008; 105(7): 2266–2270.
53. Saar G, Zhang BY, Ling W, Regatte RR, Navon G, and Jerschow A. Assessment of glycosaminoglycan concentration changes in the intervertebral disc via chemical exchange saturation transfer. *NMR Biomed.*, 2012; 25(2): 255–261.
54. Lee JS, Xia D, Parasoglou P, Chang G, Jerschow A, and Regatte RR. *Chemical Exchange Saturation Transfer Contrast by Glycosaminoglycans and Its Application for Monitoring Knee Joint Repair*. Singapore: World Scientific Publ Co Pte Ltd; 2014.
55. Muller-Lutz A, Schleich C, Schmitt B, et al. Improvement of gagCEST imaging in the human lumbar intervertebral disc by motion correction. *Skeletal Radiol.*, 2015; 44(4): 505–511.
56. Song XL, Airan RD, Arifin DR, et al. Label-free in vivo molecular imaging of underglycosylated mucin-1 expression in tumour cells. *Nat. Commun.*, 2015; 6: 6719.
57. van Zijl PCM, Jones CK, Ren J, Malloy CR, and Sherry AD. MRI detection of glycogen in vivo by using chemical exchange saturation transfer imaging (glycoCEST). *Proc. Natl. Acad. Sci. U.S.A.*, 2007; 104(11): 4359–4364.
58. Goffeney N, Bulte JW, Duyn J, Bryant LH, Jr., and van Zijl PC. Sensitive NMR detection of cationic-polymer-based gene delivery systems using saturation transfer via proton exchange. *J. Am. Chem. Soc.*, 2001; 123(35): 8628–8629.

59. van Zijl PC and Yadav NN. Chemical exchange saturation transfer (CEST): What is in a name and what isn't? *Magn. Reson. Med.*, 2011; 65(4): 927–948.
60. Terreno E, Castelli DD, and Aime S. Encoding the frequency dependence in MRI contrast media: The emerging class of CEST agents. *Contrast Media Mol. Imaging*, 2010; 5(2): 78–98.
61. Sherry AD and Woods M. Chemical exchange saturation transfer contrast agents for magnetic resonance imaging. *Annu. Rev. Biomed. Eng.*, 2008; 10: 391–411.
62. Bar-Shir A, Liu G, Greenberg MM, Bulte JW, and Gilad AA. Synthesis of a probe for monitoring HSV1-tk reporter gene expression using chemical exchange saturation transfer MRI. *Nat. Protoc.*, 2013; 8(12): 2380–2391.
63. Bar-Shir A, Liu G, Liang Y, et al. Transforming thymidine into a magnetic resonance imaging probe for monitoring gene expression. *J. Am. Chem. Soc.*, 2013; 135(4): 1617–1624.
64. Sun PZ. Simplified quantification of labile proton concentration-weighted chemical exchange rate (k(ws)) with RF saturation time dependent ratiometric analysis (QUESTRA): Normalization of relaxation and RF irradiation spillover effects for improved quantitative chemical exchange saturation transfer (CEST) MRI. *Magn. Reson. Med.*, 2012; 67(4): 936–942.
65. Randtke EA, Chen LQ, Corrales LR, and Pagel MD. The Hanes–Woolf linear QUESP method improves the measurements of fast chemical exchange rates with CEST MRI. *Magn. Reson. Med.*, 2014; 71(4): 1603–1612.
66. Woessner DE, Zhang S, Merritt ME, and Sherry AD. Numerical solution of the Bloch equations provides insights into the optimum design of PARACEST agents for MRI. *Magn. Reson. Med.*, 2005; 53(4): 790–799.
67. Chan KW, Yu T, Qiao Y, et al. A diaCEST MRI approach for monitoring liposomal accumulation in tumors. *J. Control. Release*, 2014; 180: 51–59.
68. Yu T, Chan KW, Anonuevo A, et al. Liposome-based mucus-penetrating particles (MPP) for mucosal theranostics: Demonstration of diamagnetic chemical exchange saturation transfer (diaCEST) magnetic resonance imaging (MRI). *Nanomedicine*, 2015; 11(2): 401–405.
69. Clissold SP. Aspirin and related derivatives of salicylic acid. *Drugs*, 1986; 32 Suppl 4: 8–26.
70. Rainsford KD. *Aspirin and Related Drugs*. London: Taylor and Francis Group; 2004.

71. Mock WL and Morsch LA. Low barrier hydrogen bonds within salicylate mono-anions. *Tetrahedron*, 2001; 57(15): 2957–2964.
72. Maciel GE and Savitsky GB. Carbon-13 chemical shifts + intramolecular hydrogen bonding. *J. Phys. Chem.*, 1964; 68(2): 437.
73. Yang X, Song XL, Li YG, et al. Salicylic acid and analogues as diaCEST-MRI contrast agents with highly shifted exchangeable proton frequencies. *Angew. Chem.-Int. Edit.*, 2013; 52(31): 8116–8119.
74. Hancu I, Dixon WT, Woods M, Vinogradov E, Sherry AD, and Lenkinski RE. CEST and PARACEST MR contrast agents. *Acta Radiol.*, 2010; 51(8): 910–923.
75. Liu GS, Song XL, Chan KWY, and McMahon MT. Nuts and bolts of chemical exchange saturation transfer MRI. *NMR Biomed.*, 2013; 26(7): 810–828.
76. Castelli DD, Terreno E, Longo D, and Aime S. Nanoparticle-based chemical exchange saturation transfer (CEST) agents. *NMR Biomed.*, 2013; 26(7): 839–849.
77. Song X, Yang X, Ray Banerjee S, Pomper MG, and McMahon MT. Anthranilic acid analogs as diamagnetic CEST MRI contrast agents that feature an intramolecular-bond shifted hydrogen. *Contrast Media Mol. Imaging*, 2015; 10(1): 74–80.
78. Song X, Walczak P, He X, et al. Salicylic acid analogues as chemical exchange saturation transfer MRI contrast agents for the assessment of brain perfusion territory and blood-brain barrier opening after intra-arterial infusion. *J. Cereb. Blood Flow Metab.*, 2016; pii: 0271678X16637882. [Epub ahead of print].
79. Lesniak WG, Oskolkov N, Song X, et al. Salicylic acid conjugated dendrimers are a tunable, high performance CEST MRI nanoplatform. *Nano Lett.*, 2016; 16(4): 2248–2253.
80. Fernandez-Cuervo G, Sinharay S, and Pagel MD. A catalyCEST MRI contrast agent that can simultaneously detect two enzyme activities. *Chembiochem*, 2016; 17(5): 383–387.
81. Hingorani DV, Montano LA, Randtke EA, Lee YS, Cardenas-Rodriguez J, and Pagel MD. A single diamagnetic catalyCEST MRI contrast agent that detects cathepsin B enzyme activity by using a ratio of two CEST signals. *Contrast Media Mol. Imaging*, 2016; 11(2): 130–138.
82. Sinharay S, Fernandez-Cuervo G, Acfalle JP, and Pagel MD. Detection of sulfatase enzyme activity with a catalyCEST MRI contrast agent. *Chemistry*, 2016; 22(19): 6491–6495.

83. Yang X, Song X, Ray Banerjee S, et al. Developing imidazoles as CEST MRI pH sensors. *Contrast Media Mol. Imaging*, 2016; doi: 10.1002/cmmi.1693. [Epub ahead of print].

84. Kupriyanov VV, Balaban RS, Lyulina NV, Steinschneider A, and Saks VA. Combination of ^{31}P-NMR magnetization transfer and radioisotope exchange methods for assessment of an enzyme reaction mechanism: Rate-determining steps of the creatine kinase reaction. *Biochim Biophys Acta*, 1990; 1020(3): 290–304.

85. Alger JR and Shulman RG. NMR studies of enzymatic rates in vitro and in vivo by magnetization transfer. *Q Rev Biophys.*, 1984; 17(1): 83–124.

86. Gilboa H, Chapman BE, and Kuchel PW. ^{19}F-NMR magnetization transfer between 5-FBAPTA and its complexes. An alternative means for measuring free Ca^{2+} concentration, and detection of complexes with protein in erythrocytes. *NMR Biomed.*, 1994; 7(7): 330–338.

87. Koretsky AP, Basus VJ, James TL, Klein MP, and Weiner MW. Detection of exchange-reactions involving small metabolite pools using NMR magnetization transfer techniques: Relevance to subcellular compartmentation of creatine-kinase. *Magn. Reson. Med.*, 1985; 2(6): 586–594.

88. Balaban RS and Koretsky AP. Interpretation of P-31 NMR saturation transfer experiments: What you can't see might confuse you. Focus on "Standard magnetic resonance-based measurements of the P_i → ATP rate do not index the rate of oxidative phosphorylation in cardiac and skeletal muscles." *Am. J. Physiol. Cell Physiol.*, 2011; 301(1): C12–C15.

89. Kupriyanov VV, Balaban RS, Lyulina NV, Steinschneider AY, and Saks VA. Combination of ^{31}P-NMR magnetization transfer and radioisotope exchange methods for assessment of an enzyme reaction mechanism: Rate-determining steps of the creatine-kinase reaction. *Biochim. Biophys. Acta.*, 1990; 1020(3): 290–304.

90. Bar-Shir A, Gilad AA, Chan KW, et al. Metal ion sensing using ion chemical exchange saturation transfer ^{19}F magnetic resonance imaging. *J. Am. Chem. Soc.*, 2013; 135(33): 12164–12167.

91. Bar-Shir A, Yadav NN, Gilad AA, van Zijl PC, McMahon MT, and Bulte JW. Single (19)F probe for simultaneous detection of multiple metal ions using miCEST MRI. *J. Am. Chem. Soc.*, 2015; 137(1): 78–81.

92. Tsien RY. New calcium indicators and buffers with high selectivity against magnesium and protons: Design, synthesis, and properties of prototype structures. *Biochemistry*, 1980; 19(11): 2396–2404.

93. Carter KP, Young AM, and Palmer AE. Fluorescent sensors for measuring metal ions in living systems. *Chem. Rev.*, 2014; 114(8): 4564–4601.
94. Qian X and Xu Z. Fluorescence imaging of metal ions implicated in diseases. *Chem. Soc. Rev.*, 2015; 44(14): 4487–4493.
95. Smith GA, Hesketh RT, Metcalfe JC, Feeney J, and Morris PG. Intracellular calcium measurements by ^{19}F-NMR of fluorine-labeled chelators. *Proc. Natl. Acad. Sci. U.S.A.*, 1983; 80(23): 7178–7182.
96. Metcalfe JC, Hesketh TR, and Smith GA. Free cytosolic Ca^{2+} measurements with fluorine labelled indicators using ^{19}F-NMR. *Cell Calcium*, 1985; 6(1–2): 183–195.
97. Badar-Goffer RS, Ben-Yoseph O, Dolin SJ, Morris PG, Smith GA, and Bachelard HS. Use of 1,2-bis(2-amino-5-fluorophenoxy)ethane-N,N,N',N'-tetraacetic acid (5FBAPTA) in the measurement of free intracellular calcium in the brain by ^{19}F-nuclear magnetic resonance spectroscopy. *J. Neurochem.*, 1990; 55(3): 878–884.
98. Song SK, Hotchkiss RS, Neil J, Morris PE, Jr., Hsu CY, and Ackerman JJ. Determination of intracellular calcium in vivo via fluorine-19 nuclear magnetic resonance spectroscopy. *Am. J. Physiol.*, 1995; 269(2 Pt 1): C318–322.
99. Chen W, Steenbergen C, Levy LA, Vance J, London RE, and Murphy E. Measurement of free Ca^{2+} in sarcoplasmic reticulum in perfused rabbit heart loaded with 1,2-bis(2-amino-5,6-difluorophenoxy)ethane-N,N,N',N'-tetraacetic acid by ^{19}F-NMR. *J. Biol. Chem.*, 1996; 271(13): 7398–7403.
100. Murphy E, Steenbergen C, Levy LA, Gabel S, and London RE. Measurement of cytosolic free calcium in perfused rat heart using TF-BAPTA. *Am. J. Physiol.*, 1994; 266(5 Pt 1): C1323–C1329.
101. Xia Z, Horton JW, Tang H, and Yang Y. Metabolic disorder in myocardiac intracellular free calcium after thermal injury. *Burns*, 2001; 27(5): 453–457.
102. Tirotta I, Dichiarante V, Pigliacelli C, et al. (19)F magnetic resonance imaging (MRI): From design of materials to clinical applications. *Chem. Rev.*, 2015; 115(2): 1106–1129.
103. Ahrens ET and Zhong J. In vivo MRI cell tracking using perfluorocarbon probes and fluorine-19 detection. *NMR Biomed.*, 2013; 26(7): 860–871.

Chapter 10

Evolution of Genetically Encoded CEST MRI Reporters: Opportunities and Challenges

Ethel J. Ngen, Piotr Walczak, Jeff W. M. Bulte, and Assaf A. Gilad

Division of MR Research, Russell H. Morgan Department of Radiology and Radiological Science and Cell Imaging Section, Institute for Cellular Engineering, Johns Hopkins University School of Medicine, 707 N. Broadway Ave., Baltimore, MD, 21205, USA
assaf.gilad@jhu.edu

10.1 Introduction

Molecular imaging is a rapidly growing field, which holds the promise of enabling the non-invasive visualization, characterization, and quantification of dynamic biological processes at the cellular and molecular level, in intact live subjects. With the growing need for more targeted and personalized therapeutic regimens, the development of molecular imaging strategies and tools will play a crucial role in elucidating the molecular bases of diseases, and also in permitting the timely diagnosis and treatment of pathologies [1–3].

Chemical Exchange Saturation Transfer Imaging: Advances and Applications
Edited by Michael T. McMahon, Assaf A. Gilad, Jeff W. M. Bulte, and Peter C. M. van Zijl
Copyright © 2017 Pan Stanford Publishing Pte. Ltd.
ISBN 978-981-4745-70-3 (Hardcover), 978-1-315-36442-1 (eBook)
www.panstanford.com

Molecular and cellular magnetic resonance imaging (MRI) offers several advantages over other imaging modalities, such as optical, radionuclide, and ultrasound imaging. These advantages include the ability to monitor molecular and cellular processes longitudinally at high spatial resolution, and the possibility of co-registering these dynamic processes with both functional images and high-resolution anatomical images. Although MRI requires the administration of probes at higher concentrations (10^{-3}–10^{-6} M) compared to optical imaging (10^{-10}–10^{-14} M) and radionuclide imaging (10^{-7}–10^{-9}), recent technological advances have greatly improved both signal-to-noise ratio (SNR) and contrast-to-noise ratio (CNR) [4]. These technological advances include the development of high magnetic field scanners, improved radiofrequency coils, robust gradient systems, and better image post-processing tools. In addition, the development of molecular targeting strategies and tools capable of enabling the specific visualization of molecular biomarkers will further enhance the CNR [5–7].

Genetically encoded reporter imaging (or reporter gene imaging) is a molecular imaging strategy that enables the monitoring of the transcriptional and translational regulation of genes. Reporter gene imaging is essential in a wide variety of both basic and translational research areas, including (1) monitoring the efficacy of gene therapy [8, 9]; (2) monitoring transplanted stem cell survival, migration, and differentiation [10, 11]; (3) assessing endogenous molecular events associated with various pathologies, such as enzymatic activity and receptor expression [12]; and (4) studying brain function, neural plasticity, and brain repair mechanisms following trauma [13, 14].

10.1.1 *Genetically Encoded Reporter Imaging*

A reporter gene is a gene that can be either fused to a gene of interest or cloned instead of the gene of interest, and whose product (an expressed receptor, enzyme, peptide, transporter, or protein nanostructure) can be detected with an appropriate imaging or molecular biology technique. Generally, reporter gene imaging involves the use of a reporter gene expression vector, which contains the reporter gene and a transcriptional control element (promoter-enhancer) that can initiate and regulate the reporter gene's expression.

Two types of transcriptional control elements are generally used: (1) constitutive promoters (always on), which result in a continuous expression of the reporter gene and (2) conditional promoters (on–off switchable), which respond to endogenous transcription factors and transcription-regulating complexes to regulate the expression of the reporter gene only within a given context. Examples of constitutively driven promoters include cytomegalovirus (CMV), long terminal repeat, phosphoglycerate kinase (PGK), thymidine kinase (Tk), Rous sarcoma virus, and elongation factor-1 (EF1). Examples of conditional promoters could be neuron-specific enolase, which is expressed only in neurons [15], or the promoter of the early immediate gene, *c-fos*, which is expressed in response to a specific stimulation [16, 17]. These promoters often include DNA elements that respond to a specific stimulation, such as the interferon-stimulating response element and the hypoxia response element. Overall, constitutively driven promoters are best suited to monitor gene therapy, by virtue of their ability to monitor the efficiency, location, and extent of gene expression following gene delivery. Constitutively driven promoters are also well suited to track the location and survival of transplanted stem cells in cell-based therapies. Conditional promoters, however, are best suited to monitor the dynamic processes and microenvironmental changes associated with cells, such as their differentiation status and the degree of hypoxia/normoxia in the cellular microenvironment [18].

Following the construction of the reporter gene expression vector, the reporter gene must be introduced into a cell (either a prokaryotic or a eukaryotic cell), or into an animal of interest, using either transfection or viral agents. Whereas using transfection agents leads to either a transient or stable expression of the reporter gene, using viral agents leads to stable expression of the reporter gene. Following transfection or transduction, and upon activation of the gene promoter, a gene product is generated, which can then be appropriately detected (Fig. 10.1). Traditionally, reporter genes have been classified into categories based on the nature of the gene product. These include enzyme-based, receptor-based, transporter-

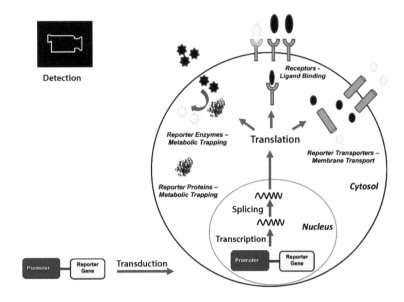

Figure 10.1 Principles of reporter gene imaging.

based, and, more recently, artificial peptide-based and gas vesicle–based reporter genes.

10.1.2 Genetically Encoded MRI Reporters

Genetically encoded MRI reporters differ from genetically encoded optical and radionuclide reporters in that they enable the repeated longitudinal visualization of gene expression at high spatial resolution without the limits of tissue depth penetration or use of ionizing radiation. Genetically encoded reporters also provide a means to co-register genetic information with functional information as well as high-resolution anatomical images.

Traditionally, genetically encoded MRI reporters have been enzyme based, receptor based, or transporter based in nature. These reporters are generally detected either following an interaction with administered paramagnetic or superparamagnetic metal-based probes or following an interaction with endogenously present metabolites or superparamagnetic iron oxides present in the tissue. Over the years, several genetically encoded MRI reporters have been developed that take advantage of different magnetic resonance (MR) techniques and contrast generation mechanisms. These include (1) multiparametric MR spectroscopy (^{1}H, ^{31}P, ^{13}C) for the detection of metabolic changes associated with enzymatic activity after the gene expression of enzymes; (2) T_2/T_2^*-weighted MRI for the detection of transverse relaxation rate changes ($\Delta R_2/\Delta R_2^*$) of water protons surrounding expressed receptors and enzymes, after the interaction of the expressed receptors with administered superparamagnetic probes or endogenously present metal ions; and (3) T_1-weighted MRI for the detection of longitudinal relaxation rate changes (ΔR_1) of water protons surrounding expressed enzymes, after the interaction of the expressed enzymes with administered paramagnetic metal chelates.

Examples of reporter genes that are detected by MR spectroscopy (^{1}H, ^{31}P) include creatine kinase and its invertebrate analogue, arginine kinase [19–21]. The metabolite concentration changes generated by these enzymes can then be detected by MR spectroscopy. Examples of genetically encoded MRI reporters detected with T_2/T_2^*-weighted MRI include ferritin proteins (responsible for iron storage) [22, 23], transferrin receptors (responsible for intracellular iron internalization) [24], and tyrosinase enzymes (responsible for melanin synthesis; melanin has a high binding affinity for metal ions) [25, 26]. Alkaline phosphatases (responsible for dephosphorylation of a variety of molecules, including metalloporphyrins) are also genetically encoded reporters that are detectable with T_2/T_2^*-weighted MRI [27]. Examples of genetically encoded MRI reporters that can be detected with T_1-weighted MRI include β-galactosidase [28, 29]. T_1 contrast generation, however, is less commonly used due to the lower sensitivity of water protons to paramagnetic metal chelates. In addition, safety concerns have been raised over the use of gadolinium-based chelates (the most commonly used T_1 contrast agents).

A major limitation of the genetically encoded reporters that exploit both the R_1 and R_2/R_2^* contrast generation mechanisms is the need for the administration of reporter probes. Upon probe interaction with the gene products, these probes either generate or amplify the MRI contrast. The administration of reporter probes for imaging gene products poses a number of challenges. Most probes are not accessible to the central nervous system (CNS) due to the presence of the blood–brain barrier. This limits their application for the study of CNS-associated pathologies. With the exception of a few activatable (on–off switchable) probes, most probes generate contrast even without interacting with the gene product. Consequently, usually a residual contrast signal is present. In addition, most probes are non-specifically taken up in vivo and have slow clearance kinetics. This usually results in delayed signal enhancement, which could generate false-negative readouts. Finally, some probes, especially gadolinium-based chelates, have low sensitivities.

To overcome the limitations associated with probe administration, a new class of genetically encoded MRI reporters has now been developed: genetically encoded chemical exchange saturation transfer (CEST) MRI reporters [30].

10.2 CEST MRI Contrast Generation Mechanism

CEST MRI is a relatively new contrast generation mechanism, which relies on magnetization transfer between a magnetized reporter of interest and molecules in its surrounding environment, after a dynamic intermolecular exchange process [31, 32]. Currently, reporter genes that exploit magnetization transfer via two main chemical exchange mechanisms have been developed: (1) proton (^{1}H) exchange (protonation/deprotonation) and (2) hyperpolarized xenon (^{129}Xe) exchange, also known as hyperCEST (exchange between protein-bound hyperpolarized ^{129}Xe/unbound hyperpolarized ^{129}Xe). HyperCEST contrast generation and genetically encoded reporters are discussed in Section 10.4.

Proton exchange with water molecules is currently the most exploited CEST mechanism, due to the abundance of water protons in biological tissues [33]. This high water concentration permits the detection of reporters at the micromolar range through signal amplification, following proton exchange and magnetization transfer to the surrounding water protons [34, 35].

In order to generate CEST contrast, an exchangeable proton on the reporter of interest is saturated with a radiofrequency pulse,

CEST MRI Contrast Generation Mechanism | 199

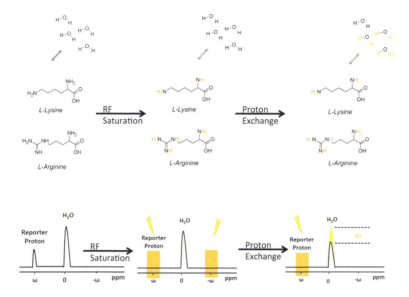

Figure 10.2 CEST contrast generation mechanism.

at a resonance frequency away from that of the surrounding water protons ($S_W^{+\Delta\omega}$). This is done to avoid direct saturation of water protons by the applied radiofrequency pulse. Another saturation pulse is applied to the opposite but equal proton resonance frequency ($S_W^{-\Delta\omega}$) to validate specific reporter proton magnetization (Fig. 10.2). For CEST contrast to be generated, two conditions need to be met. The first is that the protons on the reporter need to be saturated for sufficiently long periods of time to ensure complete saturation. The second is that exchange rates between the saturated reporter protons and those of its surrounding water molecules need to be sufficiently rapid (within the intermediate NMR timescale range) to ensure detection by MRI. The chemical exchange rates can be determined by either varying the saturation pulse time while keeping the saturation power constant (QUEST), or varying the saturation pulse power while keeping the saturation time constant (QUESP) [36].

$$MTR_{\text{asym}} = \frac{S_{CA}^{-\Delta\omega} - S_{CA}^{+\Delta\omega}}{S_{CA}^{-\Delta\omega}} \approx \frac{S_0 - S_{CA}^{+\Delta\omega}}{S_0} \quad (10.1)$$

The CEST contrast is commonly described using the term MTR_{asym} (Eq. 10.1), where $S_{CA}^{+\Delta\omega}$ is the signal intensity from the contrast agent after saturation at the appropriate exchangeable proton frequency; $S_{CA}^{-\Delta\omega}$ is the signal intensity from the contrast agent after saturation at the opposite radiofrequency frequency; and S_0 is the signal intensity from water in the absence of a saturation pulse.

10.3 Genetically Encoded CEST MRI Reporters

Genetically encoded CEST MRI reporters are a relatively new class of MRI reporter genes that exploit the CEST contrast generation mechanism. Genetically encoded CEST MRI reporters, which exploit the proton exchange mechanism, can be classified into two main categories, based on the gene products: (1) CEST-responsive peptides and (2) enzymes that generate contrast upon interaction with CEST-responsive small organic molecules or peptide-based substrates. Both types of reporter genes were developed to address two different limitations encountered with currently available reporter gene systems. Whereas CEST-responsive peptides were developed to overcome the limitations of working with administered probes with unfavorable pharmacokinetics and CNS inaccessibility, enzyme/probe systems were developed to overcome the limitations of the low sensitivity of the peptides (Table 10.1).

Genetically encoded CEST MRI reporter genes offer several advantages over other MRI reporter genes. These include (1) the ability of the CEST MRI contrast signal to be switched on and off, depending on the application of a saturation radiofrequency pulse; (2) the ability to visualize more than one gene product or molecular target simultaneously, using different saturation radiofrequency pulses (multicolor CEST) [37], although to date, this is only theoretical and has not been demonstrated experimentally (which could be particularly useful in imaging multiple gene expression or multiple cell types); and (3) the CEST signal generated does not mask out any anatomical information, and thus permits the co-registration of genetic and anatomical information.

Table 10.1 Evaluation of CEST reporter genes

Name	Description	Opportunities	Challenges
Substrate-independent			
• LRP[38]	• Lysine-rich protein	• Artificial—no endogenous counterpart. • Does not rely on administration of contrast agents. • Provides contrast at 3.6–3.7 ppm, removed from the water peak.	• Unstable DNA structure (many repetitions of AAA and AAG). • Provides contrast at 3.6–3.7 ppm where there is high contrast from endogenous proteins.
• hPRM-1[39]	• Human protamine-1	• Robust contrast (compared to all other known proteins). • No substrate is required. • Known protein, vast literature. • Human source—potentially less immunogenic. • Known to be expressed in human cells. • Expression is limited to sperm cells—lower endogenous background.	• The signal at 1.8 ppm is masked by endogenous metabolites.
• scGFP[40]	• Superpositively charged green fluorescent protein	• Fluorescent protein—good for screening and validation.	• Has not been tested in mammalian cells.

(Contd.)

Table 10.1 (Contd.)

Name	Description	Opportunities	Challenges
Enzymes			
• CD[41]	• Cytosine deaminase	• Amplification, one enzyme molecule activates multiple probes. • Relies on natural substrate. • Theranostic: the imaging probe (5-FC) is also an anti-cancer pro-drug.	• Gives signal at 2.0–2.4 ppm, high background from endogenous molecules (metabolites). • "Double negative": CEST by its nature it is a negative contrast; CD detection is via reduction in negative contrast.
• HSV1-TK[42,43]	• Thymidine kinase	• Provides contrast at 5 ppm, less contrast from endogenous molecules.	• The substrate (5-MDHT) does not cross the BBB, thus is limited to tumors in the brain.
• PKA sensor[44]	• Protein kinase A sensor	• Sensing signal transduction • A prototype for sensors of other kinases. • Switchable. The contrast is turned off by phosphorylation.	• The signal at 1.8 ppm is masked by endogenous metabolites.

10.3.1 Genetically Encoded CEST-Responsive Protein-Based Reporters

The main advantage of genetically encoded CEST-responsive, protein-based reporters is that after their expression, these reporters generate an endogenous CEST contrast, which can be visualized without the need for the administration of metal-based probes. This could be particularly useful in studying pathologies associated with the CNS, where the use of reporter probes (which might be unable to traverse the blood–brain barrier) can be avoided. Examples of such reporters include lysine-rich protein-based reporters, arginine-rich protein-based reporters, and superpositively charged green fluorescent protein reporters.

10.3.1.1 Lysine-rich protein (LRP)-based reporter genes

This was developed as the first genetically encoded CEST MRI reporter prototype and is currently the most widely used CEST reporter gene [38]. Lysine-rich protein (LRP)-based reporter genes were developed after the observation that poly-L-lysine generated an exceptionally high CEST contrast following the saturation of its exchangeable amide protons, at $\Delta\omega = 3.7$ ppm [34]. An artificial gene made of eight synthetic oligonucleotides, which encodes for a high number of lysine residues (200, 32 kDa), was designed de novo. The CEST contrast generated in transduced 9L rat gliosarcoma cells, after the expression of this transgene, was sufficiently high to distinguish 9L tumors overexpressing the transgene from control wild-type tumors not expressing the gene [38].

In addition, LRP reporter genes inserted into a herpes simplex-derived oncolytic virus (G47Δ) were later used to monitor brain cancer virotherapy, without disrupting the therapeutic effectiveness of the virus. Whereas a significant CEST signal at 3.6 ppm was observed in cell lysates and tumors expressing G47Δ-LRP, no CEST signal was observed in cell lysates and tumors expressing the empty virus (G47Δ) [46]. LRP reporter genes have also been developed under a variety of promoters as molecular-genetic tools with which to study cancer progression [47].

Although reporter gene imaging with LRP has been feasible, one of the main limitations in using LRP has been its low sensitivity. Consequently, several studies are underway to develop more sensitive LRP analogues.

10.3.1.2 Arginine-rich protein (ARP)-based reporter genes

After the success observed with LRP reporter genes, other peptides were screened to find more CEST-responsive amino acid sequences [37]. It was observed that arginine-rich peptides and proteins can also generate high CEST contrast from both their amide protons ($\Delta\omega = 3.7$ ppm) and their guanidyl protons ($\Delta\omega = 1.8$ ppm) [37]. Consequently, human protamine 1 (hPM1), an arginine-rich peptide (composed of 30% arginine residues) [48], and a less immunogenic and humanized homologue of the CEST-responsive salmon protamine sulfate, was cloned. A lentivirus that encoded the hPRM1 reporter gene under the CMV promoter was constructed and used to transduce human embryonic kidney (HEK293) cells. Using the hPM1 construct, it was possible to distinguish cell lysates of transduced HEK293 cells from non-transduced HEK293 wild-type cells [39]. Using a biophysical approach, it was demonstrated that the CEST contrast from the human protamine is affected by the pH of its environment, phosphorylation, intramolecular interactions, and interactions with other molecules, such as nucleoside triphosphate [49]. These findings may guide further design of peptide-based CEST reporters.

10.3.1.3 Superpositively charged green fluorescent proteins

Fluorescent proteins have been used extensively as reporter genes to image cellular and molecular processes [50, 51]. However, fluorescence imaging is limited for in vivo imaging due to the restricted light depth penetration. In an effort to develop dual modal MRI/fluorescent reporter genes, superpositively charged green fluorescent protein (GFP) mutants with high numbers of lysine and arginine residues were screened as potential CEST and fluorescence reporters. Briefly, *Escherichia coli*–optimized genes encoding the wild-type GFP (containing 20 lysine residues, seven arginine residues, and possessing a net charge of −7) were studied in comparison to two superpositively charged GFP variants: +36 GFP (containing 36 lysine residues, 20 arginine residues, and possessing a net charge of +36), and +48 GFP (containing 42 lysine residues, 21 arginine residues, and possessing a net charge of +48). In *E. coli* cells transformed with the respective genes, both +36 GFP and +48 GFP showed a significantly higher CEST contrast at both $\Delta\omega = 3.6$ ppm (amide protons) and $\Delta\omega = 1.8$ ppm (guanidyl protons) compared to wild-type GFP, at all saturation powers evaluated [40]. These dual modal reporters could be particularly useful in studying molecular processes because the fluorescence signal can be used for histological validation of the CEST imaging in vivo findings.

10.3.2 *Genetically Encoded Enzyme/Probe CEST MRI Reporter Systems*

Genetically encoded enzyme/probe CEST reporter systems were developed as genetic–molecular sensors to study specific and compartmentalized dynamic cellular processes, such as signal transduction, in vivo at high spatial and temporal resolution in genetically specified cell populations. These reporters function by generating an alteration in the CEST contrast following molecular processes, such as phosphorylation/dephosphorylation. These systems could be particularly helpful in understanding complex biological processes involved in diseases, as well as in the preclinical development and screening of novel therapeutics to target enzymes involved in various pathologies. One of the main advantages of genetically encoded enzyme/probe CEST reporter systems is that they enable the visualization of highly complex and specific processes at resolutions that cannot be achieved with other imaging modalities. Examples of such systems are further discussed in the following paragraphs.

10.3.2.1 Protein kinase A

Protein kinases are a class of enzymes involved in several cell signaling pathways necessary for the post-translational modification of proteins to modulate functionalization [52]. Protein kinase A (PKA) is a cyclic adenosine monophosphate (cAMP)-dependent protein kinase involved in several pathologies, and whose activity is highly dependent on its subcellular location [53]. Thus, being able to visualize the localization of this enzyme with high spatial resolution will be beneficial in targeting and inhibiting its activity in various pathologies.

PKA phosphorylates the hydroxyl group (–OH) of serine and threonine, with high specificity for the amino acid sequence, LRRAS/TLG [54]. In this context, a series of PKA peptide-based substrates were screened for their CEST activity. It was observed that, following phosphorylation of these peptides by PKA, a remarkable decrease in the CEST contrast was generated within minutes. After this observation, a CEST-responsive PKA genetic–molecular sensor was developed by cloning a synthetic gene encoding eight monomers of LRRASLG in *E. coli* [44]. A high CEST signal was observed in *E. coli* cells expressing the biosensor compared to *E. coli* cells not expressing the sensor. These CEST biosensors could be used to map PKA activity and its location for the development of targeted PKA-based therapeutics.

10.3.2.2 Herpes simplex virus type 1 thymidine kinase

Herpes simplex virus type 1 thymidine kinase (HSV1-tk) is a viral enzyme analogue of mammalian thymidine kinases, involved in the synthesis of DNA and cell division. HSV1-tk has a lower substrate specificity compared to mammalian thymidine kinases, and consequently, it has been used in nuclear medicine as a gene-directed enzyme pro-drug therapeutic strategy (Ganciclovir) and also as PET and SPECT genetic-imaging strategies [55–60]. HSV1-tk functions by catalyzing the phosphorylation of a variety of nucleosides, which, after phosphorylation, are trapped within the cells. HSV1-tk therapeutic and imaging strategies rely on this entrapment and accumulation of phosphorylated nucleosides in cells expressing HSV1-tk. Several pyrimidine-based nucleosides have been developed as PET and SPECT imaging agents [61–64].

One of the main disadvantages of the previously developed genetically encoded CEST MRI reporters is that the frequency

Genetically Encoded CEST MRI Reporters | 207

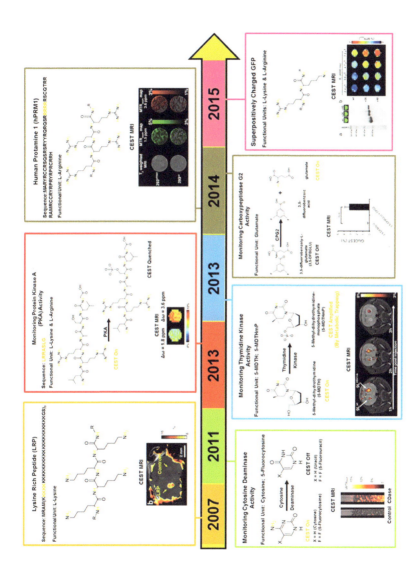

Figure 10.3 Historical time line of CEST MRI reporter gene development.

difference between the exchangeable protons on the reporter and that of water is relatively small (i.e., $\Delta\omega < 4$ ppm for amide, amine, guanidine, and hydroxyl protons). This may lead to direct saturation of water protons and, consequently, to false-positive readings. The imino proton of pyrimidine-based nucleosides has a frequency offset much farther from that of water protons ($\Delta\omega > 5$–6 ppm). Thymidine is one such pyrimidine-based nucleoside. However, since the exchange rate (k_{ex}) between the imino protons of thymidine and that of water is too fast (>3000 s^{-1}) to be detected by CEST MRI, the acid dissociation constant (pK_a) of the imino proton of thymidine was modified by rational chemical modification to reduce the k_{ex} and thus makes an optimized CEST-based contrast agent [42, 43]. Of the thymidine analogues screened for CEST activity, 5-methyl-dihydrothymidine (5-MDHT) showed excellent CEST activity and selectivity in cells expressing HSV1-tk.

10.4 Genetically Encoded Hyperpolarized Xenon (^{129}Xe) CEST MRI Reporters

One of the main limitations of proton MRI is its inherently low sensitivity (micromolar detection limits). Most recently, hyperpolarized MRI was developed to overcome the sensitivity limitation of proton MRI. Hyperpolarized MRI, as described in detail in Chapter 8, relies on the use of MRI-sensitive nuclei, which can be excited in the non-equilibrium states with an artificially high spin polarization (hyperpolarization) [65–67]. Xenon (^{129}Xe) is an example of an MRI-active nucleus that can be hyperpolarized. Hyperpolarized ^{129}Xe CEST MRI (also called hyperCEST) relies on xenon-binding constructs, such as cryptophanes [68–71], which alter the frequency offset of ^{129}Xe, when bound, compared to that when unbound. Upon saturation of bound ^{129}Xe with an appropriate radiofrequency pulse, and following magnetization transfer from saturated bound ^{129}Xe to the unbound ^{129}Xe, the signal from the unbound is quenched and this generates contrast (Fig. 10.4).

Gas vesicles occur naturally in several species of waterborne halobacteria and cyanobacteria as a means to control buoyancy [72].

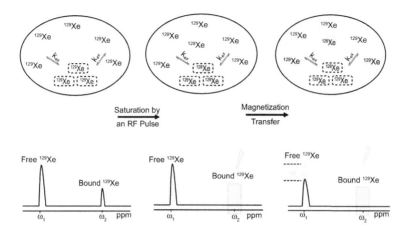

Figure 10.4 HyperCEST contrast generation mechanism.

The vesicles are permeable to gases that range in size from hydrogen to perfluorocyclobutane and are in constant equilibrium with gas molecules dissolved in the surrounding milieu. Recently, genetically encoded hyperCEST reporters, which exploit the aforementioned mechanism, were developed. This was achieved by transducing *E. coli* cells with a gene cluster that encodes 11 gas vesicle genes from *Bacillus megaterium* [73]. ^{129}Xe dissolved in culture media containing the gas vesicles showed a distinct chemical shift and a rapid exchange between the gas vesicles and the culture media. This enabled the use of these nanostructures as genetically encoded hyperCEST reporters, following saturation at an appropriate frequency. Interestingly, gas vesicles from different species, with distinct shapes and sizes, demonstrated hyperCEST at different frequency offsets. This suggests the possibility of multiplexed imaging of the different gas vesicles at different frequency offsets.

Heteronuclear CEST offers several advantages over proton CEST. These include (1) the higher sensitivity due to the lack of background signal from biological samples and (2) the large chemical shift difference (several hundred ppm for non-proton spins) between the bound and unbound gases, preventing indirect saturation of the unbound gas.

10.5 Considerations in Developing CEST MRI Genetically Encoded Reporters

In order to develop optimal genetically encoded reporters, a number of considerations must be addressed. First, it is important to identify the overall imaging objective, such as imaging stem cell survival, migration, and differentiation; or to study cancer progression and response to therapy. Next, based on the overall objective, it would be important to identify a biological process to be imaged. This could be a specific cell type, enzyme, substrate, receptor, ligand, transporter, or peptide. The most appropriate MRI-active nuclei (^{1}H, ^{19}F, ^{31}P, ^{23}Na, ^{129}Xe) to use also must be determined based on its sensitivity and specificity for the target biological process. Subsequently, a primary imaging agent associated with the target biological process must be identified and its CEST properties evaluated (peptides, enzyme substrates), usually through rational design or high-throughput screening. Upon identification of the primary CEST agent, an appropriate gene construct that encodes for the target of interest must be cloned into an expression vector and the cells transformed or transduced to express the target. Depending on the target expressed (enzyme, transporter, or gas vesicle), a CEST probe might need to be administered, which, upon interaction with the target, would generate CEST contrast. In this case, it would be necessary to evaluate the CEST properties of the interacting probe. In the event where the probe has minimal CEST activity, it would be important to modify its structure rationally to enhance the CEST activity. After this, the probe would need to be evaluated both in the absence and presence of the target in vitro. Upon sufficient activity, it could then be evaluated in vivo.

10.6 Current Challenges and Future Directions

One of the main limitations of proton-based CEST MRI reporter genes is their intrinsically low sensitivity. Hyperpolarized MRI, using heteronuclei, is currently being developed to overcome the low sensitivity encountered with proton MRI. However, the nanostructures required to entrap the polarized nuclei and induce a CEST contrast may not be suitable for studying specific pathways involved in a variety of pathologies, and expressing a large number of genes might be a challenge. Thus, it would be important to develop more sensitive proton-based reporter genes, which could be used to study protein expression with high sensitivity and specificity. High sensitivity could be achieved either by developing strategies that amplify the exchange rates of the excitable protons, or by modifying peptides to contain a higher number of excitable protons.

Another limitation encountered with proton-based CEST MRI is the low selectivity due to the small frequency offsets ($\Delta\omega$). This limitation could be overcome by developing reporters with exchangeable protons at farther frequency offsets from water protons. Alternatively, low selectivity could be improved by generating MRI-active heteronuclei reporters not endogenously expressed in the biological systems, such as fluorine-based reporters.

In addition, over the years, fluorescence and PET reporter genes have been more extensively developed and studied compared to MRI reporter genes, particularly CEST reporter genes. Fluorescence and PET reporter genes have been developed to study a variety of molecular processes involved in pathologies. Given the advantages of MRI over other imaging modalities and the amplification mechanism of CEST, developing CEST MRI reporter genes capable of imaging some of these processes would provide a way to image these processes at higher spatial resolution.

10.7 Conclusion

CEST MRI reporter genes offer several advantages for the imaging of molecular processes at high spatial and temporal resolution. However, MRI reporter genes and CEST MRI reporter genes, in particular, are at an early phase of development and still face a number of limitations that need to be addressed to make them widely accepted molecular-genetic imaging agents.

Acknowledgments

The development of genetically encoded CEST MRI reporters was supported by NIH grants: EB005252, R21EB008769, NS065284, EB018882, MSCRFII-0042, and ABTA Basic Research-117704.

References

1. Massoud TF and Gambhir SS. Integrating noninvasive molecular imaging into molecular medicine: An evolving paradigm. *Trends Mol Med*, 2007; 13(5): 183–191.
2. Weber WA, Czernin J, Phelps ME, and Herschman HR. Technology insight: Novel imaging of molecular targets is an emerging area crucial to the development of targeted drugs. *Nat Clin Prac Oncol*, 2008; 5(1): 44–54.
3. Herschman HR. Molecular imaging: Looking at problems, seeing solutions. *Science*, 2003; 302(5645): 605–608.
4. Edelman RR. The history of MR imaging as seen through the pages of radiology. *Radiology*, 2014; 273(2S): S181–S200.
5. Shazeeb MS, Sotak CH, DeLeo M, and Bogdanov A. Targeted signal-amplifying enzymes enhance MRI of EGFR expression in an orthotopic model of human glioma. *Cancer Res*, 2011; 71(6): 2230–2239.
6. Artemov D, Mori N, Ravi R, and Bhujwalla ZM. Magnetic resonance molecular imaging of the HER-2/neu receptor. *Cancer Res*, 2003; 63(11): 2723–2727.
7. Townsend TR, Moyle-Heyrman G, Sukerkar PA, MacRenaris KW, Burdette JE, and Meade TJ. Progesterone-targeted magnetic resonance imaging probes. *Bioconjugate Chem*, 2014; 25(8): 1428–1437.
8. Lears KA, Parry JJ, Andrews R, Nguyen K, Wadas TJ, and Rogers BE. Adenoviral-mediated imaging of gene transfer using a somatostatin receptor-cytosine deaminase fusion protein. *Cancer Gene Ther*, 2015; 22(4): 215–221.
9. Singh SP, Han L, Murali R, et al. SSTR2-based reporters for assessing gene transfer into non–small cell lung cancer: Evaluation using an intrathoracic mouse model. *Hum Gene Ther*, 2011; 22(1): 55–64.
10. Naumova AV, Modo M, Moore A, Murry CE, and Frank JA. Clinical imaging in regenerative medicine. *Nat Biotech*, 2014; 32(8): 804–818.

11. Kircher MF, Gambhir SS, and Grimm J. Noninvasive cell-tracking methods. *Nat Rev Clin Oncol*, 2011; 8(11): 677–688.
12. Lake MC and Aboagye EO. Luciferase fragment complementation imaging in preclinical cancer studies. *Oncoscience*, 2014; 1(5): 310–325.
13. Jouroukhin Y, Nonyane BS, Gilad A, and Pelled G. Molecular neuroimaging of post-injury plasticity. *J Mol Neurosci*, 2014; 54(4): 630–638.
14. Gilad AA and Pelled G. New approaches for the neuroimaging of gene expression. *Front Integr Neurosci*, 2015; 9: 5.
15. Vinores SA, Marangos PJ, Parma AM, and Guroff G. Increased levels of neuron-specific enolase in PC12 pheochromocytoma cells as a result of nerve growth factor treatment. *J Neurochem*, 1981; 37(3): 597–600.
16. Sheng M and Greenberg ME. The regulation and function of c-fos and other immediate early genes in the nervous system. *Neuron*, 1990; 4(4): 477–485.
17. Jouroukhin Y, Nonyane BA, Gilad AA, and Pelled G. Molecular neuroimaging of post-injury plasticity. *J Mol Neurosci*, 2014; 54(4): 630–638.
18. Serganova I, Ponomarev V, and Blasberg R. Human reporter genes: Potential use in clinical studies. *Nucl Med Biol*, 2007; 34(7): 791–807.
19. Forbes SC, Bish LT, Ye F, et al. Gene transfer of arginine kinase to skeletal muscle using adeno-associated virus. *Gene Ther*, 2014; 21(4): 387–392.
20. Renema WKJ, Kan HE, Wieringa B, and Heerschap A. In vivo magnetic resonance spectroscopy of transgenic mouse models with altered high-energy phosphoryl transfer metabolism. *NMR Biomed*, 2007; 20(4): 448–467.
21. Auricchio A, Zhou R, Wilson JM, and Glickson JD. In vivo detection of gene expression in liver by (31)P nuclear magnetic resonance spectroscopy employing creatine kinase as a marker gene. *Proc Nat Acad Sci USA*, 2001; 98(9): 5205–5210.
22. Cohen B, Ziv K, Plaks V, et al. MRI detection of transcriptional regulation of gene expression in transgenic mice. *Nat Med*, 2007; 13(4): 498–503.
23. Genove G, DeMarco U, Xu H, Goins WF, and Ahrens ET. A new transgene reporter for in vivo magnetic resonance imaging. *Nat Med*, 2005; 11(43): 450–454.
24. Liu J, Cheng ECH, Long RC, et al. Noninvasive monitoring of embryonic stem cells in vivo with MRI transgene reporter. *Tissue Eng Pt C-Meth*, 2009; 15(4): 739–747.
25. Alfke H, Stöppler H, Nocken F, et al. In vitro MR imaging of regulated gene expression. *Radiology*, 2003; 228(2): 488–492.

26. Paproski RJ, Forbrich AE, Wachowicz K, Hitt MM, and Zemp RJ. Tyrosinase as a dual reporter gene for both photoacoustic and magnetic resonance imaging. *Biomed Opt Express*, 2011; 2(4): 771–780.
27. Westmeyer Gil G, Emer Y, Lintelmann J, and Jasanoff A. MRI-based detection of alkaline phosphatase gene reporter activity using a porphyrin solubility switch. *Chem Biol*, 2014; 21(3): 422–429.
28. Louie AY, Huber MM, Ahrens ET, et al. In vivo visualization of gene expression using magnetic resonance imaging. *Nat Biotech*, 2000; 18(3): 321–325.
29. Urbanczyk-Pearson LM, Femia FJ, Smith J, et al. Mechanistic Investigation of β-galactosidase-activated MR contrast agents. *Inorg Chem*, 2008; 47(1): 56–68.
30. Liu G, Bulte JM, and Gilad A. CEST MRI reporter genes. In: Modo M, Bulte JWM, eds. *Magnetic Resonance Neuroimaging*, Vol 711: Humana Press; 2011: 271–280.
31. Wolff SD and Balaban RS. NMR imaging of labile proton exchange. *J Magn Reson*, 1990; 86(1): 164–169.
32. Ward KM, Aletras AH, and Balaban RS. A new class of contrast agents for MRI based on proton chemical exchange dependent saturation transfer (CEST). *J Magn Reson*, 2000; 143(1): 79–87.
33. Guivel-Scharen V, Sinnwell T, Wolff SD, and Balaban RS. Detection of proton chemical exchange between metabolites and water in biological tissues. *J Mag Reson Imaging*, 1998; 133(1): 36–45.
34. Goffeney N, Bulte JWM, Duyn J, Bryant LH, and van Zijl PCM. Sensitive NMR detection of cationic-polymer-based gene delivery systems using saturation transfer via proton exchange. *J Am Chem Soc*, 2001; 123(35): 8628–8629.
35. Snoussi K, Bulte JWM, Guéron M, and van Zijl PCM. Sensitive CEST agents based on nucleic acid imino proton exchange: Detection of poly(rU) and of a dendrimer-poly(rU) model for nucleic acid delivery and pharmacology. *Magn Reson Med*, 2003; 49(6): 998–1005.
36. McMahon MT, Gilad AA, Zhou J, Sun PZ, Bulte JWM, and van Zijl PCM. Quantifying exchange rates in chemical exchange saturation transfer agents using the saturation time and saturation power dependencies of the magnetization transfer effect on the magnetic resonance imaging signal (QUEST and QUESP): Ph calibration for poly-L-lysine and a starburst dendrimer. *Magn Reson Med*, 2006; 55(4): 836–847.
37. McMahon MT, Gilad AA, DeLiso MA, Cromer Berman SM, Bulte JWM, and van Zijl PCM. New "multicolor" polypeptide diamagnetic chemical

exchange saturation transfer (DIACEST) contrast agents for MRI. *Magn Reson Med*, 2008; 60(4): 803–812.

38. Gilad AA, McMahon MT, Walczak P, et al. Artificial reporter gene providing MRI contrast based on proton exchange. *Nat Biotech*, 2007; 25(2): 217–219.

39. Bar-Shir A, Liu G, Chan KWY, et al. Human PROTAMINE-1 as an MRI reporter gene based on chemical exchange. *ACS Chem Biol*, 2014; 9(1): 134–138.

40. Bar-Shir A, Liang Y, Chan KWY, Gilad AA, and Bulte JWM. Supercharged green fluorescent proteins as bimodal reporter genes for CEST MRI and optical imaging. *Chem Comm*, 2015; 51(23): 4869–4871.

41. Liu G, Liang Y, Bar-Shir A, et al. Monitoring enzyme activity using a diamagnetic chemical exchange saturation transfer magnetic resonance imaging contrast agent. *J Am Chem Soc*, 2011; 133(41): 16326–16329.

42. Bar-Shir A, Liu G, Liang Y, et al. Transforming thymidine into a magnetic resonance imaging probe for monitoring gene expression. *J Am Chem Soc*, 2013; 135(4): 1617–1624.

43. Bar-Shir A, Liu G, Greenberg MM, Bulte JWM, and Gilad AA. Synthesis of a probe for monitoring HSV1-tk reporter gene expression using chemical exchange saturation transfer MRI. *Nat Protocols*, 2013; 8(12): 2380–2391.

44. Airan RD, Bar-Shir A, Liu G, et al. MRI biosensor for protein kinase A encoded by a single synthetic gene. *Magn Reson Med*, 2012; 68(6): 1919–1923.

45. Snoussi K, Bulte JW, Gueron M, and van Zijl PC. Sensitive CEST agents based on nucleic acid imino proton exchange: Detection of poly(rU) and of a dendrimer-poly(rU) model for nucleic acid delivery and pharmacology. *Magn Reson Med*, 2003; 49(6): 998–1005.

46. Farrar CT, Buhrman JS, Liu G, et al. Establishing the lysine-rich protein CEST reporter gene as a CEST MR imaging detector for oncolytic virotherapy. *Radiology*, 2015; 275(3): 746–754.

47. Minn I, Bar-Shir A, Yarlagadda K, et al. Tumor-specific expression and detection of a CEST reporter gene. *Magn Reson Med*, 2015; 74(2): 544–549.

48. Lewis J, Song Y, de Jong M, Bagha S, and Ausió J. A walk though vertebrate and invertebrate protamines. *Chromosoma*, 2003; 111(8): 473–482.

49. Oskolkov N, Bar-Shir A, Chan KWY, et al. Biophysical characterization of human protamine-1 as a responsive CEST MR contrast agent. *ACS Macro Lett*, 2015; 4(1): 34–38.

50. Shaner NC, Campbell RE, Steinbach PA, Giepmans BNG, Palmer AE, and Tsien RY. Improved monomeric red, orange and yellow fluorescent proteins derived from *Discosoma* sp. red fluorescent protein. *Nat Biotech*, 2004; 22(12): 1567–1572.

51. Shaner NC, Steinbach PA, and Tsien RY. A guide to choosing fluorescent proteins. *Nat Meth*, 2005; 2(12): 905–909.

52. Noble MEM, Endicott JA, and Johnson LN. Protein kinase inhibitors: Insights into drug design from structure. *Science*, 2004; 303(5665): 1800–1805.

53. Zhang J, Ma Y, Taylor SS, and Tsien RY. Genetically encoded reporters of protein kinase A activity reveal impact of substrate tethering. *Proc Natl Acad Sci USA*, 2001; 98(26): 14997–15002.

54. Kemp BE, Graves DJ, Benjamini E, and Krebs EG. Role of multiple basic residues in determining the substrate specificity of cyclic AMP-dependent protein kinase. *J Biol Chem*, 1977; 252(14): 4888–4894.

55. Tjuvajev JG, Avril N, Oku T, et al. Imaging herpes virus thymidine kinase gene transfer and expression by positron emission tomography. *Cancer Res*, 1998; 58(19): 4333–4341.

56. Gambhir SS, Bauer E, Black ME, et al. A mutant herpes simplex virus type 1 thymidine kinase reporter gene shows improved sensitivity for imaging reporter gene expression with positron emission tomography. *Proc Natl Acad Sci USA*, 2000; 97(6): 2785–2790.

57. Gambhir SS, Barrio JR, Wu L, et al. Imaging of adenoviral-directed herpes simplex virus type 1 thymidine kinase reporter gene expression in mice with radiolabeled ganciclovir. *J Nucl Med*, 1998; 39(11): 2003–2011.

58. Tjuvajev JG, Hsia Chen S, Joshi A, et al. Imaging adenoviral-mediated herpes virus thymidine kinase gene transfer and expression in vivo. *Cancer Res*, 1999; 59(20): 5186–5193.

59. Tjuvajev JG, Finn R, Watanabe K, et al. Noninvasive imaging of herpes virus thymidine kinase gene transfer and expression: A potential method for monitoring clinical gene therapy. *Cancer Res*, 1996; 56(18): 4087–4095.

60. Choi SR, Zhuang Z-P, Chacko A-M, et al. Spect imaging of herpes simplex virus type 1 thymidine kinase gene expression by [123I]FIAU1. *Acad Radiol*, 2005; 12(7): 798–805.

61. Yaghoubi SS and Gambhir SS. PET imaging of herpes simplex virus type 1 thymidine kinase (HSV1-tk) or mutant HSV1-sr39tk reporter gene expression in mice and humans using [18F]FHBG. *Nat Protocols*, 2007; 1(6): 3069–3074.

62. Chacko A-M, Blankemeyer E, Lieberman BP, Qu W, and Kung HF. 5-[18F]Fluoroalkyl pyrimidine nucleosides: Probes for positron emission tomography imaging of herpes simplex virus type 1 thymidine kinase gene expression. *Nucl Med Biol*, 2009; 36(1): 29–38.
63. Müller U, Martić M, Kraljević TG, et al. Synthesis and evaluation of a C-6 alkylated pyrimidine derivative for the in vivo imaging of HSV1-TK gene expression. *Nucl Med Biol*, 2012; 39(2): 235–246.
64. Mukhopadhyay U, Soghomonyan S, Yeh HH, et al. N3-Substituted thymidine analogues V: Synthesis and preliminary PET imaging of N3-[18F]fluoroethyl thymidine and N3-[18F]fluoropropyl thymidine. *Nucl Med Biol*, 2008; 35(6): 697–705.
65. Schröder L. Xenon for NMR biosensing: Inert but alert. *Phys Medica*, 2011; 29(1): 3–16.
66. Schröder L, Lowery TJ, Hilty C, Wemmer DE, and Pines A. Molecular imaging using a targeted magnetic resonance hyperpolarized biosensor. *Science*, 2006; 314(5798): 446–449.
67. Ana-Maria O and Shah NJ. Hyperpolarized xenon in NMR and MRI. *Phys Med Biol*, 2004; 49(20): R105.
68. Bai Y, Hill PA, and Dmochowski IJ. Utilizing a water-soluble cryptophane with fast xenon exchange rates for picomolar sensitivity NMR measurements. *Anal Chem*, 2012; 84(22): 9935–9941.
69. Taratula O and Dmochowski IJ. Functionalized ^{129}Xe contrast agents for magnetic resonance imaging. *Curr Opin Chem Biol*, 2010; 14(1): 97–104.
70. Stevens TK, Ramirez RM, and Pines A. Nanoemulsion contrast agents with sub-picomolar sensitivity for xenon NMR. *J Am Chem Soc*, 2013; 135(26): 9576–9579.
71. Stevens TK, Palaniappan KK, Ramirez RM, Francis MB, Wemmer DE, and Pines A. HyperCEST detection of a ^{129}Xe-based contrast agent composed of cryptophane-A molecular cages on a bacteriophage scaffold. *Magn Reson Med*, 2013; 69(5): 1245–1252.
72. Pfeifer F. Distribution, formation and regulation of gas vesicles. *Nat Rev Micro*, 2012; 10(10): 705–715.
73. Shapiro MG, Ramirez RM, Sperling LJ, et al. Genetically encoded reporters for hyperpolarized xenon magnetic resonance imaging. *Nat Chem*, 2014; 6(7): 629–634.

Chapter 11

ParaCEST Agents: Design, Discovery, and Implementation

Mark Milne, Yunkou Wu, and A. Dean Sherry

*Advanced Imaging Research Center, UT Southwestern Medical Center,
Dallas, TX 75080, USA*
dean.sherry@utsouthwestern.edu

11.1 Introduction

11.1.1 History of paraCEST Agents

Paramagnetic chemical exchange saturation transfer (paraCEST) agents are, in the terminology of the late 1970s, aqueous nuclear magnetic resonance (NMR) shift reagents that alter the frequency of the exchanging proton resonance so that they are further away from the bulk water resonance. As the name indicates, paraCEST contrast is generated as a result of transfer of saturated proton spins originating on a paramagnetic species ($\mu_{\text{eff}} \neq 0$) into the larger pool of water proton spins and detected by magnetic resonance imaging (MRI). This spin transfer typically utilizes exchangeable protons in a paramagnetic metal ion complex in the form of –NH (amide), –NH$_2$ (amine), or –OH (hydroxyl) that slowly exchange with the bulk

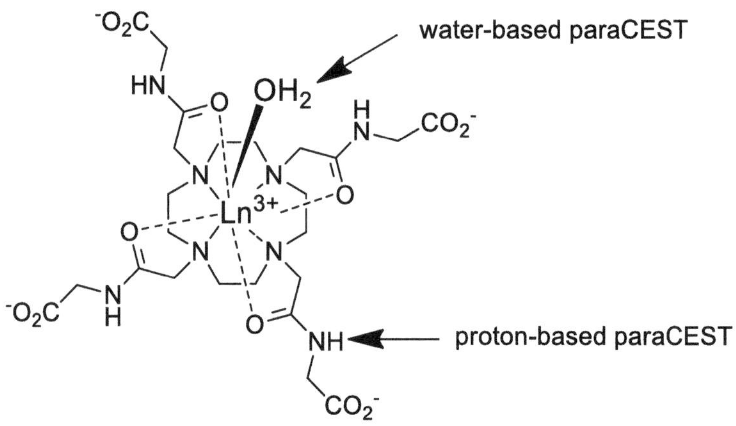

Figure 11.1 Structure of Ln DOTA-(gly)$_4^-$. Arrows indicate exchangeable protons from either amides or bound water.

water protons. Exchange of water molecules from an inner-sphere coordination position in a paramagnetic metal ion complex with the pool of bulk water molecules is also possible. In this chapter, we will refer to paraCEST agents that exchange protons as *proton-based* paraCEST agents and those that exchange entire water molecules as *water-based* paraCEST agents (Fig. 11.1).

The first paraCEST agent, reported more than a decade ago, was the water-based agent EuDOTA-(gly-OEt)$_4^{3+}$ (Fig. 11.2) [1]. Like many things in science, this molecule was an accidental discovery because no one at the time would have believed that water exchange in any Eu^{3+} complex could be so slow that a separate proton resonance could be detected for a Eu^{3+}-bound water molecule by high-resolution NMR at room temperature. Nevertheless, a single, relatively sharp ^{1}H resonance with a relative area of two protons was present at 50 ppm in the high-resolution NMR spectrum of this complex. This resonance was eventually assigned to a single, slowly exchanging, Eu^{3+}-bound water molecule. To test whether this water molecule was exchanging with solvent, a frequency-selective pre-saturation pulse was applied at 50 ppm prior to the collection of the NMR spectrum. This resulted in a substantial decrease in the intensity of the bulk water proton signal, thereby demonstrating that these two proton pools were in exchange. This was the very

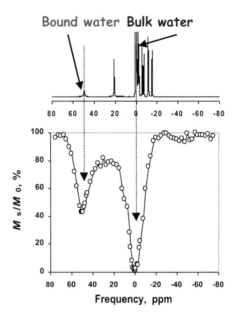

Figure 11.2 The CEST spectrum of EuDOTA-(gly-OEt)$_4$$^{3+}$ collected at 4.7 T and 25°C.

first CEST experiment performed in our lab. The full Z-spectrum (or in newer terminology, the CEST spectrum) shows the effects of applying a frequency-selective, low power pulse (∼10 μT) every 0.5 ppm from 75 to −75 ppm prior to measuring the intensity of the bulk water signal (Fig. 11.2). The "dip" seen in the CEST spectrum near 50 ppm reflects the single exchanging Eu^{3+}-bound water molecule that exchanges with bulk water at the rate of ∼4400 s^{-1}, while the "dip" observed at 0 ppm reflects the direct saturation of bulk water protons. While no other exchanging protons were apparent in this spectrum, this agent does have four exchanging – NH amide protons, but they resonate at a frequency quite close to the bulk water protons (∼4 ppm) and so they are not easily detected under these experimental conditions.

According to the exchange theory, to generate a CEST signal from this bound water, the water exchange rate must be slow enough to meet the slow-to-intermediate exchange requirement, $\Delta\omega \geq k_{ex}$, where $\Delta\omega$ is the chemical shift difference between the exchangeable protons and the bulk water protons, and k_{ex} is the water exchange rate. For the paraCEST agent shown in Fig. 11.2, $\Delta\omega = 2\pi \times 10{,}000$ Hz $= 62{,}832$ s^{-1}, so clearly $\Delta\omega \ggg k_{ex}$ for this complex. The $\Delta\omega$ values of many protons in LnDOTA-based complexes are much larger than the $\Delta\omega$ values of typical diamagnetic CEST molecules (<5 ppm in most cases), so the proton or water molecule exchange rates in paraCEST agents can be much faster while still meeting the basic requirement, $\Delta\omega \geq k_{ex}$.

11.2 Lanthanide-Induced Shifts

When involved in the design and development of paraCEST agents, it is important to understand some fundamentals about paramagnetic NMR shifts in these complexes. Why, for example, are the chemical shifts of water-based exchanging species so much larger (typically 50–600 ppm) than the chemical shifts of proton-based exchanging species (3–80 ppm) in the same paramagnetic complex? The answer to this lies in understanding some basic geometrical relationships between dipolar NMR shifts and chemical structure.

The lanthanides became popular in the NMR world in 1969 with the first report by Hinckley of "lanthanide shift reagents" for NMR [2]. This discovery led to a flurry of activities in the use of lanthanide-induced NMR shift (LIS) for two main purposes: (1) to spread out or amplify the proton chemical shifts in molecules (essentially turning a 60 MHz NMR spectrometer into a 400 or 500 MHz NMR instrument) and (2) to use LIS data to predict the solution structures of molecules. Today, high-field spectrometers are ubiquitous in most chemistry labs, so the addition of a shift reagent is no longer required. However, the second application, i.e., the use of LIS to evaluate the solution structure of a molecule, remains of interest to numerous lanthanide chemists around the world [3] and, more recently, to structural biologists interested in evaluating solution structures of proteins and other macromolecules [4]. For paramagnetic lanthanide complexes, the chemical shift of a ligand nucleus (Δ_{obs}) consists of a diamagnetic (Δ_{dia}) and a paramagnetic (Δ_p) component. The diamagnetic component simply refers to the chemical shift of a proton in the corresponding diamagnetic complex, using either La^{3+} or Lu^{3+} as the paramagnetic substitute. The paramagnetic (Δ_p) component may be further separated into contact (Δ_c) and pseudocontact (Δ_{pc}) shift components, although for most of the complexes and exchange sites we regard as paraCEST agents, it is the pseudocontact (Δ_{pc}) shift component that is of most interest because of its direct relationship to the complex structure (Eq. 11.1) [5].

$$\Delta_{pc} = D_1(3\cos^2\theta - 1)/r^3 + D_2(\sin^2\theta \cos 2\phi)/r^3 \quad (11.1)$$

Here D_1 and D_2 are terms containing Bleaney's constant (Cj) [6] characteristic of each Ln^{3+} ion (see Table 11.1) plus a ligand field term (LF), which is related to the amount of electron density donated by each ligand donor atom to the Ln^{3+} ion; θ, ϕ, and r are the polar coordinates of each ligand nucleus relative to the highest fold symmetry axis of the complexed Ln^{3+}. For Ln^{3+} complexes having three-fold or higher axial symmetry, Eq. 11.1 simplifies to Eq. 11.2.

$$\Delta_{pc} = D_1(3\cos^2\theta - 1)/r^3 \quad (11.2)$$

This geometrical function is illustrated in Fig. 11.3.

Figure 11.4 illustrates that for an Ln^{3+} complex having C$_4$ symmetry, the largest dipolar or pseudocontact NMR shifts will be observed for the bound water molecule (positioned along the C$_4$ axis of symmetry) and all ethylene ligand protons positioned on the opposite side of the molecule from water. The magnitude of those paramagnetic shifts (Δ_{pc}) is largely determined by any differences in angle from the C$_4$ symmetry axis (θ) plus the distance between the Ln^{3+} and those individual protons (r^{-3}). Given this diagram, it is not surprising to observe that the bound water molecule in such LnDOTA-tetraamide chelates displays the largest Δ_{pc} by a factor of ~2 over the next nearest shifted proton. The acetate and amide protons, by virtue of lying ~90° relative to the principle axis of symmetry, will always display much smaller Δ_{pc} values, typically by a factor of 2 or more. In the example shown here, the two primary amide protons are magnetically different with one positioned near the magic angle (54.7°). In this case, the two protons would have different Δ_{pc} values (perhaps in some cases even different signs) even though they are attached to the same nitrogen atom. This has been observed in proton-based paraCEST agent, Yb-DOTAM, where

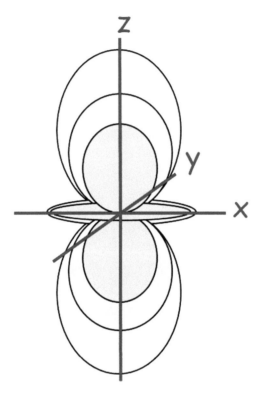

Figure 11.3 A plot of Eq. 11.2 in the Cartesian coordinate system. With a paramagnetic Ln^{3+} ion situated at the origin, the magnitude of a dipolar Δ_{pc} shift for any atom in a complex will be determined by the distance and angle of that atom relative to the paramagnetic center.

the two amide protons differ in chemical shift by about 3 ppm and, interestingly, also display differences in proton exchange rates [7].

It is clear from the aforementioned discussion that the lanthanide-induced paramagnetic shifts (Δ_{pc}) are quite sensitive to the geometry of the molecule under study. A second factor in Eq. 11.2 is the magnitude of the D_1 term. As indicated earlier, this term consists of a magnetic constant characteristic of each lanthanide ion (Bleaney's C_j values) and a second constant that reflects the strength of the ligand field (LF) surrounding the lanthanide ion (how tightly the ligand holds onto the Ln^{3+}). For comparison purposes, assume that the ligand field term is constant for the moment and

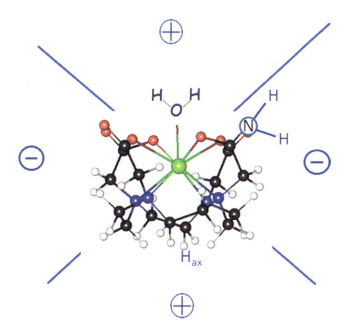

Figure 11.4 Using an LnDOTA-amide molecule as an example, one would predict that the largest dipolar shifts (Δ_{pc}) would be seen for those atoms positioned along the highest fold symmetry axis of the complex (where θ is closest to zero). In this case, the H_2O molecule and the axial ligand proton H_{ax} would be shifted in the same direction (assume + is a shift to higher frequency (downfield) while the $-NH_2$ protons would be shifted in the opposite direction (−, a shift to lower frequency (upfield)).

simply compare theoretical shift values for each lanthanide, given by the Bleaney C_j values (see Table 11.1). A comparison of C_j values suggests that Dy^{3+} complexes would always be preferred as a basic platform for building paraCEST agents simply because, all other things being equal, this ion will always produce the largest pseudocontact shifts, Δ_{pc}. This would be followed closely by Tb^{3+} and Tm^{3+}. While this order of Δ_{pc} is indeed preserved in isostructural complexes of Dy^{3+}, Tb^{3+}, and Tm^{3+}, there are other factors that make Dy^{3+} complexes less desirable as paraCEST complexes.

The other factor in determining the magnitudes of D_1 and D_2 is the term involving the strength of the ligand field. Quantitative

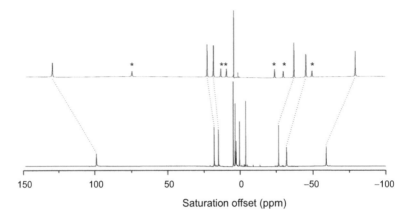

Figure 11.5 High-resolution ^{1}H NMR spectra of YbDOTA$^-$ (top) and YbDOTA-(gly)$_4$$^-$ (bottom) recorded at 25°C. The resonances labeled with an asterisk (*) reflect a smaller population of the minor twisted square antiprism (TSAP) isomer in YbDOTA$^-$ that is barely detectable in the spectrum of YbDOTA-(gly)$_4$$^-$. Dotted lines are to indicate which proton signals corresponded to one another between the two compounds.

comparisons of this effect are not as readily available unless one performs more elaborate theoretical estimates using a bonding theory such as density functional theory (DFT). The simplest way to compare the effects of ligand field is to compare high-resolution NMR spectra of any given Ln^{3+} within complexes with similar symmetry. For example, the NMR spectra of YbDOTA$^-$ and YbDOTA-(gly)$_4^-$ are compared in Fig. 11.5. Both complexes exist in solution largely as square antiprism (SAP) complexes with C$_4$ symmetry and both have an overall negative charge. Nevertheless, the observed Δ_{pc} for each proton in the spectrum of YbDOTA$^-$ is ∼25% larger than the corresponding proton in the spectrum of YbDOTA-(gly)$_4$$^-$. This reflects the greater combined electron donor strength of the N and O donor atoms of DOTA compared to those in DOTA-(gly)$_4$. Although this is most easily attributed to differences in electron donation from the four negatively charged carboxylate oxygen atoms of DOTA compared to the four neutral amide oxygen atoms of DOTA-(gly)$_4$, differences in donor strength of the macrocyclic nitrogen donor atoms also contribute because of slight differences in their pK_a values. Thus, one can conclude that DOTA provides a stronger ligand field to all Ln^{3+} ions than does DOTA-(gly)$_4^-$. As we shall see, the field strength also play an important role in affecting water exchange rates and hence the CEST signal.

11.3 T_1 and T_2 Considerations in the Design of paraCEST Agents

It is well known from Bloch theory that the most favorable condition for CEST generation is when the T_1 values of bulk water protons are as long as possible. T_2 has a different effect on CEST from T_1 but follows a similar trend in which longer is beneficial in MR imaging. If one compares the bulk paramagnetic effect of each of lanthanide ion on the T_1 and T_2 of bulk water protons (see Table 11.1), then one would conclude that Dy^{3+} is the least favorable lanthanide for CEST because it has the largest influence on shortening the T_1 of bulk

Table 11.1 Bleaney's constants (C_j) for relative paramagnetic shifts for a series of isostructural Ln^{3+} complexes plus paramagnetic shifts measured for the bound water resonances in CEST spectra of the LnDOTA-(gly-OEt)$_4^{3+}$ complexes. Experimental water proton T_1 and T_2 values were measured for 10 mM samples of LnCl$_3$ in water at 25 °C and 400 MHz. The paraCEST index was calculated as $|\delta|/T_{1or2P}^{-1}$ (see text for details)

Ln^{3+} Complexes	C_j[a]	δ, ppm (bound water)[b]	T_1(s)[c]	$1/T_{1P}$	ParaCEST Index (T_1)	T_2(s)[c]	$1/T_{2P}$	ParaCEST Index (T_2)
Pr^{3+}	−11.0	−60	2.18	0.18	333	1.08	0.58	103
Nd^{3+}	−4.2	−32	1.73	0.29	110	1.70	0.25	130
Sm^{3+}	−0.7	−4	2.75	0.08	50	1.79	0.22	18
Eu^{3+}	4.0	+50	2.76	0.08	625	2.35	0.08	609
Tb^{3+}	−86	−600	0.17	5.60	107	0.13	7.12	84
Dy^{3+}	−100	−720	0.10	9.33	77	0.08	12.8	56
Ho^{3+}	−39	−360	0.17	5.56	65	0.16	5.95	61
Er^{3+}	33	+200	0.17	5.56	36	0.16	5.95	34
Tm^{3+}	53	+500	0.20	4.72	106	0.18	5.21	96
Yb^{3+}	22	+200	1.08	0.65	308	0.55	1.49	135

[a] Bleaney (1972) [6]
[b] Zhang and Sherry (2003) [9]
[c] T_1 values were measured by inversion recovery sequence; T_2 values were measured using a CPMG pulse sequence. LaCl$_3$ was used as the diamagnetic reference, where T_1 is 3.52 s and T_2 is 2.91 s. The errors in these measurements were ±10 ms.

Table 11.2 Chemical shifts and exchange lifetimes for water molecule and amide proton exchange for the series of LnDOTA-(gly)$_4{}^-$ complexes

Ln^{3+} Complexes	δ, ppm (Bound Water)[a]	τ_m^{298} μs Ln^{3+}- bound H$_2$O[a]	δ, ppm (NH)[b]	τ_m^{298} μs Amide Protons[b]
Pr^{3+}	−60	20	13	—
Nd^{3+}	−32	80	11	—
Sm^{3+}	−4	320	—	—
Eu^{3+}	+50	382	−4	—
Tb^{3+}	−600	31	61	—
Dy^{3+}	−720	17	77	—
Ho^{3+}	−360	19	39	290
Er^{3+}	+200	9	−22	300
Tm^{3+}	+500	3	−51	400
Yb^{3+}	+200	3	−16	400

[a] Zhang and Sherry (2003) [9]
[b] Aime et al. (2002) [8]

water protons. This was observed experimentally by Aime et al., in comparing CEST spectra of 30 mM solutions of the LnDOTA-(gly)$_4{}^-$ series. In this case, they observed nearly complete quenching of the amide-based CEST signal in the Dy^{3+} complex, and the largest amide-based CEST signal was from the Yb^{3+} complex (the Eu^{3+} complex was not included in this comparison because the −NH signal in this complex is too close to water to make a direct comparison with the other ions) [8]. Hence, the most favorable lanthanides for CEST are those that produce the largest Δ_{pc} shifts but also have the smallest influence on bulk water T_1. Thus, a comparison of the potential of using the different lanthanide ions for CEST can be made by a paraCEST index, defined as $|\delta|/T_{1or2P}{}^{-1}$ where $T_{1or2P}{}^{-1}$ is the paramagnetic contribution to the relaxation rate ($T_{1or2P}{}^{-1} = 1/T_{1or2\,para} - 1/T_{1or2\,dia}$). These values are summarized in Table 11.1.

From these, one quickly concludes that the most favorable lanthanide ions for paraCEST applications lie in the approximate order, Eu^{3+} >> Pr^{3+} > Yb^{3+} > Nd^{3+} ∼ Tb^{3+} ∼ Tm^{3+} > Dy^{3+} > Ho^{3+} > Sm^{3+} > Er^{3+}. The T_2 paraCEST index order is similar.

So far we have described the effects of T_1 and T_2 relaxation time on paraCEST agents by simply stating that longer values are better. To get a deeper understanding of why this is true, one can refer to the paper by Woessner et al. [10], which described CEST quantitatively using numerical solutions to the Bloch equations. The Bloch equations predict that, at steady state, the net Z magnetization of water protons after a long presaturation pulse (M_z) relative to the Z magnetization in the absence of such a pulse (M_0) is given by Eq. 11.3.

$$\frac{M_Z}{M_0} = \frac{\tau_a}{T_{1a} + \tau_a} \quad (11.3)$$

Here τ_a is the lifetime of a proton in the water pool, and T_{1a} is the longitudinal relaxation time of bulk water. Figure 11.6 (top) shows simulated CEST spectra for a simple two-pool exchange model where T_{1a} (the T_1 of bulk water protons) is varied while holding the other parameters constant. Clearly, the intensity of the CEST exchange peak at 50 ppm in this simulation heavily depends on T_{1a} and longer is better. Figure 11.6 (bottom) also illustrates the impact of altering the T_{2a} of pool A (water) on the CEST signal of pool B with all other variables held constant. Clearly, the impact of T_{2a} on the CEST intensity of pool B was minimal, even though the bulk water line width ($\propto 1/T_{2a}$) broadened considerably with the T_{2a} decreasing. Although it would be easy to conclude from this illustration that T_{2a} is unimportant in terms of the CEST intensity, a broad water proton line width has other consequences in terms of detection and imaging of paraCEST agents in vivo. We will return to this point in more detail below. Nevertheless, the impact of paraCEST agents on water proton T_1 and T_2 is important to keep in mind when designing new agents. Referring to the Bleaney constants in Table 11.1 again, one would predict that it would be most difficult to design a viable Dy^{3+}-based paraCEST agent because its influence on the T_1 of bulk water is the greatest among the lanthanide ions.

Woessner et al. [10] also demonstrated the effects of adding a third pool of exchanging protons to a simple two-pool model. Given that many water-based paraCEST agents have exchanging –NH or –OH protons in addition to an exchanging water molecule, it is important to understand the influence of these other exchanging species on the intensities of other CEST peaks. Figure 11.7 illustrates the influence of the third pool of exchanging protons with a very short T_1 value (pool C) on the intensity of the pool B CEST peak. Clearly, the intensity of the pool B CEST peak is significantly

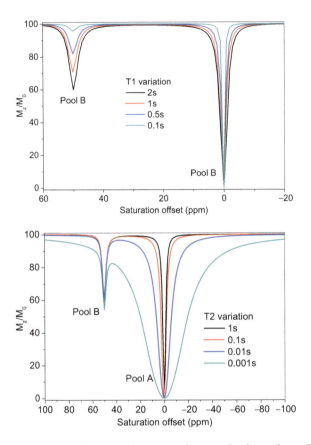

Figure 11.6 Simulated two-pool Z-spectra showing the dependency T_{1a} and T_{2a} on M_z/M_0. The top graph is for T_{2a} held constant at 0.1 s, while T_{1a} was varied from 0.1 to 2 s as indicated in the graph. The bottom graph shows simulations where T_{1a} was fixed at 2 s, while T_{2a} was varied. The remaining parameters included $C_{pool\,A} = 111$ M, $C_{pool\,B} = 40$ mM, $T_{1b} = 0.1$ s, $T_{2b} = 0.05$ s, $B_1 = 200$ Hz, and saturation time = 2 s.

influenced by the short T_{1c} protons in pool C when they exchange rapidly with water protons. In the example shown, it was assumed that no direct proton exchange occurred between protons in pool B and pool C, only exchanges between B ↔ A and C ↔ A. The net effect of this C ↔ A exchange is that the short T_{1c} protons in pool C reduce T_{1a} in proportion to the exchange rate. If the rate of proton exchange between C ↔ A is slow, then pool C has

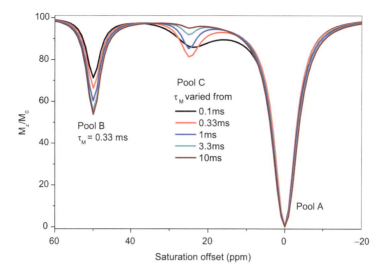

Figure 11.7 A three-pool Bloch simulation where protons exchange only between pools B ↔ A and pools C ↔ A. The water lifetime in pool B was fixed, $\tau_M = 330$ μs typical of a water-based paraCEST agent, and the proton lifetime in pool C was varied as shown. The remaining parameters included $C_{pool\ A} = 111$ M, $C_{pool\ B} = 20$ mM, $C_{pool\ C} = 20$ mM, $T_{1a} = 2$ s, $T_{2a} = 0.2$ s, $T_{1b} = 0.1$ s, $T_{2b} = 0.1$ s, $T_{1c} = 0.3$ ms, $T_{2c} = 0.3$ ms, $B_1 = 512$ Hz, and saturation time = 2 s.

little influence on the intensity of the pool B CEST peak. However, as the pool C protons begin exchanging much more rapidly, then the intensity of the pool B peak is reduced considerably, ~50% in this example. It is important to understand the impact of other exchanging proton species when one begins to translate paraCEST agents in vivo. In tissues, there are a very large number of different types of protons all exchanging with tissue water. This means that the sensitivity of a paraCEST agent in vivo could be a small fraction of its measured sensitivity in an NMR tube, depending on the T_{1c} of those endogenous exchanging protons. One might consider using this phenomenon to gain some perspective about the local chemical environment of a paraCEST agent in vivo. Perhaps one could design an experiment to differentiate between an agent exposed to a large number of rapidly exchanging, short T_1 endogenous protons or an agent sequestered in a chemical environment in tissue where the number of exchanging protons is limited. The fact that paraCEST agents can be detected in vivo indicates that this effect may be small in most cases, but it could certainly play a role in limiting the sensitivity of paraCEST agents. It is important to keep this in mind as more in vivo applications are reported.

The impact of a reduction in water T_{1a} and T_{2a} on the detection of paraCEST agents in vivo has been demonstrated experimentally. The first example is a redox-sensitive paraCEST probe containing two stable free-radical nitroxides (Fig. 11.8) designed to reduce T_{1a} and thereby quench the CEST signal [11]. The measured water T_{1a} of a 10 mM sample of the bis-nitroxide Eu^{3+} complex was only 0.2 s, which reduced the water-based CEST signal of this complex to less than 2% at 51 ppm. Chemical reduction of this complex by ascorbate resulted in an increase in the water proton T_1 to 2.6 s and full recovery of the CEST signal. The unique feature of this agent may ultimately prove useful as a sensor of tissue redox state in vivo. In a healthy mouse, the nitroxide agent is filtered by the kidneys without biological reduction as evidenced by the lack of CEST signal in the bladder (after sufficient time for the agent to be fully cleared). Once the agent is in the bladder, the injection of ascorbic acid (IV) resulted in chemical reduction of the agent in the bladder and full CEST activation. Subsequent unpublished experiments using this CEST-based redox sensor in tumor-bearing mice indicate that the agent can undergo biological reduction in diseased animals, so it may ultimately serve useful as a biomarker of tissue redox state in vivo (Ratnakar, unpublished observations).

As discussed, T_{2a} has little direct impact on the intensity of a CEST signal as demonstrated in Fig. 11.6, and a short T_{2a} certainly does have an impact on the detection of paraCEST agents in vivo using standard MRI sequences. This was first illustrated in experiments where the water-based paraCEST agent EuDOTA-(gly)$_4^-$ was injected into mice and after a few minutes, as the agent concentrated in the kidney by normal glomerular filtration, the kidney images turned dark [12]. A similar imaging darkening is widely observed for Gd^{3+}-based agents when used in high concentrations. For Gd^{3+}, this effect is simply due to the paramagnetic shortening of T_{2a}, but for EuDOTA-(gly)$_4^-$, which is only weakly paramagnetic, the line broadening seen here in kidney

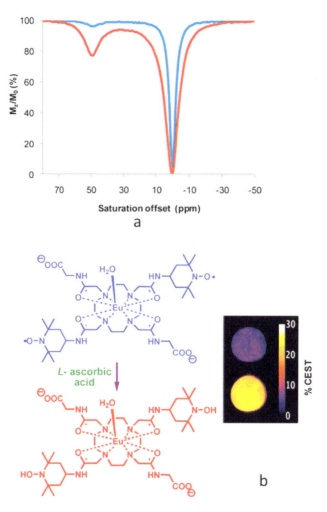

Figure 11.8 (a) CEST spectra of a 10 mM aqueous solution of redox sensor recorded at 9.4 T, 298 K, pH 7, $B_1 = 10$ µT, and irradiation time of 5 s before (blue) and after (red) reduction of the complex with L-ascorbic acid. (b) The inverted difference CEST images of phantoms containing 10 mM redox sensor before (blue) and after (red) reduction. The image shown was created by acquiring spin echo CEST images (9.4 T, 298 K, TR = 10 s, irradiation time = 5 s, $B_1 = 10$ µT) at −51 ppm (off resonance) and at +51 ppm (on resonance) and subtracting the on-resonance image from the off-resonance image. Reprinted with permission from Ref. [11], Copyright 2013, American Chemical Society.

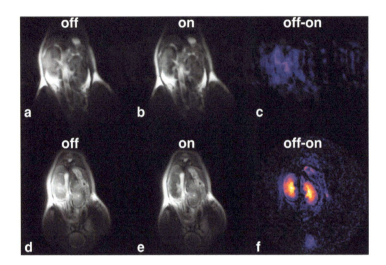

Figure 11.9 Images were collected using a standard FSEMS-CEST sequence after administration of a 1.0 mmol/kg intravenous dose of EuDOTA-(gly)$_4^-$. No obvious off–on CEST signal was evident in the kidneys (a–c). The same coronal slice was acquired using SWIFT-CEST (d–f). The presence of agent in the kidneys was clearly seen in the resulting off–on CEST image. Reprinted with permission from Ref. [12], Copyright 2011, John Wiley and Sons.

images was traced to a different phenomenon—T_2 broadening due to slow-to-intermediate exchange between a highly shifted Eu^{3+}-bound water molecule (at 50 ppm) and bulk water. This $T_{2\text{exch}}$ effect was described by Swift and Connick for paramagnetic systems over 50 years ago [13]. For the example shown in Fig. 11.9, most of the water signal from kidney is not detected when using a standard fast spin echo imaging sequence but can be observed using SWIFT, a newer imaging technique designed to detect short T_2 components. The CEST image collected by SWIFT (Fig. 11.9f) shows the expected gradient of paraCEST agent as it was filtered through the kidneys.

The above two examples serve to illustrate the importance of considering both T_{1a} and T_{2a} in the design and implementation of new paraCEST agents in vivo. If one wants to design an amide-based paraCEST agent to measure a specific biological activity, for example, the best approach would be to choose a lanthanide ion having the largest possible paraCEST index (Eu^{3+}, Yb^{3+}, Tb^{3+}, or Tm^{3+} preferred, see Table 11.1) and form a complex between that ion and a ligand that yields a complex with either very fast water exchange, to limit T_{2ex} shortening, or, better yet, no water exchange because an inner-sphere position is not available for water to limit both T_1 and T_2 shortening. This approach was recently followed in the design of an amide-based paraCEST agent intended to reduce the contribution of T_{2exch} to the water line width by limiting water access to the paramagnetic center [14].

11.4 Water Molecule Exchange, Proton Exchange, and CEST Contrast

The rate of water exchange in lanthanide complexes plays a significant role in all types of MRI contrast agents, including Gd^{3+}-based T_1 agents, Dy^{3+}-based T_{2exch} agents, and paraCEST agents. An optimal Gd^{3+}-based T_1 agent must have a relatively short water exchange lifetime (τ_M), roughly 24 ns at 3 T [15]. At the other extreme, the water exchange lifetime in an optimal paraCEST agent must be quite long, on the order of 100 μs or longer. This dramatic difference in lifetime means that lanthanide complexes having water exchange rates varying by 10,000-fold or more have been described [16]. Using ligand design principles to control water exchange rates in lanthanide complexes has become a central theme in our lab and others over the past 20 years or more. This flexibility offers a wide opportunity to use chemistry to create novel types of MRI agents that are exquisitely sensitive to biological functions.

There are many more examples of "functional" paraCEST agents based on alterations in either water or proton exchange rates. If water or proton exchange is too fast, the resulting CEST signal will be small because the exchanging protons cannot be efficiently saturated before exchange occurs. The opposite is also true if exchange is too slow. In this case, a CEST signal will be limited because the saturated spins are not efficiently transferred into the much larger pool of bulk water molecules. Consequently, the lifetime ($\tau_M = k_{ex}^{-1}$) is one of the most important variables to consider when designing new paraCEST agents. For diaCEST agents, the proton exchange rate at pH 7 is largely determined by the chemical structure of each compound containing an exchangeable

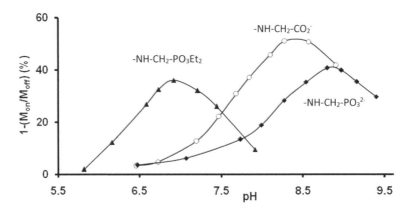

Figure 11.10 pH dependence of CEST intensity $[1 - (M_{on}/M_{off})]$ for three YbDOTA-tetraamide complexes (the chemical formula of each amide side chain is indicated in the figure). The CEST spectra were recorded on 20 mM samples at 9.4 T, 298 K, and $B_1 = 11.75$ μT.

–NH (amide), –NH$_2$ (amine), or –OH (alcohol) proton. This is largely true for amide- or amine-based paraCEST agents as well although at least one example has been published showing that the charge and identity of a functional group near an amide proton in paraCEST complexes can influence the –NH proton exchange rates [17]. As shown in Fig. 11.10, the optimal CEST intensity differs for each of these four complexes, and this reflects differences in –NH proton exchange rates for each complex. For the complex containing the carboxylate substituent (–NH–CH$_2$–CO$_2^-$), the amide proton lifetime was 3.3 ms, while for the complex containing the uncharged phosphonate diester substituent (–NH–CH$_2$–PO$_3$Et$_2$), the amide proton lifetime was much shorter, 0.11 ms, a 30-fold difference at a pH of 7.5. This illustrates that the pH profile of any –NH group can be readily controlled by altering the chemistry near the amide functionality.

For water-based paraCEST agents, one can use basic coordination chemistry principles to alter the rate of water exchange in these complexes over several orders of magnitude [16]. For example, bound water lifetimes, τ_M, have been reported that span from 3.7 ns for DyDOTA [18] to 416 μs for a EuDOTA derivative with mixed keto/amide chelating arms [19]. This reflects a 10^6-fold difference

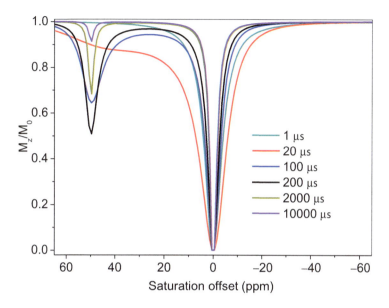

Figure 11.11 Simulated CEST profiles for a two-pool exchange model showing the effects of various pool B (50 ppm) water lifetimes (τ_M). The remaining parameters included $C_{pool\,A} = 111$ M, $C_{pool\,B} = 20$ mM, $T_{1a} = 2$ s, $T_{1a} = 2.5$ s, $T_{2a} = 1.2$ s, $T_{1b} = 0.2$ s, $T_{2b} = 0.1$ s, $B_1 = 9.4$ μT, and saturation time = 2 s.

in water exchange rates for water-based paraCEST agents compared to the relatively modest 30-fold difference observed for the amide-based paraCEST complexes illustrated in Fig. 11.10. Figure 11.11 shows the effect of different exchange rates or τ_M on the CEST intensity of a water-based paraCEST agent simulated using a fixed $B_1 = 9.4$ μT. Although this was simulated for a model Eu^{3+}-based agent with a bound water chemical shift of 50 ppm, these curves would look similar for any other Ln^{3+} complex with a bound water peak at another frequency. Only the shapes of the curves would differ. For a fast water exchange complex having a $\tau_M \leq 1$ μs, essentially no CEST is detected using 9.4 μT for activation. One could, of course, continue to increase B_1 until CEST is detected, but eventually the amount of power used in any experiment will be limited by sample heating. For a complex having a $\tau_M = 20$ μs, CEST would then be detected as a broad shoulder on the high-frequency side of the water peak. The bound water peak continues to grow in intensity and sharpen for complexes as τ_M moves toward 100–200 µs, then once again decreases in intensity as τ_M gets even longer. For a complex having a $\tau_M = 10$ ms, the intensity of the bound water peak at 50 ppm is again relatively small. This illustrates that there is an optimal exchange rate associated with the largest CEST signal. Quantitatively, Bloch theory predicts that a maximum CEST signal will be observed when the exchange rate matches the applied power according to $k_{ex} = 2\pi B_1$ [10]. In this example, a $B_1 = 9.4$ µT corresponds to 400 Hz at 9.4 T, so the optimal exchange rate is predicted to be $2\pi \cdot 400$ Hz $= 2513$ s^{-1} or a water lifetime of $\tau_M = 399$ µs. This means that if one is limited to a B_1 of 10 µT or less for an in vivo application using a 9.4 T animal scanner, one will need to identify water-based paraCEST agents having bound water lifetimes on the order of 374 µs to have the best chance of detecting the CEST signal in vivo. If one is restricted to even lower B_1 values, then even slower exchanging systems are required for optimal CEST.

11.5 Modulation of Inner-Sphere Water Exchange Rates

Some of the chemical features of ligands that influence water exchange rates in lanthanide complexes were recently reviewed [20]. In the section that follows, we review some recent results from our laboratory to illustrate a few of the important chemical features that can be used to modulate water exchange rates in paraCEST agents. One may notice that all of the paraCEST complexes described below are amide derivatives of DOTA. The reasons for this may not be obvious at this point, but one can highlight two important properties of DOTA-type ligands. First, they form thermodynamically favorable and kinetically inert complexes with the lanthanide ions, clearly important for potential in vivo applications. Second, DOTA-type ligands are considered "strong field" ligands for the lanthanides, so the paramagnetic shifts of all protons in such complexes tend to be large (the term D_1 in Eq. 11.2 is large and therefore $\Delta\omega$ is large; see Table 11.2). Given the CEST requirement, $\Delta\omega \geq k_{ex}$, the goal then is to make $\Delta\omega$ as large as possible (using DOTA derivatives) and reduce k_{ex} so that it is less than $\Delta\omega$ but no slower (see Fig. 11.11). This requires knowledge of the water exchange mechanism and the features of ligand design that can be used to modulate the rate of water exchange between an inner-sphere coordination position and bulk water.

The ligand design features we will illustrate can be divided into four categories: (1) electron donor strength of the ligand donor atoms; (2) charge on ligand side chains; (3) steric hindrance or bulkiness of side-chain amide groups; (4) the presence of coordination isomers (SAP/TSAP). The difference between a ligand having "strong" versus "weak" electron donor atoms can have a dramatic effect on water exchange rates. This was illustrated by theory and experiment in Green et al. [19]. The simplest explanation of why ligand donor atoms affect water exchange rates is to consider the overall charge on a lanthanide ion. In considering DOTA-type ligands, the chelated lanthanide ion will typically accept nine donor atoms, four nitrogen atoms from the macrocyclic ring, four oxygen atoms from the carboxylate or amide oxygen atoms, and one water molecule. The mechanism of water exchange in these complexes is known to occur via a dissociative pathway [21]. This means that the single water molecule must leave the inner-sphere of the lanthanide ion before another water molecule can replace it, so the rate at which the water molecule leaves will depend on the total amount of electron density donated by the other eight ligand donor atoms. The more electron density contributed by the other atoms, the easier it is for a water molecule to escape the grasp of the lanthanide ion. Figure 11.12 compares the amount of negative charge on the oxygen donor atoms for three different types of ligand donor atoms as estimated by density functional theory (DTF) calculations [19]. Theory predicts that the amount of negative charge on the oxygen donor atoms in ligands such as these falls in the order, carboxylate > amide > ketone. This is entirely consistent with the observation that LnDOTA complexes display fast water exchange, while LnDOTA-tetraamide complexes display much slower water exchange rates. This calculation also predicts that water exchange would be slowest if the oxygen donor atoms are ketone like. This prediction was then verified experimentally in the ligand series shown in Fig. 11.13.

While the amide/keto example demonstrates how electron density can modulate water exchange rates, there are other published

240 | ParaCEST Agents

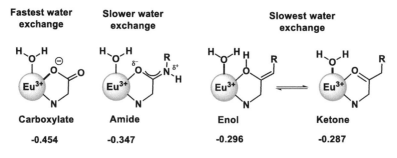

Mulligen charge of the O-atom

Figure 11.12 Resonance contributions in EuDOTA-type complexes containing a carboxylate oxygen atom, amide oxygen atom, or ketone oxygen atom. In those complexes having oxygen atoms that contribute the least amount of negative charge to the Eu^{3+}, one would anticipate that the Eu^{3+} expect to gain more electrons from the bound water oxygen (as indicated by the "thicker" bond) and this would translate to slower rates of water dissociation.

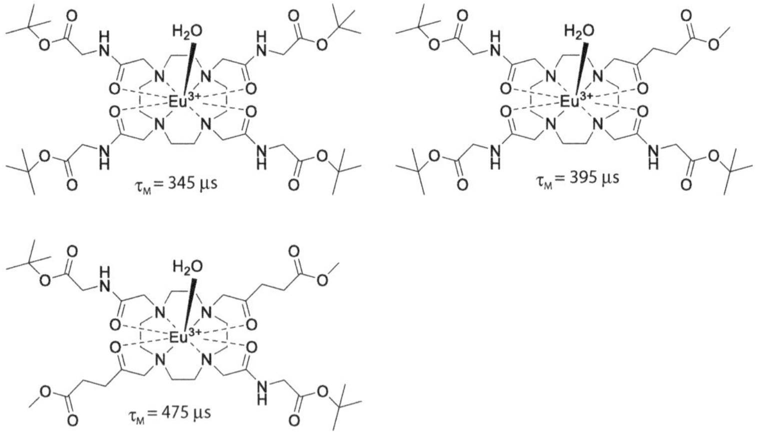

τ_M = 345 μs

τ_M = 395 μs

τ_M = 475 μs

Figure 11.13 Structures of three different EuDOTA-type ligands having a mixture of amide and ketone oxygen atoms. As shown, water exchange is much slower (longer τ_M) in those systems with the most ketone-type bonds as predicted by theory.

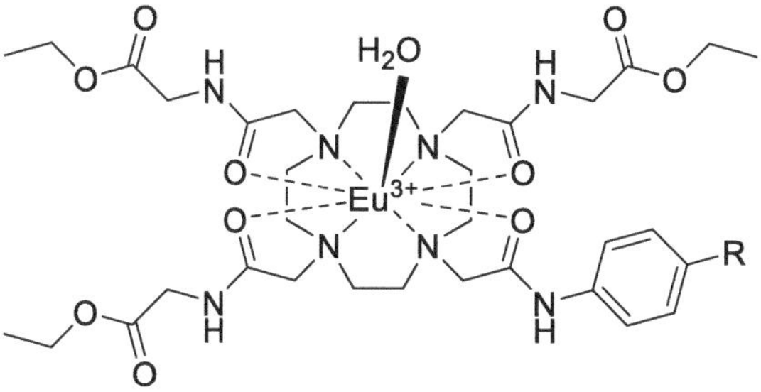

Figure 11.14 EuDOTA-tetraamide complexes with differing mesomeric electron-withdrawing groups (CO$_2^t$Bu, CN), mesomeric electron-donating groups (OMe), inductive electron-withdrawing groups (F), and inductive electron-donating groups (Me). The water exchange rates varied [τ_M values ranging from 144 μs (F) to 352 μs (CO$_2^t$Bu)] as anticipated based on the amount of negative charge on the single amide oxygen donor atom holding the variable substituent group.

examples where the amount of negative charge on a ligating oxygen atom of the ligand has been put to practical use. For example, Ratnakar et al. [22] compared water exchange rates in a simple series of EuDOTA-tetraamide complexes containing one aromatic amide side chain with different electron-donating/withdrawing substituents (Fig. 11.14). The relationship between water exchange rates and predicted electron density on the single variable amide oxygen donor atom indicated that this correlation is quite powerful and may be used as a strategy in the design of "responsive" paraCEST agents.

Equipped with this background knowledge, Ratnakar et al. [23] then designed a redox-sensitive paraCEST agent based on the reduction of a single electron-withdrawing methylquinolinium group (Fig. 11.15). In this example, the positively charged species Compound 1 displayed rather rapid water exchange ($\tau_M = 78$ μs), almost too fast for CEST. However, reduction of this species by the addition of β-NADH to produce Compound 2 resulted in an almost two-fold slowing of the water exchange ($\tau_M = 130$ μs), which was sufficient to "turn on" the CEST signal [23]. This interesting simple example of a "functional" CEST agent led others in the lab to think about other chemical platforms for creating functional agents based on changes in electron density on a single oxygen donor atom.

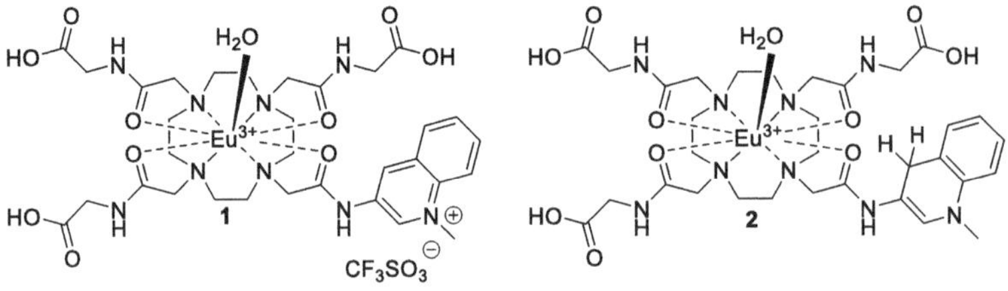

Figure 11.15 Reduction of Compound 1 to Compound 2 using NADH turns on the CEST signal by slowing water exchange.

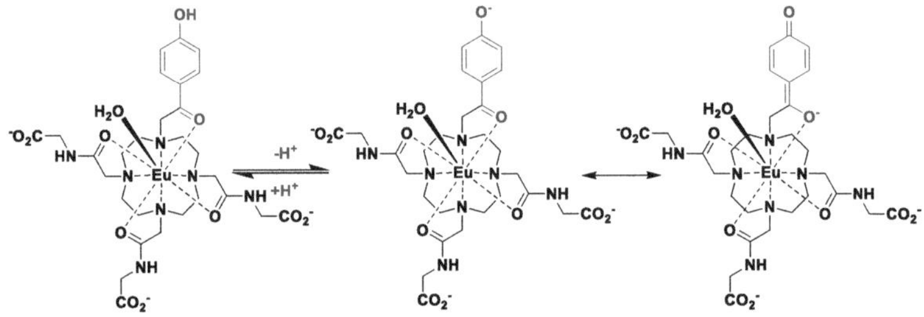

Figure 11.16 The structure of the pH-responsive paraCEST agent. Deprotonation of the phenolic proton results in conjugation of the resulting quinone-like structure with the acetyl oxygen atom coordinated to the Eu^{3+}.

One of the more interesting and perhaps practical examples is the unique paraCEST agent designed by Wu et al. [24] for reporting tissue pH (Fig. 11.16). In this construct, the charge on the single ketone-type oxygen atom of the ligand is altered through resonance stabilization as the phenolic group is deprotonated. This seemingly small change in electron donor strength gave the surprising result that the frequency of the Eu^{3+}-bound water exchange peak shifted upfield in proportion to the degree of deprotonation of the phenolic group. This important discovery provided a very simple way to measure pH using ratiometric imaging principles without knowing the exact concentration of the agent in solution. Concentration-independent agents such as this are required in the design of functional reporters of pH and other physiological parameters because the agent concentration in tissue is not well known and can be quite dynamic (changing as an agent is cleared by kidney filtration). In many paraCEST and all diaCEST agents, the amplitude of the CEST signal can be quite sensitive to pH [14, 25, 26], but the amplitude of the signal also depends on concentration and so one cannot get a direct readout of pH without knowing the agent concentration. In this unique pH sensor, the amplitude of the CEST signal changes very little with pH because one is not deprotonating the bound water molecule. Rather, the chemical shift of the water exchange peak changes with pH as a result of changes in ligand field strength so the chemical shift becomes the direct readout for signaling a change in pH (Fig. 11.17). If one then measures the intensity of the CEST water exchange peak at two frequencies (for example, 55 and 49 ppm), the resulting water intensity ratio changes linearly with pH.

The change in chemical shift observed in Fig. 11.17 is due to the differences in ligand field effects generated by switching from a weak field keto-type donor atom at lower pH values (e.g., pH 6, 50 ppm) to a stronger field quinone-like structure at higher pH values (e.g., pH 7.6, 54 ppm). Surprisingly, the sensitivity of chemical shift to changes in pH was quite large, about 2 ppm/pH unit. We have successfully used this agent in vivo in mice to image pH. As the CEST spectrum in Fig. 11.18 illustrates, the water exchange rate in this complex appears to be quite amenable for CEST detection in tissues.

Two other basic chemical factors are important in determining water exchange rates in paraCEST complexes. The first is overall charge of a complex. If one compares amide side chains having negatively charged carboxyl groups versus carboxylate ester groups, one always finds that complexes containing the ester groups display significantly slower water exchange. For example, if one compares Compound 3 in Fig. 11.19 (τ_M = 156 µs) with its corresponding ester, Compound 4 (τ_M = 247 µs), water exchange is slower in the ester. The explanation for this is complicated, but one simple rule of thumb is that the ester complex also has an overall positive charge and this makes it more difficult for a water molecule to leave the coordination sphere of the Eu^{3+} ion via a dissociative mechanism. Effectively, one can consider that Eu^{3+} in the ester complex has greater net positive charge than the Eu^{3+} ion in the carboxylate complex where the overall net charge is minus one [27]. This same trend was also seen in a comparison of water exchange rates in Compound 5 (τ_M = 104 µs) versus Compound 6 (τ_M =

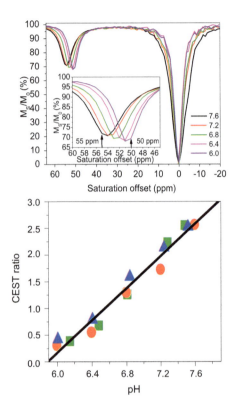

Figure 11.17 (Top) pH dependence of CEST spectra recorded at 9.4 T and 298 K. Insert: expanded view of the water exchange peak as a function of pH. [Eu^{3+}] = 10 mM, B_1 = 14.1 µT, saturation time = 2 s. (Bottom) pH ratiometric plot created by measuring the intensity of the CEST signal at 55 and 49 ppm. CEST spectra were recorded at three different concentrations 5, 10, and 20 mM to demonstrate that this ratio is independent of concentration.

210 µs) [27]. In this case, both lifetimes are shorter than those measured for the simple glycinate and glycinate ester derivatives. This demonstrates the other important factor here; the extra hydrophobicity introduced by the two methyl groups in Compounds 5 and 6 disrupts the water hydration layer above the Eu^{3+}-bound water molecule thereby allowing the bound water to exchange faster. This same effect is observed whenever a bulky hydrophobic group is introduced as a substituent on the amide. Compare, for

Modulation of Inner-Sphere Water Exchange Rates | 245

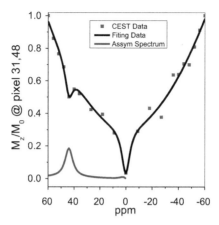

Figure 11.18 In vivo CEST spectrum from a signal pixel (pixel size = 0.5 mm × 0.5 mm) within a mouse kidney 6 min post intravenous injection of the pH-responsive paraCEST agent shown in Fig. 11.16. The CEST spectrum shows the typical broad MT signal from tissue and the sharper CEST water exchange peak from the agent. A simple asymmetry analysis of the CEST spectrum reveals the agent peak at ~45 ppm. This indicates that the pH in this particular pixel was rather low [27].

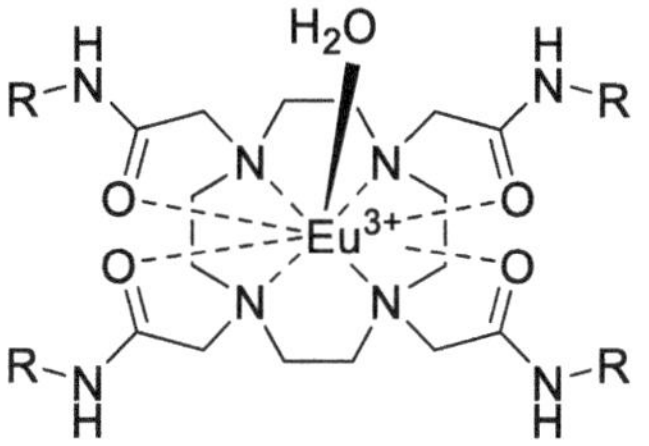

Figure 11.19 Structures of a series of EuDOTA-tetraamide complexes with differing overall charge and hydrophobicity. Both factors have a major influence on the rate of water exchange. Reprinted with permission from Ref. [28], Copyright 2009, John Wiley and Sons.

example, Compound 5 with Compound 7, the replacement of the single carboxylate group in Compound 5 ($\tau_M = 104$ μs) with a methyl group in Compound 7 ($\tau_M = 10$ μs) increases the water exchange rate by one order of magnitude. Thus, the chemical characteristics of an amide side chain have a significant influence on the rate of water exchange in these complexes.

The bulkiness of a ligand side chain is the second structural variation that can also impact the rate of water exchange in paraCEST agents. Aime et al. calculated the surface area of accessible water for a series of different lanthanide complexes and reported a nice correlation between this term and the bound water lifetime [29]. The bulkiness of ligand side chains can also impact the rates of water exchange by forcing a change in geometry of the complex. There are two favorable coordination geometries for nine-coordinate LnDOTA-type complexes, a square antiprismatic structure (SAP), and a twisted square antiprismatic structure (TSAP). These coordination isomers differ only in the twist angle between the plane of the four oxygen atoms and the four nitrogen atoms, so their energy difference is quite small. Consequently, both coordination isomers are typically present and in equilibrium in solution. Given this small difference in energy, the size or bulkiness of the side-chain groups can alter this equilibrium quite easily so that one isomer becomes favored over the other. In general, SAP isomers are favored for ligands with less bulky side chains, while bulkier groups tend to favor the TSAP structure. While both geometries are octa-coordinate, the Ln-O bond lengths are slightly longer in the TSAP isomers and this allows for faster water exchange in the more open TSAP geometry. Consequently, SAP structures are always more favorable for CEST than TSAP structures. Figure 11.20 shows a schematic crystal representation of the SAP and TSAP geometries with distances for both geometries from previously reported crystal structures [14, 30]. One can see quite easily from this figure that the SAP structure is more compact, so the water molecule (not shown) would lie closer to the lanthanide ion. The TSAP structure is more expanded such that the ligand's oxygen atoms are further away from the Ln^{3+} ion. This forces the water molecule further away from Ln^{3+} as well so that it becomes easier for it to dissociate from Ln^{3+}. This is the reason why water exchange is so much faster in TSAP complexes. To our knowledge, an Ln^{3+}-bound water exchange peak has not been firmly identified in any complex known to exist largely in the TSAP geometry. These differences in water exchange rates between isomers may prove useful in creating other types of paraCEST agents wherein fast water exchange is considered an asset [14].

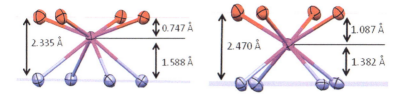

Figure 11.20 SAP (left) and TSAP (right) crystal representations with atoms omitted for clarity [14, 30]. Distances between planes are reported and are measured from the planes of the coordinating atoms: O4 – red, N4 – blue, and the lanthanide center in purple.

11.6 Techniques to Measure Exchange Rates

As demonstrated by the many examples illustrated above, optimization of water exchange rates is vital to the continued development of paraCEST agents. Several techniques are available to measure water exchange rates, depending on the exchange rate regime under study. The measurement of very fast water exchange rates, 10^7–10^9 s^{-1}, requires more elaborate methods, including fitting of temperature-dependent proton T_1 and T_2 measurements, ^{17}O line widths, or often a complete multi-parameter fitting of NMRD curves. These can be rather complex and are often best done through collaborations with labs that specialize in such analyses. However, in complexes where water exchange rates are much slower, 10^3–10^5 s^{-1}, a typical exchange regime for paraCEST agents, then more direct methods are available to anyone developing paraCEST complexes. The goal in this final section is not to review all methods available for all water and proton exchange regimes but to highlight those commonly available to investigators. The techniques used to measure water or proton exchange rates should be simple and straightforward such that agents with a large range of τ_M values can be evaluated quickly and accurately. Whenever possible, techniques should be complementary so that similar concentrations and temperature ranges can be used for all comparisons. In the following sections, we discuss three techniques used in our lab for measuring water exchange rates in paraCEST complexes.

11.6.1 Direct Measurement of ^{1}H NMR Resonance Line Widths

The most direct method for measuring proton or water exchange rates is by careful measurements of NMR line widths over the widest range of temperatures as permissible. An early investigation into the contribution of proton exchange to the r_1 relaxivity of GdDOTA-tetraamide complexes illustrates two very important points in this regard [31]. First, if one uses a ^{1}H NMR technique to evaluate exchange rates, one cannot distinguish between exchanging water molecules versus exchanging protons. Second, it is important to understand that proton exchange between an Ln^{3+}-coordinated water molecule and bulk solvent is typically *slow* compared to the rates of whole water molecule exchange except at the extremes of low or high pH [31]. Thus, proton exchange between pools of water molecules that differ only in chemical shift is relatively slow, similar to proton exchange between an amide –NH proton and water protons. So if one finds that water exchange is slow enough in a paraCEST complex to detect separate resonances by ^{1}H NMR, then monitoring the ^{1}H NMR line shape of either the Ln^{3+}-coordinated water molecule or the bulk water resonance over a range of temperatures allows a direct measure of the water molecule exchange rate using classical NMR fitting routines [32]. An example of one such data set is shown in Fig. 11.21 [33]. From data such as these, τ_M can be derived by fitting the line shapes recorded at each temperature to NMR exchange models using either self-written Matlab programs or commercial NMR software such as ACD Labs. The bound water lifetime in EuDOTA-(gly-OEt)$_4$$^{3+}$ obtained using this direct NMR line shape analysis was 225 µs at 25°C. A similar value was also obtained by fitting a single CEST spectrum of this compound at the same temperature to the Bloch equations modified for exchange (described in more detail below). This gives one confidence that the values obtained by fitting CEST spectra are quite reliable as long as the experimental conditions are well controlled and B_1 value is carefully calibrated. Interestingly, the high-resolution NMR spectrum of the acid derivative of this same compound, EuDOTA-(gly)$_4$$^{-1}$, does not show a resolved Eu^{3+}-coordinated water peak except at lower temperatures, so this

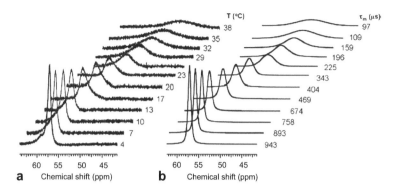

Figure 11.21 (a) High-resolution ^{1}H NMR spectra of EuDOTA-(gly-OEt)$_4^{3+}$ in water at pH 7 (showing only the downfield region). The ^{1}H resonance near 57 ppm is the Eu^{3+}-bound water molecule. (b) Calculated Eu^{3+}-bound water resonance at each temperature, based on line-fitting of the bound and bulk water resonances at each temperature. The bound water lifetimes at each temperature are shown. Reprinted with permission from Ref. [33], Copyright 2010, John Wiley and Sons.

method could not be used to evaluate the water exchange rate in the acid derivate of this complex. Nevertheless, Bloch fitting of the CEST spectrum of EuDOTA-(gly)$_4^{-1}$ collected at 25°C gave a water molecule exchange lifetime of 156 µs for the hydrolyzed derivative. This demonstrates that the direct NMR line-fitting method is only applicable to paraCEST agents having bound water lifetimes $\geq$200 µs at a particular temperature using a 400–500 MHz NMR spectrometer, whereas fitting of CEST spectra can be done reliably on faster exchanging systems, perhaps at lifetimes as low as 10 µs. Of course, if one collects high-resolution NMR spectra and CEST spectra at ultra-high magnetic fields (15 T or greater), then faster water exchange systems are amenable to either of these techniques.

11.6.2 Omega Plots

In 2010 Dixon et al. reported a new technique to estimate water exchange rates in paraCEST agents in which the intensity of any CEST exchange peak recorded at steady state is measured as a function of applied saturation power (B_1) [33]. The nice feature of this method is that a plot of $M_z/(M_0 - M_z)$ versus saturation power

($1/\omega^2$; rad/s) is linear over a wide range of power levels and the x-axis intercept of such a plot provides a direct estimate of the water exchange rate. While the theory behind the Omega plotting method may not be as obvious as the direct ^{1}H line shape analysis, the linearity of the Omega methods makes it convenient for anyone to use. The method also offers two additional advantages. The first is that the Omega plot method does not rely on knowing the agent concentration. While in phantom studies, the agent concentration is typically known, the Omega plot method would be most useful for the determination of water exchange rates in vivo where the agent concentration can be variable in different tissues and often unknown. The second benefit is that the Omega plot method is amenable to complexes having water exchange lifetimes shorter than 200 μs. Figure 11.22 shows an Omega plot for EuDOTA-(gly)$_4$$^{-1}$ in water. As indicated earlier, a fit of the CEST spectrum of this compound to Bloch equations at a known agent concentration yielded a τ_M = 156 μs, while here without including the agent concentration in the calculation, the Omega plot yielded an identical value (158 μs).

11.6.3 Bloch Fitting

Fitting of the entire CEST spectrum to Bloch equations has become one of the most widely used methods to measure water or proton exchange rates [10, 34]. One of the major advantages of this method is that one can fit the experimental CEST spectra to multiple pools of exchanging proton species and include experimental parameters such as the T_1 and T_2 of bulk water and the chemical shift estimates of all exchanging species. This technique, while robust, requires the user to have a priori estimates of the relaxation rates and chemical shifts of each species and an appropriate Bloch model–dependent fitting routine. Often times, exchangeable proton or water molecule resonances cannot be observed by ^{1}H NMR, so their relaxation rates can only be estimated. Although the chemical shifts of many exchanging protons can be estimated from the CEST spectrum itself, in some cases, exchanging –NH or –OH protons may lie too close to the bulk water resonance to be observed but nevertheless need to be included in the model. Given that Bloch fitting often requires

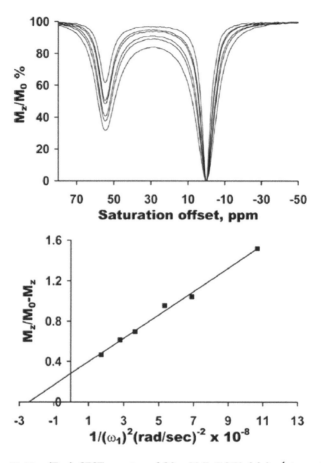

Figure 11.22 (Top) CEST spectra of 20 mM EuDOTA-(gly)$_4{}^{-1}$ in water at pH 7, 9.4 T, collected at 25°C with varying B_1 = 487, 606, 689, 833, 952, and 1212 Hz, with a presaturation time of 10 s. (Bottom) An omega plot ($M_z/(100 - M_z)$ versus $1/\omega_1{}^2$) derived from spectroscopic data for 20 mM EuDOTA-(gly)$_4{}^{-1}$ in water at pH 7 and 25°C, using a 10 s saturation pulse at the indicated power levels. Reprinted with permission from Ref. [33], Copyright 2010, John Wiley and Sons.

a number of initial estimations, it is important to collect multiple CEST spectra at different agent concentrations and applied B_1 values to obtain reliable estimates of exchange rates. For a more detailed description of the equations used in fitting of the CEST spectra, we refer the reader to Woessner et al. [10].

11.7 Summary

Since the first report of paraCEST 15 years ago, a large number of new agents have been described and tested in vitro. Most have been based on paramagnetic lanthanide complexes, but more recently, a number of paramagnetic transition metal ion complexes have also appeared [35–39]. Unfortunately, only a limited number of in vivo applications of paraCEST agents have been described, so it is worthwhile considering why this might be the case. One limitation may be that many chemists who think about new agent designs do not necessarily have easy access to animal MR scanners and the opportunity to test their compounds in vivo. Second, these agents can behave quite differently in vivo than they do in an NMR tube. Differences in temperature and the presence of other exchanging protons species in tissues can all influence the observed signals from paraCEST agents. Third, and perhaps most important, is the question of sensitivity. How much agent must be given to an animal to provide an acceptable in vivo CEST signal? Is the toxicity of these agents acceptable at these levels? Referring once again to Eq. 11.3, one could ask how much agent is required to produce a 5% decrease in intensity of the bulk water after complete saturation of a single water molecule or proton exchange site. Equation 11.3 shows that τ_a must be on the same order of magnitude or somewhat smaller than T_{1a} to impact M_z/M_0. Given the relationship between $\tau_a = \tau_M/([\text{agent}]/55M)$ [40], substitution into Eq. 11.3 yields the following relationship:

$$\frac{M_z}{M_0} = \frac{1}{1 + ([\text{agent}]*T_{1a}/55M*\tau_M)} \quad (11.4)$$

If one accepts a 5% decrease in the water intensity to be sufficient for detecting paraCEST agents in vivo, then the term in the denominator, $[\text{agent}]*T_{1a}/55M*\tau_M$, should have a value of ~0.1. If one assumes a value of 2 s for T_{1a} in vivo, then Bloch theory predicts that a water-based paraCEST agent having a bound water lifetime (τ_M) of 363 μs should be detectable at an in vivo concentration of 1 mM. This concentration is similar to that used clinically for Gd^{3+}-based imaging agents. This, of course, assumes that the agent can be fully saturated using an acceptable amount of power (B_1) without sample heating. Thus, while there may remain challenges to applying paraCEST agents in vivo to answer important clinical questions, important advancements have been made in both agent construct (optimizing water/proton exchange rates) and pulse sequence development for full activation of paraCEST agents. Many labs around the world are making excellent contributions toward this goal, and we look forward to participating in the advances needed to bring paraCEST agents to the clinical setting.

Acknowledgments

We wish to thank our many collaborators over the years for contributing to our collective knowledge of paraCEST agents and thank the NIH (CA115531, EB015908, and EB004582) and the Robert A. Welch Foundation (AT-584) for financial support during the writing of this chapter.

References

1. Zhang S, Winter P, Wu K, and Sherry AD. A novel europium(III)-based MRI contrast agent. *Journal of the American Chemical Society*, 2001; 123(7): 1517–1518.
2. Hinckley CC. Paramagnetic shifts in solutions of cholesterol and dipyridine adduct of trisdipivalomethanatoeuropium(3). A shift reagent. *Journal of the American Chemical Society*, 1969; 91(18): 5160.
3. Geraldes CFGC. Lanthanides: Shift reagents. In *Encyclopedia of Inorganic and Bioinorganic Chemistry*. New Jersey: John Wiley & Sons; 2012.
4. Pintacuda G, John M, Su X-C, and Otting G. NMR structure determination of protein–ligand complexes by lanthanide labeling. *Accounts of Chemical Research*, 2007; 40(3): 206–212.
5. Sherry AD and Geraldes CFGC. *Lanthanide Probes in Life, Chemical and Earth Sciences*. Amsterdam: Elsevier; 1989.
6. Bleaney B. Nuclear magnetic resonance shifts in solution due to lanthanide ions. *Journal of Magnetic Resonance*, 1972; 8(1): 91–100.
7. Zhang SR, Michaudet L, Burgess S, and Sherry AD. The amide protons of an ytterbium(III) dota tetraamide complex act as efficient antennae for transfer of magnetization to bulk water. *Angewandte Chemie-International Edition*, 2002; 41(11): 1919.

8. Aime S, Barge A, Delli Castelli D, et al. Paramagnetic lanthanide(III) complexes as pH-sensitive chemical exchange saturation transfer (CEST) contrast agents for MRI applications. *Magnetic Resonance in Medicine*, 2002; 47(4): 639–648.
9. Zhang SR and Sherry AD. Physical characteristics of lanthanide complexes that act as magnetization transfer (MT) contrast agents. *Journal of Solid State Chemistry*, 2003; 171(1–2): 38–43.
10. Woessner DE, Zhang S, Merritt ME, and Sherry AD. Numerical solution of the Bloch equations provides insights into the optimum design of PARACEST agents for MRI. *Magnetic Resonance in Medicine*, 2005; 53(4): 790–799.
11. Ratnakar SJ, Soesbe TC, Lumata LL, et al. Modulation of CEST images in vivo by T_1 relaxation: A new approach in the design of responsive PARACEST agents. *Journal of the American Chemical Society*, 2013; 135(40): 14904–14907.
12. Soesbe TC, Togao O, Takahashi M, and Sherry AD. SWIFT-CEST: A new MRI method to overcome T_2 shortening caused by PARACEST contrast agents. *Magnetic Resonance in Medicine*, 2012; 68(3): 816–821.
13. Swift TJ and Connick RE. NMR-relaxation mechanisms of O17 in aqueous solutions of paramagnetic cations and the lifetime of water molecules in the first coordination sphere. *Journal of Chemical Physics*, 1962; 37(2): 307–320.
14. Milne M, Lewis M, McVicar N, Suchy M, Bartha R, and Hudson RHE. MRI paraCEST agents that improve amide based pH measurements by limiting inner sphere water T_2 exchange. *RSC Advances*, 2014; 4(4): 1666–1674.
15. Caravan P. Strategies for increasing the sensitivity of gadolinium based MRI contrast agents. *Chemical Society Reviews*, 2006; 35(6): 512–523.
16. Sherry AD and Wu Y. The importance of water exchange rates in the design of responsive agents for MRI. *Current Opinion in Chemical Biology*, 2013; 17(2): 167–174.
17. Opina ACL, Wu Y, Zhao P, Kiefer G, and Sherry AD. The pH sensitivity of –NH exchange in LnDOTA–tetraamide complexes varies with amide substituent. *Contrast Media and Molecular Imaging*, 2011; 6(6): 459–464.
18. Soesbe TC, Ratnakar SJ, Milne M, et al. Maximizing T_2-exchange in Dy^{3+} DOTA-(amide)X chelates: Fine-tuning the water molecule exchange rate for enhanced T_2 contrast in MRI. *Magnetic Resonance in Medicine*, 2014; 71(3): 1179–1185.

19. Green KN, Viswanathan S, Rojas-Quijano FA, Kovacs Z, and Sherry AD. Europium(III) DOTA-derivatives having ketone donor pendant arms display dramatically slower water exchange. *Inorganic Chemistry*, 2011; 50(5): 1648–1655.
20. Siriwardena-Mahanama BN and Allen MJ. Strategies for optimizing water-exchange rates of lanthanide-based contrast agents for magnetic resonance imaging. *Molecules*, 2013; 18(8): 9352–9381.
21. Helm L and Merbach AE. Inorganic and bioinorganic solvent exchange mechanisms. *Chemical Reviews*, 2005; 105(6): 1923–1960.
22. Ratnakar SJ, Woods M, Lubag AJM, Kovács Z, and Sherry AD. Modulation of water exchange in europium(III) DOTA-tetraamide complexes via electronic substituent effects. *Journal of the American Chemical Society*, 2007; 130(1): 6–7.
23. Ratnakar SJ, Viswanathan S, Kovacs Z, Jindal AK, Green KN, and Sherry AD. Europium(III) DOTA-tetraamide complexes as redox-active MRI sensors. *Journal of the American Chemical Society*, 2012; 134(13): 5798–5800.
24. Wu Y, Soesbe TC, Kiefer GE, Zhao P, and Sherry AD. A responsive europium(III) chelate that provides a direct readout of pH by MRI. *Journal of the American Chemical Society*, 2010; 132(40): 14002–14003.
25. McVicar N, Li AX, Suchý M, Hudson RHE, Menon RS, and Bartha R. Simultaneous in vivo pH and temperature mapping using a PARACEST-MRI contrast agent. *Magnetic Resonance in Medicine*, 2013; 70(4): 1016–1025.
26. Suchy M, Li AX, Bartha R, and Hudson RHE. Analogs of Eu^{3+} DOTAM-Gly-Phe-OH and Tm^{3+} DOTAM-Gly-Lys-OH: Synthesis and magnetic properties of potential PARACEST MRI contrast agents. *Bioorganic and Medicinal Chemistry*, 2008; 16(11): 6156–6166.
27. Wu Y, Zhang S, Soesbe TC, et al. pH imaging of mouse kidneys in vivo using a frequency-dependent paraCEST agent. *Magnetic Resonance in Medicine*, 2016; 75(6): 2432–2441.
28. Mani T, Tircso G, Togao O, et al. Modulation of water exchange in Eu(III) DOTA-tetraamide complexes: Considerations for in vivo imaging of PARACEST agents. *Contrast Media and Molecular Imaging*, 2009; 4(4): 183–191.
29. Aime S, Barge A, Batsanov AS, et al. Controlling the variation of axial water exchange rates in macrocyclic lanthanide (III) complexes. *Chemical Communications*, 2002; 0(10): 1120–1121.

30. Milne M, Chicas K, Li A, Bartha R, and Hudson RHE. ParaCEST MRI contrast agents capable of derivatization via "click" chemistry. *Organic and Biomolecular Chemistry*, 2012; 10(2): 287–292.
31. Aime S, Barge A, Bruce JI, et al. NMR, relaxometric, and structural studies of the hydration and exchange dynamics of cationic lanthanide complexes of macrocyclic tetraamide ligands. *Journal of the American Chemical Society*, 1999; 121(24): 5762–5771.
32. Zhang S, Wu K, Biewer MC, and Sherry AD. ^{1}H and ^{17}O NMR detection of a lanthanide-bound water molecule at ambient temperatures in pure water as solvent. *Inorganic Chemistry*, 2001; 40(17): 4284–4290.
33. Dixon WT, Ren J, Lubag AJM, et al. A concentration-independent method to measure exchange rates in PARACEST agents. *Magnetic Resonance in Medicine*, 2010; 63(3): 625–632.
34. Li AX, Hudson RHE, Barrett JW, Jones CK, Pasternak SH, and Bartha R. Four-pool modeling of proton exchange processes in biological systems in the presence of MRI-paramagnetic chemical exchange saturation transfer (PARACEST) agents. *Magnetic Resonance in Medicine*, 2008; 60(5): 1197–1206.
35. Dorazio SJ, Tsitovich PB, Siters KE, Spernyak JA, and Morrow JR. Iron(II) PARACEST MRI contrast agents. *Journal of the American Chemical Society*, 2011; 133(36): 14154–14156.
36. Olatunde AO, Dorazio SJ, Spernyak JA, and Morrow JR. The NiCEST approach: Nickel(II) paraCEST MRI contrast agents. *Journal of the American Chemical Society*, 2012; 134(45): 18503–18505.
37. Dorazio SJ, Olatunde AO, Spernyak JA, and Morrow JR. CoCEST: Cobalt(II) amide-appended paraCEST MRI contrast agents. *Chemical Communications*, 2013; 49(85): 10025–10027.
38. Jeon I-R, Park JG, Haney CR, and Harris TD. Spin crossover iron (II) complexes as PARACEST MRI thermometers. *Chemical Science*, 2014; 5(6): 2461–2465.
39. Olatunde AO, Cox JM, Daddario MD, Spernyak JA, Benedict JB, and Morrow JR. Seven-coordinate Co-II, Fe-II and six-coordinate Ni-II amide-appended macrocyclic complexes as ParaCEST agents in biological media. *Inorganic Chemistry*, 2014; 53(16): 8311–8321.
40. Zhang SR, Merritt M, Woessner DE, Lenkinski RE, and Sherry AD. PARACEST agents: Modulating MRI contrast via water proton exchange. *Accounts of Chemical Research*, 2003; 36(10): 783–790.

Chapter 12

Transition Metal paraCEST Probes as Alternatives to Lanthanides

Janet R. Morrow and Pavel B. Tsitovich

Department of Chemistry, University at Buffalo, the State University of New York, Amherst, NY 14260, USA
jmorrow@buffalo.edu

12.1 Introduction

The application of paramagnetic metal ion complexes as MRI contrast agents has an interesting history. Complexes developed initially as T_1 relaxivity agents contained paramagnetic Fe(III) and Mn(II), followed by studies on Gd(III) complexes [1]. More recently, complexes of lanthanide ions other than Gd(III) have been developed as MRI contrast agents that operate by chemical exchange saturation transfer (CEST) [2–4]. In particular, Eu(III), Tm(III), and Yb(III) have been extensively used as CEST or paramagnetic CEST (paraCEST) agents as described in detail in Chapter 11. Why would one then want to develop transition metal ion paraCEST agents based on different elements? In this chapter, we show that the unique coordination chemistry, tunable redox potentials as well

---

Chemical Exchange Saturation Transfer Imaging: Advances and Applications
Edited by Michael T. McMahon, Assaf A. Gilad, Jeff W. M. Bulte, and Peter C. M. van Zijl
Copyright © 2017 Pan Stanford Publishing Pte. Ltd.
ISBN 978-981-4745-70-3 (Hardcover), 978-1-315-36442-1 (eBook)
www.panstanford.com

as suitable paramagnetic properties of transition metal ions make them excellent candidates for the development as paraCEST agents.

All paraCEST agents are paramagnetic metal ion complexes that produce highly shifted ligand proton resonances. In addition, paraCEST agents have ligand protons, generally H_2O, NH, or OH groups that are in chemical exchange with bulk water protons. As mentioned in detail in Chapter 6, exchange rate constants must be slow on the NMR timescale so that distinct proton resonances for the agent and for bulk water are observed [5]. In order to be able to efficiently saturate the proton magnetization with the presaturation pulse, the paramagnetically shifted proton resonances should be relatively sharp. Thus, the paramagnetic metal ion center should ideally produce small proton relaxation enhancements, but large paramagnetic proton shifts [3].

The line broadening of the ligand proton resonance in a complex is closely connected to the electronic relaxation rate constant of the bound metal ion [6]. Slowly relaxing electronic states give rise to highly broadened proton resonances (e.g., Mn(II) or Gd(III)), whereas paramagnetic metal ions with more rapidly relaxing states (e.g., Eu(III) or Yb(III)) typically give sharper proton resonances. Among the first row transition metal ions, high-spin Fe(II) and Co(II) have the most rapidly relaxing electronic states and, consequently, favorably narrow line widths [6, 7]. Certain coordination geometries of high-spin Ni(II) complexes have favorable electronic relaxation rate constants and sharp proton resonances. Unfortunately, the most common geometries for Ni(II), such as pseudo-octahedral, typically do not have optimal paramagnetic properties as shift agents.

Paramagnetic induced proton shifts (hyperfine shifts) have two contributions, one from dipolar or through–space interactions of the electron and proton spins and the other from delocalization of spin density on the ligand protons from through–bond contributions [7, 8]. Protons that are more than two bonds away from a paramagnetic Ln(III) center show predominantly dipolar contributions to the hyperfine shift. This is because Ln(III) ions do not have substantial covalent character to their coordination bonds, making contact contributions much less important than for transition metal ions [8]. The more substantial contact contributions for the transition metal ions are both a blessing and a curse for the coordination chemist.

The advantage of having both contact and dipolar contributions to the hyperfine shift is that the exchangeable protons get two contributions and this may produce highly shifted resonances if the two contributions reinforce each other. The disadvantage is that it is difficult to predict the magnitude of the hyperfine shift and to parse the two contributions to the shift [9]. The difficulty in predicting the hyperfine shifts for transition metal ion complexes gives a large trial and error component to their development.

Unlike Ln(III) complexes, transition metal paraCEST agents have multiple electronic spin states that are readily accessible. These spin states depend on complex geometry and on the ligand donor groups. For example, pseudo-octahedral Fe(II) (d_6) complexes may be paramagnetic with two paired and four unpaired d-electrons or may be diamagnetic with all six d-electrons paired. The spin state of the Fe(II) center may be modulated by simple substituent changes on the ligands. This has been shown for simple pyridine pendents attached to 1,4,7-triazacyclononane that coordinate Fe(II) to produce a strong ligand field and a diamagnetic complex [10], whereas the addition of a methyl group adjacent to the coordinating nitrogen as in Fe(L2) gives a weak crystal field and a paramagnetic complex (Scheme 12.1) [11, 12]. Thus, the nature of the coordination sphere of the metal ion can be varied to influence the spin state. Spin state changes may also be produced as a function of temperature. These so-called "spin crossover" complexes, when based on Fe(II), may go from diamagnetic ($S = 0$) to paramagnetic ($S = 2$) upon increase in temperature [13]. Such temperature-dependent switches may be powerful tools for the registration of temperature through MRI [14].

Another important difference is that transition metal paraCEST agents may have multiple accessible oxidation states, unlike Ln(III) agents that are present in the trivalent state under most conditions. In particular, Fe(II)/Fe(III) and Co(II)/Co(III) oxidation states are readily accessible and their stability depends on the nature of the ligands in the coordination sphere [15, 16]. These redox potentials can be tuned to favor the paramagnetic divalent state to produce stable paraCEST agents based on Fe(II), Co(II), or Ni(II) [17]. Alternatively, redox potentials can be tuned to match the negative redox potentials observed in cells so that the trivalent state is more favorable [18]. As will be discussed in the following paragraphs, this makes it feasible to switch magnetic properties by oxidizing or reducing the metal ion center.

Finally, all of the three metals featured here—iron, cobalt, and nickel—are biologically essential elements. In particular, iron is of interest for the development of contrast agents because iron is readily handled by the human body [19]. Metabolically recyclable MRI contrast agents based on iron are a long-term goal.

12.2 Coordination Chemistry of Iron(II), Cobalt(II), and Nickel(II)

The coordination chemistry of these three divalent transition metal ions with macrocyclic ligands is rich. One challenge for Fe(II) and Co(II) azamacrocycle coordination chemistry is the necessity of preventing oxidation to higher states. In order to stabilize the divalent state over the trivalent state, it is useful to consider differences in their coordination chemistry. Fe(II) has a larger ionic radius compared to Fe(III) by 18% for high-spin six-coordinate complexes, while six-coordinate high-spin Co(II) is 29% larger than low-spin six-coordinate Co(III) [20]. Thus, ligands that dictate higher coordination geometries are preferable. Fe(II) and Co(II) are softer Lewis acids than are Fe(III) or Co(III). Nitrogen donors, aromatic nitrogens, and neutral oxygen donor groups are desirable for the stabilization of the divalent state [21]. Macrocycles are the ligands of choice in order to optimize stability and kinetic inertness toward dissociation. The addition of pendent groups to the macrocycle completes the coordination sphere and encapsulates the metal ion. Encapsulation prevents direct innersphere coordination to adventitious ligands in biological systems such as carbonate and phosphate.

Several macrocyclic complexes of paramagnetic Fe(II), Co(II), and Ni(II) are shown in Scheme 12.1. Notably, complexes may be six, seven, or eight coordinate [22–24]. Four different macrocyclic frameworks have been utilized in the Morrow laboratory for paraCEST agents, including diaza-15-crown-5, CYCLAM (1,4,8, 11-tetraazacyclotetradecane), CYCLEN (1,4,7,10-tetraazacyclodo-

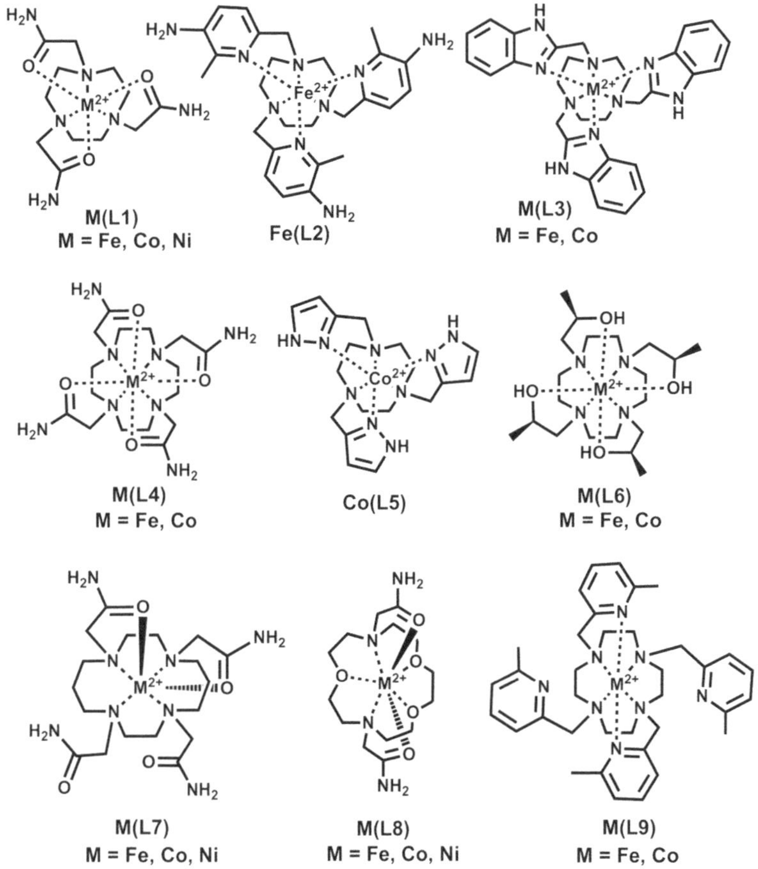

Scheme 12.1

decane), and TACN (1,4,7-triazacyclononane) [17]. The nitrogens in these macrocycles serve to coordinate the metal ion and as a site of attachment of the pendent groups. The metal ion is roughly in the plane of the macrocycle for complexes of the diaza-crown and for CYCLAM, but above the plane of the macrocycle in complexes of CYCLEN and TACN.

There is a preponderance of amide pendents in paraCEST agents reported to date [17]. Amides generally coordinate through the oxygen atom and are neutral donors, fitting the criteria for favoring binding to divalent metal ions. Amide pendents are sterically efficient pendent groups and synthetically straightforward to add to macrocycles. The NH protons are readily exchangeable with bulk water. As described later, exchange rate constants for amide NH protons are pH dependent and range from 200 to 800 s^{-1} at near neutral pH [12, 22, 25–27]. Notably, there is a partial double bond between the carbonyl carbon and nitrogen atom that gives rise to two chemically inequivalent NH protons for each pendent.

The other neutral donor group commonly used is an alcohol pendent, such as found in L6 [25]. Data from pH-potentiometric titrations are consistent with the binding of the metal ion to the alcohol groups without ionization at neutral pH. Notably, the alcohol pendents contain a methyl group on the carbon adjacent to the alcohol. The chirality of this group arises from the alkylation of the macrocycle with a single enantiomer of propylene oxide. This methyl group imparts rigidity to the macrocyclic complex, which inhibits dissociation of the metal ion.

Certain pendents containing heterocyclic imine donors are excellent candidates for the stabilization of the divalent states of Co(II) and Fe(II) [12, 17, 18, 24, 28]. The heterocyclic groups may be derivatized to further expand the properties of these pendent groups. Pyridine derivatives are especially useful for the stabilization of the divalent state. Reported paraCEST complexes have a methyl group placed adjacent to the coordinating nitrogen to increase the metal to nitrogen bond length and promote the formation of high-spin complexes [11, 12]. In the TACN-based macrocycle, this methyl group is necessary for the formation of a high-spin Fe(II) complex. The methyl groups are oriented such that they act as splines that influence the pendent group coordination to the metal ion, providing a rigid macrocyclic complex. An amino group may be added to give an exchangeable proton for CEST imaging as in Fe(L2) [12]. Other pendent groups contain five-membered heterocycles with exchangeable NH groups such as pyrazole and benzimidazole [18, 28]. These heterocycles, when used with the TACN framework, facilitate the oxidation of the metal ion center of Co(II) and Fe(II) complexes.

Recently reported crystal structures show diverse coordination chemistry for the macrocyclic complexes shown in Scheme 12.1. CYCLEN-based macrocycles give eight-coordinate Fe(II) in Fe(L4),

NMR Spectra, CEST Spectra, and Imaging | 263

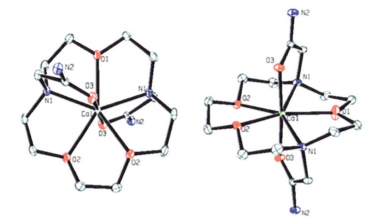

Figure 12.1 Crystal structure of complex cation of [Co(L8)]Cl$_2$. Reprinted with permission from Ref. [22], Copyright 2014 American Chemical Society.

which contains the sterically efficient amide pendent groups [23]. The smaller Co(II) ion gives rise to the seven-coordinate metal center in Co(L4) [23]. CYCLAM derivatives have the metal ion in the center of the macrocycle with two of the pendents coordinated axially, including Co(L7) and Ni(L7) [23]. The diaza-crown derivative forms seven-coordinate Co(L8), and Fe(L8) (Fig. 12.1) [22]. The smaller Ni(II) ion forms a six-coordinate complex with L8. Both Fe(II) and Co(II) form six-coordinate complexes with the CYCLEN-based ligand L9 due to the greater size and bulkiness of the picolyl pendent groups in comparison to ligands with amide (L4) or alcohol (L6) pendents [24].

12.3 NMR Spectra, CEST Spectra, and Imaging

Several of the paramagnetic Fe(II), Co(II), and Ni(II) complexes shown in Scheme 12.1 produce highly shifted proton resonances with relatively narrow line widths. Co(II) complexes, in particular, have large hyperfine proton shifts (−150 to 320 ppm) with macrocyclic resonances as sharp as 50 Hz [18, 24]. Fe(II) complexes generally give sharp proton resonances, barring dynamic processes in solution [11, 12, 28]. However, most of the Ni(II) macrocyclic complexes in Scheme 12.1 have broad proton resonances [26]. The notable exception is the Ni(II) complex of L8, which has narrow and highly shifted proton resonances (Fig. 12.2) [22]. This is the only such case we have found for Ni(II) complexes that typically give broad proton resonances due to the relatively long electronic relaxation times that are characteristic of six-coordinate Ni(II). The Co(II) and Fe(II) complexes of the L8 ligand also give sharp proton resonances (Fig. 12.2).

Several of the complexes in Scheme 12.1 have dynamic processes that interconvert between isomers in solution and this affects the appearance of their proton NMR spectra, generally leading to extremely broad resonances. These dynamic processes most commonly occur for complexes containing amide pendents and either TACN or CYCLEN macrocycles, including the Fe(II) and Co(II) complexes of L1 or L4 [12, 25, 27]. In contrast, Fe(II) or Co(II) macrocycles with amide-appended CYCLAM or 4,10-diaza-15-crown macrocycles (L7, L8) [22, 23, 26] lack dynamic processes occurring on the NMR timescale and have sharp proton resonances. Notably, complexes of heterocyclic pendents such as amino or methyl-pyridines on TACN or CYCLEN frequently produce very sharp proton resonances due to the rigid nature of the pendent [12]. Incorporation of rigidity into the macrocyclic complex is important because dynamic processes that broaden the proton resonances may detract from the application of the complex as a paraCEST agent. Interestingly, several of the amide-appended complexes that show dynamic processes still produce CEST peaks, albeit typically broader than complexes that do not have dynamic processes [12, 25]. However, these complexes have amide groups that are bound tightly to the metal center, such that the NH amide protons are not broadened as severely as the macrocyclic CH protons. Thus, if part of the ligand containing exchangeable protons is rigid, the dynamic process in the rest of the ligand may not substantially influence the CEST effect.

There are substantial contributions from contact shifts in the hyperfine shifted proton resonances of complexes in Scheme 12.1, especially in pendent donor groups that have delocalized bonding. For example, amide groups have a delocalized multiple bond over the oxygen, carbon, and nitrogen atoms. This also leads to the

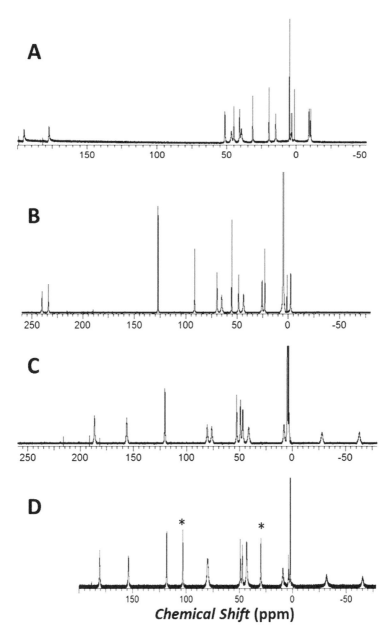

Figure 12.2 ^{1}H NMR spectra in D$_2$O of (A) Ni(L8), (B) Co(L8), (C) Fe(L8) in D$_2$O, or (D) Fe(L8) in CD$_3$CN showing amide protons. Reprinted with permission from Ref. [22], Copyright 2014 American Chemical Society.

two magnetically distinct protons. The two different protons on a single amide group are typically separated by 60–80 ppm, as shown in Fig. 12.2D [17]. This separation is most likely due to each of the protons experiencing different unpaired spin density. Notably, amides bound to Ln(III) do not show such large chemical shift differences for the two distinct amide NH protons [29]. The protons of heterocyclic pendents with delocalized pi-systems, including pyrazoles, benzimidazoles, or aminopyridines, are also likely to have substantial through–bond contributions to the hyperfine shift. This is illustrated by various chemical shifts of aromatic proton resonances in Co(L5) [18]. While both aromatic CH protons are located at approximately similar distances with respect to the metal center, their chemical shifts differ by 68 ppm, indicative of different contact shift contributions. Interestingly, work to date shows that these pendents have hyperfine shifted resonances that vary substantially based on both bonding considerations and also on through–space differences of the structurally distinct macrocycle frameworks. Moreover, the delocalization of the aromatic pi-system may be sensitive to changes of pH due to the presence of ionizable groups in the aromatic ring.

There are many challenges associated with the assignment of proton resonances in paramagnetic transition metal complexes. It is especially difficult if proton resonances are broadened and overlap due to structural fluxionality. For complexes that give sharp proton resonances, the number of resonances can be used to determine the symmetry of the complex in solution. However, methods that are generally useful in making assignments of protons, such as 2D COSY NMR, experience significant limitations due to fast relaxation of protons. In these cases, T_1 measurements of certain protons often help to identify the proximity of the proton to the paramagnetic metal ion center.

12.3.1 CEST Spectra

The exchangeable NH or OH protons of the complexes in Scheme 12.1 give rise to CEST peak positions that vary from 6.5 ppm for Fe(L2) [12] to 135 ppm for Co(L5) versus bulk water [18]. For paraCEST agents with amide pendents, two CEST peaks, one for

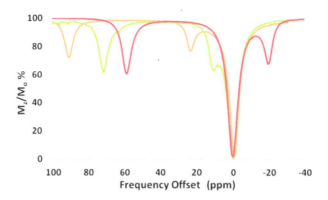

Figure 12.3 CEST spectra (11.7 T) of 10 mM Fe(L8) (yellow), Ni(L8) (green), Co(L8) pink at pH 7.4, 100 mM NaCl, 20 mM buffer, RF presaturation pulse of 2 s, $B_1 = 24$ µT. Reprinted with permission from Ref. [22], Copyright 2014 American Chemical Society.

each inequivalent amide proton, may be observed. For example, Ni(II), Fe(II), and Co(II) complexes of L8 all show two CEST peaks separated by 66, 74, and 82 ppm, respectively (Fig. 12.3) [22]. One peak is shifted downfield, and the other is closer to water. Not all amide-based paraCEST agents show multiple peaks. Fe(II) and Co(II) complexes of L1 and L4 as well as Ni(L1) show a single CEST peak ranging from 45 to 77 ppm versus water [12, 26, 27]. Presumably, the additional CEST peak is under the water signal or is not observed due to exchange rates that are not optimized to produce CEST peaks. The heterocyclic pendents with exchangeable NH protons form Fe(II) or Co(II) complexes with unpredictable and variable hyperfine shifts. Thus, Co(L5) produces a CEST peak at 135 pm, most likely from the pyrazole NH. The Fe(L3) CEST peak falls at 53 ppm, and Fe(L2) falls at 6.5 ppm (Fig. 12.4) [17].

The intensity of the CEST peaks for the complexes with exchangeable protons in Scheme 12.1 is quite variable [17]. The simplified expression for the intensity of the CEST signal, which is based on the assumption of steady-state attainment of the proton exchange with irradiation, predicts that the intensity of the CEST effect will increase with the number of exchangeable protons (n), the rate constant for exchange (k_{ex}), the concentration of the paraCEST agent (C), and the time constant for bulk water

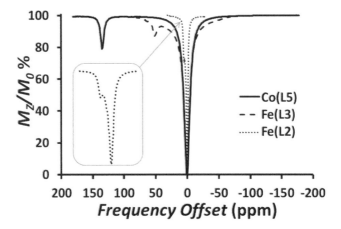

Figure 12.4 CEST spectra (11.7 T) of complexes containing heterocyclic pendents. Reprinted from Ref. [17], Copyright 2013, with permission of Springer.

proton relaxation (T_1) [2]. Also of importance is $\Delta\omega$, the difference in Hz between the paramagnetically shifted exchangeable proton resonance and the bulk water proton resonance. A large $\Delta\omega$ allows for a large proton exchange rate constant, which is generally beneficial. Overall, a paraCEST agent that has a large number of magnetically equivalent exchangeable protons, poor relaxivity, a far-shifted proton resonance, and an optimal proton exchange rate would give the most intense CEST signal. Also of note is that the more effectively the exchangeable proton resonance is saturated, the stronger the CEST effect. Effective saturation correlates to an increase in the saturation power and time allotted for saturation. A decrease in CEST may result from the broadening of the resonance by T_2-based factors and by dynamic processes. Notably, there is an optimal power for the presaturation pulse (B_1), which is based on the rate constant (k_{ex}) for the exchange of the proton ($k_{ex} = 2\pi B_1$) [30]. The presaturation power is limited by the necessity of keeping power deposition in tissue relatively low, typically at 100–200 Hz. Thus, the optimal rate constant for this presaturation pulse power is on the order of 1000 s^{-1} [31]. We discuss these factors later. The number of equivalent exchangeable protons depends on the overall symmetry of the complex. The most highly symmetric complexes are Fe(L4) (which has local C_4 axis of symmetry) that leads to two sets of four amide NH protons, Fe(L1), Co(L1), and Ni(L1), which have two sets of three NH protons, Fe(L2) which has six NH protons, Co(L3) and Co(L5) which have three NH protons. M(L8) complexes have two sets of two amide protons, whereas the Ni(L7) and Co(L7) complexes each have four different NH amide protons. Yet the intensity of the CEST peaks does not correlate well to the number of equivalent protons alone. For the amide complexes, intensity is largest for complexes that are rigid, such as M(L7) or M(L8), despite their lower number of exchangeable protons [17]. This is attributed, in part, to their narrower CEST peak intensities due to the lack of macrocycle dynamics. These complexes also have poor T_1 relaxivities, a desirable property given that water proton relaxation may compete with CEST. For example, T_1 relaxivities for the effective paraCEST agents Ni(L8) and Co(L8) are 0.012 mMs^{-1} and 0.038 mMs^{-1}, respectively, compared with 0.21 mMs^{-1} and 0.125 mMs^{-1} for less effective agents Ni(L1) and Co(L1), respectively [17]. Rate constants for proton exchange are also important contributors to the intensity of the CEST peaks. Rate constants for the paraCEST agents vary from 200 s^{-1} to 900 s^{-1} for amides at near neutral pH in buffered solution [17]. The large rate constant of 12,400 s^{-1} for pyrazoles in Co(L5) at pH 7.5 and 37°C produces a relatively intense CEST peak for this complex [18].

12.3.2 CEST Imaging

Solutions of the paraCEST agents are imaged on an MRI scanner by using information about the CEST peak position obtained from CEST NMR experiments. Experiments using MRI involve production of an image upon irradiation at the frequency of the exchangeable proton, followed by subtraction of an image produced upon irradiation at a frequency that should not produce CEST. For example, the images in Fig. 12.5 were produced by using a presaturation pulse at +59 ppm and then at −59 ppm from the bulk water signal. These experiments are initially carried out on tubes of contrast agent and are referred to as phantoms. CEST MR images of several of the transition metal ion paraCEST agents in Scheme 12.1 were acquired on a 4.7 T preclinical MR scanner. These images were made by using

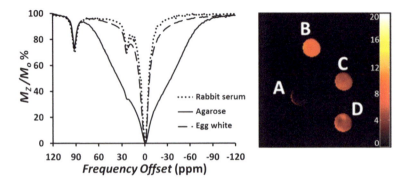

Figure 12.5 Left: CEST spectrum (11.7 T) of 10 mM Fe(L8) at pH 7.3, with 2 s RF presaturation pulse, $B_1 = 24$ µT. Right: CEST images of phantoms (4.7 T) created with five pulses of 12 µT for 1 s each at (±59 ppm). All samples contained 20 mM HEPES pH 7.4, 100 mM NaCl. (A) Buffer only; (B) 4 mM Co(L8); (C) Co(L8) in rabbit serum; (D) Co(L8) in 4% agarose gel, all at 37°C. Reprinted with permission from Ref. [22], Copyright 2014 American Chemical Society.

a pulse train (presaturation power of 12 µT for 1 s each) applied symmetrically about the bulk water resonance [12, 18, 22, 27]. Either conventional CEST spin-echo acquisition or CEST-fast imaging with steady-state free precession technique (CEST-FISP) was used. In the CEST-FISP method [32], 20 repetition image datasets were averaged together. Experiments using CEST-FISP permitted us to image Co(L8) at concentrations as low as 250 µM in phantoms [27]. Phantom imaging provides valuable initial information about the detection limits of transition metal complexes of different relaxivity and CEST contrast prior to conducting in vivo studies.

12.4 Responsive Agents

12.4.1 pH-Responsive Agents

The magnitude of the CEST effect is generally pH dependent because proton exchange with water may be acid or base catalyzed. The pH dependence of paraCEST agents with NH groups on amides, OH groups of alcohols, or NH groups of aniline has been reported [33–35]. One would expect a correlation of the rate constants with the pK_a's of the exchangeable protons, but unfortunately there are few pK_a values that have been reported for these complexes. Most commonly observed is base-catalyzed proton exchange, which produces an increase in the rate constant for exchange with increasing pH values. An increase in the rate constant generally leads to an increase in the CEST peak as long as the saturation B_1 is appropriate. Typically, there is a pH region that gives a linear dependence of CEST on pH [36].

The CEST effect of the exchangeable NH or OH protons on the paraCEST agents in Scheme 12.1 shows a dependence on pH. Agents with amide protons show an increase in CEST intensity with increasing pH over the range of pH 6.0 to 8.0, consistent with base-catalyzed exchange [22, 25–27]. Alcohol protons on Fe(L6) show an initial increase in intensity with pH from pH 5 to 7 followed by a decrease above pH 7 as increasing rate constants lead to peak broadening [25]. The pyrazole protons of Co(L5) also show an increase with pH at acidic pH values up to neutral pH, with a maximal signal at pH 6.9–7.0 followed by a decrease at higher pH values as the CEST peak broadening occurs due to an increase in proton exchange rate constants [18].

Complexes that demonstrate multiple CEST peaks have the advantage that the pH-dependent change in intensity of each of the two peaks, if different, can be used to develop ratiometric contrast agents. For example, Co(L7) has four CEST peaks. The two most highly shifted peaks show a distinct pH-dependent change in CEST intensity so that a ratio of the CEST peak intensity can be monitored. The ratio is used to produce a pH-responsive agent, which gives contrast independent of the probe concentration (Fig. 12.6) [27].

12.4.2 Redox-Responsive Agents

An advantage of working with Fe(II) and Co(II) paraCEST agents is that the ease of oxidation of the divalent to the trivalent state can be tuned over a wide range, typically by more than 1.5 V. Many of the complexes in Scheme 12.1 are highly stabilized in the divalent state. For example, Fe(L1) has a redox potential of 860 mV versus NHE, designating an Fe(II) center, which is not very reducing [11]. Similarly, the Fe(II) complexes of L4 and L6 have

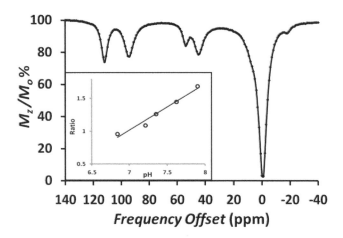

Figure 12.6 CEST spectrum (11.7 T) of 10 mM Co(L7) at pH 7.4, 37 °C. The inset shows a plot of the ratio of the $[1/(M_z/M_0)]$ at 112 ppm to 95 ppm over the pH range of 6.8 to 7.8.

redox potentials that are >800 mV versus NHE [11, 25]. The Co(II) complexes of L1, L4, L7, and L8 are very stable in air, consistent with high redox potentials [27]. However, the redox potential can be decreased to be more favorable for the trivalent state. Tuning the redox potential to a value that is close to the redox potential of cells or extracellular space provides the opportunity to switch between metal ion oxidation states based on the changing redox status of the cells [37, 38]. For the Co(II)/Co(III) couple in particular, this is an attractive proposition because the high-spin Co(II) state, which is ideal for paraCEST, oxidizes to the diamagnetic Co(III) state, which is MRI silent [18]. Thus, these transition metal paraCEST agents are magnetic state switches that depend on redox potential of their surroundings.

The Co(L5) complex is well-suited for reporting on redox status under physiological conditions, given its redox potential of −107 mV versus NHE for the Co(III) to Co(II) couple [18]. The divalent Co(II) paraCEST agent oxidizes in air to form the paraCEST-silent Co(III) complex (Fig. 12.7). Reduction of the Co(III) complex of L5 is accomplished with dithionate or cysteine to give the Co(II) complex. This reversible oxidation state switch can be followed by proton NMR spectroscopy, as well as by CEST spectra or CEST imaging.

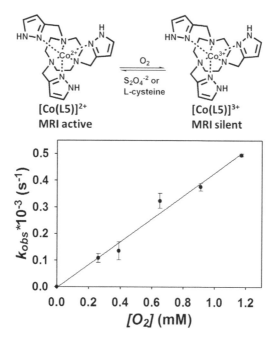

Figure 12.7 A redox-activated paraCEST agent that cycles between Co(II) and Co(III) oxidation states. A plot of first-order rate constants for the oxidation of $[Co(L5)]^{2+}$ versus oxygen concentration. Reprinted with permission from Ref. [18], Copyright 2013, John Wiley and Sons.

Interestingly, the rate constants for oxidation of the Co(II) complex predict an approximate half-life of 2.6 h in arterial blood with a pO_2 of 100 mm Hg, but 26 h under conditions similar to hypoxic tissue at pO_2 of 10 mm Hg (Fig. 12.7).

One feature of the Co(II)/Co(III) couple that may be useful in the application of these complexes as hypoxia probes is that the rate constants for electron self-exchange are tunable over a range of 10^6 to 10^{16} s^{-1}. The rate constants for redox depend on the flexibility of the ligand in adjusting to the two different-sized metal ions in the divalent versus trivalent oxidation states. One could imagine that the kinetics of redox could be optimized for the length of time that the contrast agent circulates and the region of interest for imaging.

The development of redox-activated transition metal paraCEST agents that undergo redox at the metal ion center adds a new dimension to responsive MRI contrast agents [39]. Currently, there are multiple examples of Gd(III) contrast agents that undergo redox on a ligand appendage with a change in relaxivity, but these organic groups are arguably less easily tunable, either kinetically or thermodynamically, than metal ions [40, 41]. The Mn(II)/Mn(III) couple has been reported as a redox probe [42–44], but both oxidation states produce T_1 contrast agents, making it more difficult to distinguish changes in signal due to metal ion oxidation state switches. Recently, MRI contrast agents based on Eu(II) complexes have been reported as effective T_1 agents [45, 46]. These complexes may be switched upon oxidation to give Eu(III) paraCEST agents.

12.4.3 Temperature-Responsive Agents

Paramagnetic complexes are useful as temperature sensors, given the strong dependence of the hyperfine shifted proton resonances on temperature. The dipolar contribution to the hyperfine shift has an inverse T^{-2} relationship to temperature, whereas the contact shift is T^{-1} [6]. Yet over a relatively small temperature range, the hyperfine shifted proton resonances may demonstrate a linear change with temperature [47–50]. Thus, the chemical shift of the exchangeable proton on the paraCEST agent correlates with temperature, and the corresponding CEST peak position is also a function of temperature. A linear plot of the CEST peak position as a function of temperature gives a slope with a temperature coefficient or CT value. This so-called paraCEST thermometry has been used to map temperature in phantoms and in tissue by using Ln(III) complexes [48, 51–53].

Transition metal paraCEST agents also show promise for thermometry. Shown in Fig. 12.8 is the CEST spectrum of Co(L5) as a function of temperature. An increase in temperature leads to a downfield shift in the CEST peak, due to the temperature dependence of the hyperfine shift. The approximate CT coefficient for Co(L5) is -0.70 ppm/°C. An increase in the intensity of the CEST peak is observed, as anticipated for an increase in the rate constant with temperature.

An interesting enhancement in the temperature-dependent CEST peak shift is obtained by using complexes that change spin state

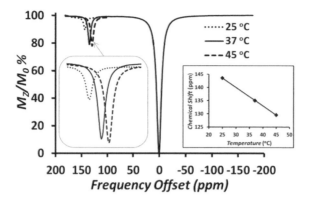

Figure 12.8 CEST spectrum (11.7 T) of Co(L5) at pH 7.0 and three temperatures. Inset: plot of the CEST peak position as a function of temperature.

upon change in temperature [14]. A Fe(II) complex of a tripodal ligand with pyridine pendents and a bound water is a paraCEST agent with an enhanced CT value. This spin crossover complex goes from diamagnetic ($S = 0$) to paramagnetic ($S = 2$) upon heating the complex over the temperature range of 20–60 °C. The exchangeable water on this complex presumably gives rise to the CEST effect. This new paraCEST thermometer has a CT of 1.0 ppm/°C, the highest known to date.

Related to paraCEST agents are paraSHIFT agents for mapping temperature [47–50]. The hyperfine shifted proton resonances in these paramagnetic complexes are monitored as a function of temperature by magnetic resonance spectroscopy (MRS). In this method, multiple spectra are taken over each pixel. Given that there are multiple proton resonances, procedures that are free of the dependence of the concentration of agent have been developed by taking the ratio of the temperature dependence of two peaks [48]. The significant temperature dependence of proton resonances in paraSHIFT examples of transition metal complexes is also promising [18, 24]. In order to design highly temperature-dependent paraSHIFT or paraCEST agents, the protons should have large hyperfine shifts and a large number of equivalent protons. For paraSHIFT agents as well as paraCEST agents, the proton resonances should be relatively narrow to maximize signal-to-noise ratios.

12.5 Toward In Vivo Studies

To date, fewer in vivo studies of paraCEST contrast agents are reported than desired. There are likely several reasons for this, including the relatively high concentrations of the agent required for contrast especially if there is interference from magnetization transfer (MT) effects that arise in tissue and from T_2^* water exchange effects that interfere with CEST imaging [3]. One way to decrease the MT interference is by obtaining complexes that have CEST peaks that are >100 ppm from bulk water. In general, this has been challenging without using the highly paramagnetic Ln(III) such as Dy(III), which also broaden proton resonances [3]. Recently, a series of Tm(III) complexes containing amide NH protons that are highly shifted (>150 ppm) have been reported [53]. Other Tm(III) agents with amide protons have shown CEST contrast in mice thigh muscle [54]. Other approaches involve using special MRI methods to overcome the T_2 shortening caused by Eu(III) paraCEST agents [55] or to develop Tm(III) complexes that have limited inner sphere water T_2 exchange [56]. Additional methods use a WALTZ-16 pulse to generate on-resonance paramagnetic chemical exchange effects (OPARACHEE) for Ln(III) complexes that have relatively rapid rates of water exchange [57, 58].

As the first step toward in vivo studies, several of the complexes in Scheme 12.1 have been studied as paraCEST agents in blood serum. Incubation of the Co(II) complex of L5 in rabbit serum under nitrogen to prevent oxidation gives a CEST peak that is undiminished after 48 h [18]. Studies of Fe(II), Co(II), or Ni(II) complexes of L8 in blood serum show that the reactivity of the complexes changes across the series. The seven-coordinate Fe(II) shows unchanged CEST signal in the presence of the serum (Fig. 12.5 (left)) as does the Co(II) complex. In contrast, the Ni(L8) complex shows a diminished CEST peak, a decreased magnetic moment, and an increase in relaxivity, consistent with the loss of the Ni(II) ion from the ligand. A similar effect is observed for the Ni(L8) complex in solutions of albumin, suggesting reaction of the complex with this blood protein [22].

Agarose was used to mimic the MT effect in tissue. CEST spectra in agarose have a broad band centered about the water peak. This broad band envelopes the Co(L8) CEST peak and partially that of Ni(L8). Fe(L8) alone has a CEST peak shifted sufficiently to experience less interference from the MT band (Fig. 12.5 (left)). The further shifted the CEST peak, the less interference there is from the MT effect [22]. CEST images of Co(L8) showed decreased contrast in agarose on a 4.7 T MRI scanner (Fig. 12.5 (right)).

12.6 Summary

Transition metal ion paraCEST agents are a new class of compounds that show great promise as MRI contrast agents. The versatile coordination chemistry of transition metal ions allows for the development of new pendent groups with exchangeable protons that have not been used in the past for paraCEST agents. Furthermore, the existence of multiple oxidation and spin states for transition metal ions makes them uniquely suitable for their development as responsive contrast agents to map changes in redox status or in temperature. There are many potentially rewarding avenues of research for the future, including incorporation of analogs into nanoparticles and liposomes as well as additional work to develop successful agents for in vivo studies.

Acknowledgments

We thank the Bruce Holm Catalyst Fund, the NIH (CA-173309) and NSF (CHE-1310374) for support.

References

1. Lauffer RB. Paramagnetic metal complexes as water proton relaxation agents for NMR imaging: Theory and design. *Chem. Rev.*, 1987; 87(5): 901–927.
2. Zhang S, Merritt M, Woessner DE, Lenkinski RE, and Sherry AD. PARACEST agents: Modulating MRI contrast via water proton exchange. *Acc. Chem. Res.*, 2003; 36(10): 783–790.

3. Viswanathan S, Kovacs Z, Green KN, Ratnakar SJ, and Sherry AD. Alternatives to gadolinium-based metal chelates for magnetic resonance imaging. *Chem. Rev.*, 2010; 110(5): 2960–3018.
4. Terreno E, Delli Castelli D, and Aime S. Encoding the frequency dependence in MRI contrast media: The emerging class of CEST agents. *Contrast Media Mol. Imaging*, 2010; 5(2): 78–98.
5. van Zijl PCM and Yadav NN. Chemical exchange saturation transfer (CEST): What is in a name and what isn't? *Magn. Reson. Med.*, 2011; 65(4): 927–948.
6. Bertini I and Luchinat C. *NMR of Paramagnetic Molecules in Biological Systems.* Menlo Park: The Benjamin/Cummings Publishing Company, Inc.; 1986.
7. Bertini I, Luchinat C, Parigi G, and Pierattelli R. NMR spectroscopy of paramagnetic metalloproteins. *ChemBioChem*, 2005; 6(9): 1536–1549.
8. Geraldes CFGC and Luchinat C. Lanthanides as shift and relaxation agents in elucidating the structure of proteins. In: Sigel A, Sigel H, eds., *Metal Ions in Biological Systems. Lanthanides and Their Interrelations with Biosystems*, Vol 40. New York: Marcel Dekker Inc.; 2003: 513–588.
9. Bertini I, Turano P, and Vila AJ. Nuclear magnetic resonance of paramagnetic metalloproteins. *Chem. Rev.*, 1993; 93(8): 2833–2932.
10. Christiansen L, Hendrickson DN, Toftlund H, Wilson SR, and Xie CL. Synthesis and structure of metal-complexes of triaza macrocycles with three pendant pyridylmethyl arms. *Inorg. Chem.*, 1986; 25(16): 2813–2818.
11. Dorazio SJ, Tsitovich PB, Gardina SA, and Morrow JR. The reactivity of macrocyclic Fe(II) paraCEST MRI contrast agents towards biologically relevant anions, cations, oxygen or peroxide. *J. Inorg. Biochem.*, 2012; 117: 212–219.
12. Dorazio SJ, Tsitovich PB, Siters KE, Spernyak JA, and Morrow JR. Iron(II) PARACEST MRI contrast agents. *J. Am. Chem. Soc.*, 2011; 133(36): 14154–14156.
13. Gütlich P, Hauser A, and Spiering HG. Thermal and optical switching of iron(II) complexes. *Angew. Chem. Int. Ed.*, 1994; 33: 2024–2054.
14. Jeon I-R, Park JG, Haney CR, and Harris TD. Spin crossover iron(II) complexes as PARACEST MRI thermometers. *Chem. Sci.*, 2014; 5(6): 2461–2465.
15. Bernhardt PV, Chen K-I, and Sharpe PC. Transition metal complexes as mediator-titrants in protein redox potentiometry. *J. Biol. Inorg. Chem.*, 2006; 11(7): 930–936.

16. Kumar K, Rotzinger FP, and Endicott JF. Oxidation-reduction reactions with macrocyclic ligands. Dependence of the rate advantage for the inner-sphere electron-transfer pathway on electronic structure for low-spin cobalt(III)-(II), nickel(III)-(II), and copper(III)-(II) couples. *J. Am. Chem. Soc.*, 1983; 105: 7064–7074.

17. Dorazio SJ, Olatunde AO, Tsitovich PB, and Morrow JR. Comparison of divalent transition metal ion paraCEST MRI contrast agents. *J. Biol. Inorg. Chem.*, 2014; 19(2): 191–205.

18. Tsitovich PB, Spernyak JA, and Morrow JR. A redox-activated MRI contrast agent that switches between paramagnetic and diamagnetic states. *Angew. Chem. Int. Ed.*, 2013; 52(52): 13997–14000.

19. Theil EC and Goss DJ. Living with iron (and oxygen): Questions and answers about iron homeostasis. *Chem. Rev.*, 2009; 109(10): 4568–4579.

20. Shannon RD. Revised effective ionic radii and systematic studies of interatomic distances in halides and chalcogenides. *Acta Cryst.*, 1976; A32: 751–767.

21. Hancock RD and Martell AE. Ligand design for selective complexation of metal ions in aqueous solution. *Chem. Rev.*, 1989; 89(8): 1875–1914.

22. Olatunde AO, Cox JM, Daddario MD, Spernyak JA, Benedict JB, and Morrow JR. Seven-coordinate CoII, FeII and six-coordinate NiII amide-appended macrocyclic complexes as paraCEST agents in biological media. *Inorg. Chem.*, 2014; 53: 8311–8321.

23. Olatunde AO, Bond CJ, Dorazio SJ, et al. Six, seven or eight coordinate FeII, CoII or NiII complexes of amide-appended tetraazamacrocycles for paraCEST thermometry. *Chem. Eur. J.*, 2015; 21(50): 18290–18300.

24. Tsitovich PB, Cox JM, Benedict JB, and Morrow JR. Six-coordinate Iron(II) and Cobalt(II) paraSHIFT agents for measuring temperature by magnetic resonance spectroscopy. *Inorg. Chem.*, 2016; 55(2): 700–716.

25. Dorazio SJ and Morrow JR. Iron(II) complexes containing octadentate tetraazamacrocycles as paraCEST magnetic resonance imaging contrast agents. *Inorg. Chem.*, 2012; 51(14): 7448–7450.

26. Olatunde AO, Dorazio SJ, Spernyak JA, and Morrow JR. The NiCEST approach: Nickel(II) paraCEST MRI contrast agents. *J. Am. Chem. Soc.*, 2012; 134: 18503–18505.

27. Dorazio SJ, Olatunde AO, Spernyak JA, and Morrow JR. CoCEST: Cobalt(II) amide-appended paraCEST MRI contrast agents. *Chem. Commun.*, 2013; 49: 10025–10027.

28. Tsitovich PB and Morrow JR. Macrocyclic ligands for Fe(II) paraCEST and chemical shift MRI contrast agents. *Inorg. Chim. Acta.*, 2012; 393: 3–11.

29. Zhang S, Michaudet L, Burgess S, and Sherry AD. The amide protons of an ytterbium(III) dota tetraamide complex act as efficient antennae for transfer of magnetization to bulk water. *Angew. Chem. Int. Ed.*, 2002; 41(11): 1919–1921.

30. Woessner DE, Zhang S, Merritt ME, and Sherry AD. Numerical solution of the Bloch equations provides insights into the optimum design of PARACEST agents for MRI. *Magn. Reson. Med.*, 2005; 53(4): 790–799.

31. Sherry AD and Wu Y. The importance of water exchange rates in the design of responsive agents for MRI. *Curr. Opin. Chem. Biol.*, 2013; 17(2): 167–174.

32. Shah T, Lu L, Dell KM, Pagel MD, Griswold MA, and Flask CA. CEST-FISP: A novel technique for rapid chemical exchange saturation transfer MRI at 7 T. *Magn. Reson. Med.*, 2011; 65(2): 432–437.

33. Aime S, Barge A, Delli Castelli D, et al. Paramagnetic lanthanide(III) complexes as pH-sensitive chemical exchange saturation transfer (CEST) contrast agents for MRI applications. *Magn. Reson. Med.*, 2002; 47(4): 639–648.

34. Sheth VR, Liu G, Li Y, and Pagel MD. Improved pH measurements with a single PARACEST MRI contrast agent. *Contrast Media Mol. Imaging*, 2012; 7(1): 26–34.

35. Opina AC, Wu Y, Zhao P, Kiefer G, and Sherry AD. The pH sensitivity of –NH exchange in LnDOTA-tetraamide complexes varies with amide substituent. *Contrast Media Mol. Imaging*, 2011; 6(6): 459–464.

36. Delli Castelli D, Terreno E, and Aime S. Yb^{III}-HPDO3A: A dual pH- and temperature-responsive CEST agent. *Angew. Chem. Int. Ed.*, 2011; 50(8): 1798–1800.

37. Banerjee R. Redox outside the box: Linking extracellular redox remodeling with intracellular redox metabolism. *J. Biol. Chem.*, 2012; 287(7): 4397–4402.

38. Khramtsov VV and Gillies RJ. Janus-faced tumor microenvironment and redox. *Antioxid. Redox Signal.*, 2014; 21(5): 723–729.

39. Tsitovich PB, Burns PJ, McKay AM, and Morrow JR. Redox-activated MRI contrast agents based on lanthanide and transition metal ions. *J. Inorg. Biochem.*, 2014; 133: 143–154.

40. Do QN, Ratnakar JS, Kovacs Z, and Sherry AD. Redox- and hypoxia-responsive MRI contrast agents. *ChemMedChem*, 2014; 9(6): 1116–1129.
41. Tu C and Louie AY. Strategies for the development of gadolinium-based 'q'-activatable MRI contrast agents. *NMR Biomed.*, 2013; 26(7): 781–787.
42. Loving GS, Mukherjee S, and Caravan P. Redox-activated manganese-based MR contrast agent. *J. Am. Chem. Soc.*, 2013; 135(12): 4620–4623.
43. Aime S, Botta M, Gianolio E, and Terreno E. A p(O(2))-responsive MRI contrast agent based on the redox switch of manganese(II/III)-porphyrin complexes. *Angew. Chem. Int. Ed.*, 2000; 39(4): 747–750.
44. Rolla GA, Tei L, Fekete M, Arena F, Gianolio E, and Botta M. Responsive Mn(II) complexes for potential applications in diagnostic magnetic resonance imaging. *Bioorg. Med. Chem.*, 2011; 19(3): 1115–1122.
45. Garcia J and Allen MJ. Interaction of biphenyl-functionalized Eu^{2+}-containing cryptate with albumin: Implications to contrast agents in magnetic resonance imaging. *Inorg. Chim. Acta.*, 2012; 393: 324–327.
46. Garcia J and Allen MJ. Developments in the coordination chemistry of europium(II). *Eur. J. Inorg. Chem.*, 2012; 2012(29): 4550–4563.
47. Aime S, Botta M, Fasano M, et al. A new ytterbium chelate as contrast agent in chemical shift imaging and temperature sensitive probe for MR spectroscopy. *Magn. Reson. Med.*, 1996; 35(5): 648–651.
48. Coman D, Kiefer GE, Rothman DL, Sherry AD, and Hyder F. A lanthanide complex with dual biosensing properties: CEST (chemical exchange saturation transfer) and BIRDS (biosensor imaging of redundant deviation in shifts) with europium DOTA-tetraglycinate. *NMR Biomed.*, 2011; 24(10): 1216–1225.
49. James JR, Gao Y, Miller MA, Babsky A, and Bansal N. Absolute temperature MR imaging with thulium 1,4,7,10-tetraazacyclododecane-1,4,7,10-tetramethyl-1,4,7,10-tetraacetic acid (TmDOTMA-). *Magn. Reson. Med.*, 2009; 62(2): 550–556.
50. Harvey P, Blamire AM, Wilson JI, et al. Moving the goal posts: Enhancing the sensitivity of PARASHIFT proton magnetic resonance imaging and spectroscopy. *Chem. Sci.*, 2013; 4(11): 4251–4258.
51. Zhang S, Malloy CR, and Sherry AD. MRI thermometry based on PARACEST agents. *J. Am. Chem. Soc.*, 2005; 127(50): 17572–17573.

52. Li AX, Wojciechowski F, Suchy M, et al. A sensitive PARACEST contrast agent for temperature MRI: Eu^{3+}-DOTAM-glycine (Gly)-phenylalanine (Phe). *Magn. Reson. Med.*, 2008; 59(2): 374–381.
53. Stevens TK, Milne M, Elmehriki AA, Suchy M, Bartha R, and Hudson RH. A DOTAM-based paraCEST agent favoring TSAP geometry for enhanced amide proton chemical shift dispersion and temperature sensitivity. *Contrast Media Mol. Imaging*, 2013; 8(3): 289–292.
54. McVicar N, Li AX, Suchy M, Hudson RHE, Menon RS, and Bartha R. Simultaneous in vivo pH and temperature mapping using PARACEST-MRI contrast agent. *Magn. Reson. Med.*, 2013; 70(4): 1016–1025.
55. Soesbe TC, Togao O, Takahashi M, and Sherry AD. SWIFT-CEST: A new MRI method to overcome T2 shortening caused by PARACEST contrast agents. *Magn. Reson. Med.*, 2012; 68(3): 816–821.
56. Milne M, Lewis M, McVicar N, Suchy M, Bartha R, and Hudson RHE. MRI paraCEST agents that improve amide based pH measurements by limiting inner sphere water T2 exchange. *RSC Adv.*, 2014; 4(4): 1666–1674.
57. Vinogradov E, Zhang S, Lubag A, Balschi JA, Sherry AD, and Lenkinski RE. On-resonance low B1 pulses for imaging of the effects of PARACEST agents. *J. Magn. Reson.*, 2005; 176(1): 54–63.
58. Jones CK, Li AX, Suchy M, Hudson RH, Menon RS, and Bartha R. In vivo detection of PARACEST agents with relaxation correction. *Magn. Reson. Med.*, 2010; 63(5): 1184–1192.

Chapter 13

Responsive paraCEST MRI Contrast Agents and Their Biomedical Applications

Iman Daryaei and Mark D. Pagel

Department of Chemistry and Biochemistry, Department of Medical Imaging, University of Arizona Tucson, Arizona 85724, USA
mpagel@u.arizona.edu

13.1 Introduction

Paramagnetic chemical exchange saturation transfer (paraCEST) MRI contrast agents typically consist of a paramagnetic metal ion and an organic chelate to form a metal chelator. ParaCEST agents have typically used a macrocyclic tetraazacyclododecane as the base structure due to the high kinetic stabilities of these metal chelates. The ligands of the metal chelates can include amide, amine, or hydroxyl groups that slowly exchange protons with bulk water. In addition, a water molecule may be tightly bound to the metal chelate, causing the chemical exchange of the water molecule with bulk solvent to be slow. Thus, this bound water molecule is considered a part of the structure of the paraCEST agent. Selective radio

---

Chemical Exchange Saturation Transfer Imaging: Advances and Applications
Edited by Michael T. McMahon, Assaf A. Gilad, Jeff W. M. Bulte, and Peter C. M. van Zijl
Copyright © 2017 Pan Stanford Publishing Pte. Ltd.
ISBN 978-981-4745-70-3 (Hardcover), 978-1-315-36442-1 (eBook)
www.panstanford.com

frequency (RF) saturation of a slowly exchanging proton in a ligand or tightly bound water, followed by exchange with bulk water, can generate the CEST signal.

The exchangeable protons of paraCEST agents can have exceptionally large chemical shifts (MR frequency relative to the MR frequency of water) due to the hyperfine shift of the proximal metal ion. This large range of chemical shifts improves the specificity of RF saturation of the intended proton relative to inadvertently saturating the bulk water. The chemical shift is highly dependent on the bond properties of the chemical group, as well as the location of the proton with respect to the atomic orbitals of the metal ion. Molecular interactions that change the chemical group of the paraCEST agent can cause changes in the MR chemical shift of the proton. Similarly, molecular or environmental changes can change the conformation of the agent, which can alter the location of the proton relative to the metal ion and change the proton's chemical shift. This change in chemical shift can be measured by recording the shift in saturation frequency that generates the greatest CEST effect from the paraCEST agent. Therefore, paraCEST agents can detect molecular biomarkers such as enzyme activities, nucleic acids, metabolites, and ions and can also detect environmental biomarkers such as pH, redox state, and temperature.

The exchangeable protons of paraCEST agents can also have a wide range of chemical exchange rates, typically about 10^2–10^4 Hz. The amplitude of the CEST signal recorded in a CEST spectrum (CEST amplitude) increases with increasing chemical exchange rate. A larger chemical shift permits the paraCEST agent to have a higher chemical exchange rate, because the chemical shift should be higher than the exchange rate to avoid MR coalescence of the MR resonance of the agent's exchangeable proton with the MR resonance of bulk water. The chemical exchange rate is dependent on the type of chemical group on the agent, the electronegativities of other chemical groups of the agent, and access of the exchangeable proton to bulk water. Each of these characteristics can be altered by molecular biomarkers, such as enzymes that catalyze a change in chemical group with the exchangeable proton or other chemical groups that change the electronegativity of the agent. Environmental biomarkers such as pH and temperature can inherently alter the chemical exchange rate of the paraCEST agent. Therefore, monitoring the CEST amplitude can be used to detect molecular and environmental biomarkers.

As a profound advantage for molecular imaging with MRI, the large range of chemical shifts of paraCEST agents greatly facilitates the detection of two CEST effects from the same agent. If one CEST effect is responsive to a biomarker while the other CEST effect is an unresponsive "control" signal, then the ratiometric comparison of both CEST effects can improve the specificity for detecting the intended biomarker. Most importantly, each CEST effect is dependent on the concentration of the agent, but the ratio of the two CEST effects is concentration independent. Similarly, both CEST effects may be dependent on the T_1 or T_2 relaxation time constants of a chemical sample or in vivo tissue, but the ratio of the two CEST effects is largely or entirely independent of T_1 and T_2 relaxation times. The concentration of bulk water that is accessible to the agent within in vivo tissues can also influence each CEST effect, but the bulk water concentration does not alter the ratio of these CEST effects. This ratiometric approach with a responsive and unresponsive "control" signal has been exploited to quantitatively measure enzyme activity and to improve the detection of metabolites and pH.

Unfortunately, paraCEST agents have limited detection sensitivity and require concentrations in the single-millimolar range or higher for adequate detection [1]. To boost sensitivity, many paraCEST agents have been packaged in a liposome or an adenovirus, and many paraCEST agents have been covalently conjugated to a dendrimer or protein carrier. However, this "packaging" of many paraCEST agents can hinder interactions or responsiveness to molecular and environmental biomarkers. As an alternative solution, a larger voxel size can compensate for poor detection sensitivity, but at the expense of creating a coarse spatial resolution. Creating coarse spatial resolution compromises one of the great strengths of in vivo MRI relative to other imaging modalities. As a partial solution to this problem, some paraCEST agents have been designed to irreversibly alter their CEST effect after interacting with a biomarker [2]. This strategy of an irreversibly responsive paraCEST agent allows a population of altered agents to accumulate, so that the sensitivity is dictated by the agent's concentration rather than the concentration of the biomarker. The accumulation of a stable, high concentration of altered agent can then be detected using CEST MRI methods which overcome the limited sensitivity of paraCEST agents. In effect, the sensitivity becomes "contrast agent-limited" rather than "molecular target-limited."

To date, responsive paraCEST agents have been designed to detect enzyme activities, proteins, nucleic acids, metabolites, and ions, and to quantitatively measure redox state, pH, and temperature. The following summaries describe paraCEST agents that detect each of these molecular and environmental biomarkers. An emphasis is placed on describing the detection mechanism (change in chemical shift versus change in chemical exchange rate), the use of a ratiometric approach to improve the biomarker detection, issues regarding reversible versus irreversible agents, and demonstrations in solutio and in vivo.

13.2 ParaCEST Agents That Detect Enzyme Activities

The direct detection of a protein with MRI contrast agents has not been successfully accomplished with paraCEST agents. For comparison, T_1 and T_2^* MRI contrast agents have been used to detect highly abundant proteins such as collagen and fibrinogen [3–6], artificially overexpressed cell receptors, and receptors with high cell membrane turnover. Yet these relaxation-based agents cause less than a 20% change in image contrast during in vivo studies, requiring major efforts to generate reliable MRI results for even these most highly expressed protein systems. ParaCEST agents have even lower detection sensitivity relative to relaxation-based agents and, therefore, are a poor choice for directly detecting proteins.

Alternatively, many paraCEST agents have been developed that detect enzyme activity. The rapid catalysis of a single enzyme molecule can alter many paraCEST agent molecules in a practical timeframe, which can amplify the detection of the enzyme. The enzyme catalysis causes an irreversible change to the paraCEST agent that allows for the accumulation of a large concentration of agent and reduces the need for rapid temporal sampling,

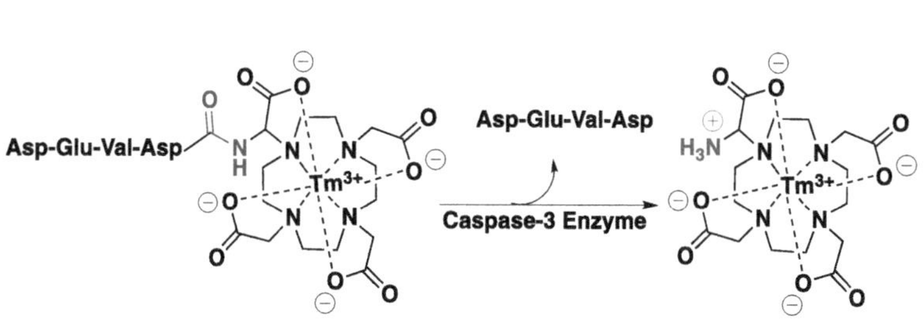

Figure 13.1 Detection of caspase-3 enzyme by a derivative of Tm-DOTA. Adapted from Refs. [2] and [7].

which further facilitates the enzyme detection. The ratiometric comparison of an irreversibly responsive CEST signal and an enzyme-unresponsive "control" signal has been used to improve the detection of enzyme activity. The catalytic efficiencies of enzymes and chemical exchange rates of the agents have been quantitatively measured, which shows that these agents can produce quantitative imaging results. Finally, a variety of paraCEST agents that have been developed show that this approach is a platform technology that can be employed to detect many types of enzymes.

The seminal example of an enzyme-responsive paraCEST agent consists of a Tm(III) macrocyclic chelate that is conjugated to the C-terminal end of a peptide (Fig. 13.1) [2, 7]. The caspase-3 protease was shown to efficiently cleave the C-terminal amino acid of the selected peptide sequence, which caused the amide group of this amino acid to be converted to an amine. This cleavage changed the chemical shift of the CEST effect from −51 ppm to +8 ppm and also accelerated the chemical exchange rate to create a lower CEST amplitude. Importantly, as little as 3.44 nM of caspase-3 enzyme catalyzed 5.2 mM of the agent that created a 5% CEST amplitude, showing that enzyme catalysis can increase the CEST amplitude to levels for practical detection. To improve this analysis, the CEST signals of an enzyme-responsive Tm(III) paraCEST agent were compared to CEST of a Yb(III) macrocyclic chelate that was unresponsive to caspase-3 enzyme activity. Finally, the Michaelis–Menten enzyme kinetics of this peptide cleavage were measured using CEST MRI. Surprisingly, the catalytic efficiency of this process was discovered to be greater than the efficiency of cleaving a

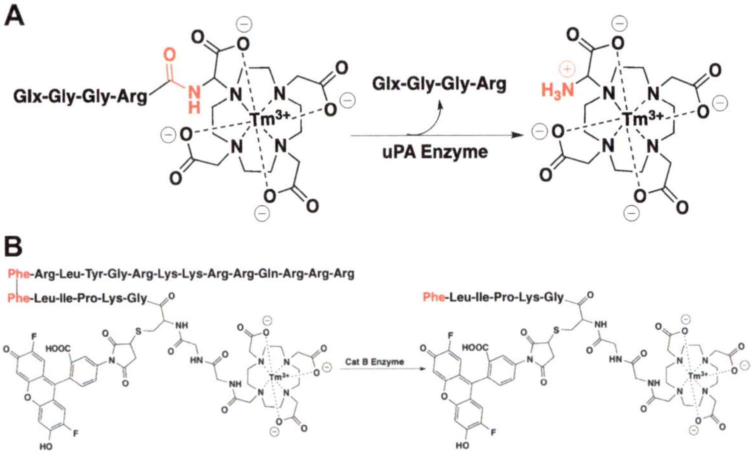

Figure 13.2 (A) Detection of uPA enzyme by Tm-DOTA conjugated to a peptide. Glx is an amino acid with either a glutamate or glutamine side chain [8]. (B) Detection of Cat D enzyme using a derivative of Tm-DO3A. The conjugated peptide contains a TAT sequence for cell uptake purposes and a peptide sequence that can be cleaved by the enzyme [9].

commercially available fluorescence agent, demonstrating that a metal chelate does not necessarily hinder enzyme catalysis.

This approach has also been used to detect the protease activity of cathepsin D and urokinase Plasminogen Activator (uPA), simply by changing the peptide sequence conjugated to a Tm(III) chelate to match a well-known peptide substrate for each agent (Fig. 13.2) [8, 9]. These additional paraCEST agents demonstrate that this approach is a platform technology that can be used to detect many protease enzymes. Only 3 mM of agent was needed to detect 0.5 nM of cathepsin D, using a variation of CEST MRI known as On-resonance PARAmagnetic CHemical Exchange Effect (OPARACHEE) MRI, which is described as follows. A concentration of 50 mM of agent was used to detect 4.1 nM of uPA by monitoring the disappearance of the CEST signal at −52 ppm, relative to an enzyme-unresponsive CEST signal from a Yb(III) chelate at −16 ppm. Furthermore, the uPA-sensitive paraCEST agent was used to detect the activity of this protease enzyme within in vivo tumor

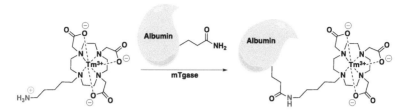

Figure 13.3 A derivative of Tm-DO3A that detects enzyme activity by the formation of a covalent bond. Reprinted with permission from Ref. [10], Copyright 2013, American Chemical Society.

tissues, which shows that this approach can affect in vivo biomedical diagnostics.

More recently, 25 mM of a paraCEST agent was shown to detect 327 nM of transglutaminase (TGase), an "anti-protease" that links an amine to the side chain of glutamine to create a new amide group (Fig. 13.3) [10]. The contrast agent consisted of a Tm(III) macrocyclic chelate with an aliphatic amine ligand, which was linked to the glutamine side chains in albumin to create an amide group through transglutaminase enzyme activity. This conversion of the amine to an amide caused the CEST signal to change from -12.7 ppm to -11.6 ppm, which was an almost negligible difference. A Hanes Wolff-QUantification of Exchange using Saturation Power (HW-QUESP) CEST analysis showed that the chemical exchange rate changed from 2533 Hz before TGase catalysis to 1260 Hz after catalysis. This large change demonstrated the value of quantitative imaging. This transglutaminase-responsive paraCEST agent should be compared with an enzyme-unresponsive CEST signal to ensure that the change in chemical exchange rate is due to transglutaminase activity and not due to other changes such as pH or temperature.

This platform technology has been extended by using ligands that are cleaved by other enzymes besides proteases. Importantly, these paraCEST agents use a ligand that separates the macrocyclic chelate from the enzyme-active functional group by using a spontaneously disassembling linker (Fig. 13.4). This separation creates a modular approach where the functional group can be easily modified to detect new types of enzymes without affecting the macrocyclic chelate. As a first example, 20 mM of a ligand with a galactose

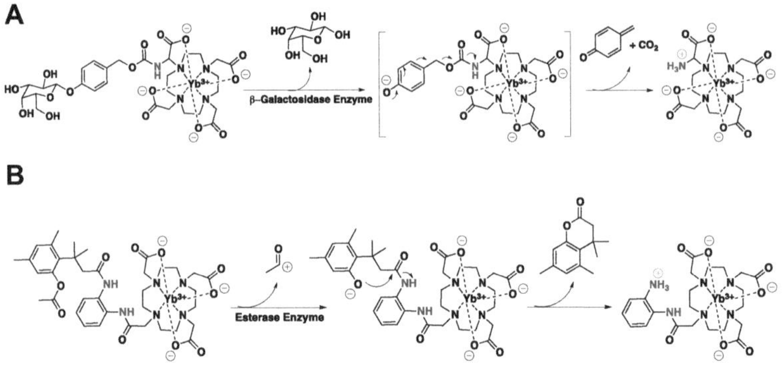

Figure 13.4 Linker with self-disassembly properties upon activation can be used to detect enzyme activity. (A) Detection of β-galactosidase enzyme by a derivative of Yb-DOTA [11, 12]. (B) Detection of esterase enzyme activity with Yb-DO3A-oAA. The ester group is cleaved by esterase enzyme to create an intermediate compound with a very fast rate for an intramolecular reaction to release the contrast agent [13].

sugar can be cleaved by approximately 200 µM β-galactosidase, which causes spontaneous cleavage of a benzyloxycarbamate spacer, which in turn converts an amide to an amine that is proximal to a Yb(III) macrocyclic chelate [11, 12]. The new amine can generate CEST signals at −16.7 and −20.5 ppm. As a second example, an ester group of a Yb(III) macrocyclic chelate at 25 mM concentration can be cleaved by 617 nM of esterase, causing an intermolecular lactonization that releases the ligand from the chelate, which causes an amide group on the chelate to covert to an amine [13]. This conversion creates a new CEST signal at +9 ppm due to the proximity of the amine to the Yb(III) ion. This CEST signal was compared with a second enzyme-unresponsive CEST signal at −10 ppm from the same agent, which further improved the detection of the esterase enzyme activity.

13.3 ParaCEST Agents That Detect Nucleic Acids

The direct detection of deoxyribonucleic acid is difficult to achieve with paraCEST agents due to the poor sensitivity of CEST MRI and the relatively low concentration of nucleic acids within in vivo

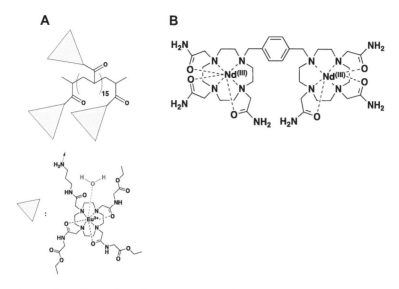

Figure 13.5 Examples of contrast agents that detect nucleic acids. (A) A polymeric contrast agent with 17 contrast agent units in each chain [14]. (B) A dinuclear complex of Nd for the detection of specific types of DNA hairpins [15].

systems. Yet paraCEST agents may possibly detect exceptional cases of high concentrations of nucleic acids, such as the delivery of high payloads of gene therapies (Fig. 13.5). For example, a polymeric Eu(III) chelate reduced its CEST amplitude by ~33% after binding to 600 nM of DNA from salmon testes, although the chemical shift remained invariant at 52 ppm [14]. Also a dimeric Nd(III) macrocyclic chelate increased its CEST amplitude by ~15% after binding to 0.5 mM of a DNA hairpin, while the chemical shift at 12 ppm did not change [15]. Therefore, the interactions of each agent with DNA altered the chemical exchange rate of the bound water and amide protons of each agent, respectively.

Further developments of these studies could improve the detection of nucleic acids with paraCEST agents. For example, the CEST amplitudes of both DNA-responsive agents are dependent on pH, which could be ratiometrically compared to a pH-responsive, DNA-unresponsive "control" agent to improve DNA detection. In addition, the quantitative evaluation of chemical exchange rates may improve the quantitative detection of specific DNA sequences. However, both of these agents directly bind to DNA and, therefore, are limited to detecting high DNA concentrations. Furthermore, both agents reversibly bind to DNA, so that rapid changes in DNA concentration cannot be accurately monitored with these paraCEST agents. Due to these limitations, paraCEST agents have not yet detected nucleic acid oligomers during in vivo studies.

13.4 ParaCEST Agents That Detect Metabolites

Metabolites can often be present at concentrations greater than 1 mM, which provides great opportunities for detection with paraCEST agents at similar concentrations (Fig. 13.6). For example, 5 mM of glucose has been detected with a similar concentration of a Eu(III) chelate that has phenylbornate ligands [16–18]. The 1:1 binding of glucose to the agent slowed the chemical exchange of the bound water, causing the peak at 50 ppm in the CEST spectrum to become broader, which was used to quantitatively measure a 383 M

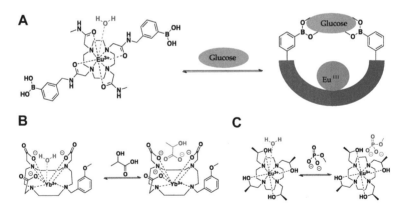

Figure 13.6 (A) Phenylbornate ligands have high affinity to glucose. The interaction of phenylbornate ligands with glucose blocks the interaction of water molecules with the Eu-complex, causing a change in CEST of this agent [16–18]. (B) The interaction of lactate with a Yb-complex blocks the accessibility of water molecules to Yb^{III} [19]. (C) A methyl phosphate interacts with Eu^{III}-complex in the same fashion as previous examples that hinders accessibility of water to the complex [20–22].

association constant between glucose and the agent. Similarly, 5 mM of lactate has been detected with a heptadentate Yb(III) chelate. The binding of lactate to the agent caused a peak in the CEST spectrum to shift from −29.1 ppm to −15.5 ppm, which was used to determine a 8000 M association constant between lactate and the agent [19]. As another example, the pendant alcohol groups of a Eu(III) chelate generated CEST at 6 ppm, which then shifted to 8 ppm upon binding to a methylphosphate metabolite at 5 mM concentration [20–22]. This change in shift was used to estimate an association constant of 100 M when the methylphosphate replaced the bound water in the complex of the agent.

As additional advantages, these three examples showed that evaluating changes in chemical shift or a ratio of two CEST amplitudes at different chemical shifts can quantitatively monitor metabolite concentrations within ∼10% accuracy. Furthermore, each example exploits reversible binding of the metabolite to the agent, which may be appropriate for these metabolites that have high, stable concentrations. As a disadvantage, good specificity for detecting the intended metabolite is difficult to achieve for each of these agents. The glucose-binding agent was also shown to bind to fructose, and the methylphosphate-binding agent also bound to diethylphosphate. The lactate-binding agent relied on the affinity of this agent for carboxylate-containing metabolites and, therefore, is also likely to bind to other metabolites in addition to lactate. This detection specificity is an inherent problem with metabolites that have very similar structures, which may possibly contribute to the lack of in vivo demonstrations of this class of agent.

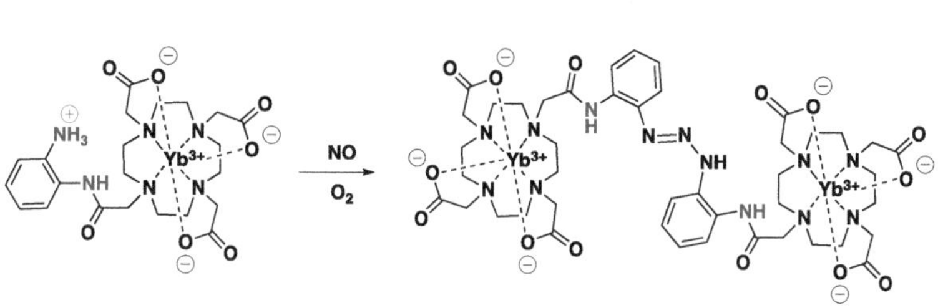

Figure 13.7 Dimerization of two Yb-DO3A-oAA molecules through a nitrogen bridge diminishes CEST signals from Yb-DO3-oAA [23].

As an alternative approach, an irreversibly responsive paraCEST agent that detects nitric oxide has been developed (Fig. 13.7), which is an important metabolite in many pathologies, including cancer [23]. A Yb(III) chelate showed 7% and 4% CEST signals at 8 and −11 ppm, respectively, which were assigned to the amide and amine of the agent's ortho-aminoanilide ligand. This agent experienced an irreversible alteration to its covalent structure after treatment with NO and O_2, causing both CEST effects to disappear. The CEST effect of an NO-unreponsive Tm(III) chelate was measured at −51 ppm and compared with the CEST effects of the Yb(III) chelate to improve the detection methodology. Using an irreversible paraCEST agent was critical for detecting NO because this metabolite has a fleeting 0.1–5 s lifespan in vivo and, therefore, cannot be transiently detected using relatively slow CEST MRI acquisition methods. This irreversibly responsive agent could accumulate a sufficient 5 mM concentration for detection over time, even though NO was not present at this same concentration at any instant (a sum of 5 mM of NO was present over 1 h). As with reversibly responsive agents, this irreversibly responsive agent has not yet been used during in vivo studies.

13.5 ParaCEST Agents That Detect Ions

The development of paraCEST agents that detect ions has been a natural result of the strong understanding of metal chelation among MRI contrast agent researchers (Fig. 13.8). For example, a Eu(III) macrocyclic chelate with pyridine ligands can chelate Zn^{2+} [24]. The Zn^{2+} ion is coordinated to a hydroxide ion that catalyzes the exchange of protons between the Eu(III)-bound water molecule and the bulk solvent. This acceleration of the chemical exchange of the bound water caused a loss of CEST amplitude with a Zn^{2+} concentration as low as 2 mM. The chemical shift of this paraCEST agent at 50 ppm did not change when Zn^{2+} was chelated by the paraCEST agent. As another example, the coordination of bis-carboxylate ligands to Ca^{2+} caused an electronic redistribution in the Yb(III) macrocyclic chelate, causing a 10-fold slower chemical exchange from the amide groups of the agent, resulting in a 60% loss of CEST at 200 mM

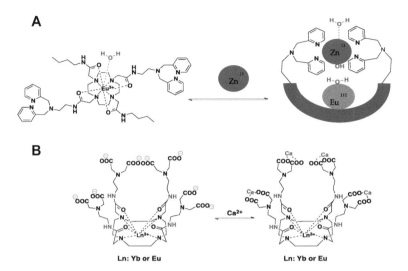

Figure 13.8 (A) Interaction of Zn^{2+} with side arms of the Eu-complex causes a change in CEST signal [24] (B) Ca^{2+} can interact with the carboxylate arms of a derivative of Yb-DOTA or Eu-DOTA complexes [25].

Ca^{2+} concentration [25] Again, the chemical shift of -13 ppm did not change when Ca^{2+} was added to this paraCEST agent.

These examples demonstrate advantages and disadvantages when using paraCEST agents to detect ions. As an advantage, the design of these agents may represent a platform technology, where the types of ligands may be changed to chelate other ions. Also the in vivo concentration of some ions can be high, which reduces concern about the inherent insensitivity of paraCEST agents. Both of these paraCEST agents contain multiple labile protons that can further boost the CEST effect on a per-ion basis. As a disadvantage, both these paraCEST agents change their CEST signal in a reversible manner, while ion flux can be more rapid than a CEST MRI acquisition protocol, which causes the CEST signal to reflect a non-linear weighted average of ion concentration during the CEST MRI study. Moreover, some ligands may have low specificity for chelating the intended ion relative to other similar ions, which may further compromise the quantification of the intended ion concentration. As a testament, the Yb(III) macrocyclic chelate that detects Ca^{2+} was also shown to lose CEST when bound to Mg^{2+}.

In addition, both paraCEST agents have CEST amplitudes that are dependent on pH and temperature. The ratiometric comparison with an ion-unresponsive CEST agent that is dependent on pH and temperature may potentially improve on detection, but this has not yet been experimentally demonstrated. These disadvantages should be overcome before using these paraCEST agents during in vivo studies.

13.6 ParaCEST Agents That Detect Redox State

The redox state of tissue environments can be an important biomarker for many pathologies that have disregulated concentrations of reducing agents. To detect oxidative versus reductive environments, a paraCEST agent has been developed, which consists of a Eu(III) chelate with two quinolinium ligands and a bound water molecule (Fig. 13.9A) [26]. The oxidized form generates a weak CEST signal at 43 ppm at 20 mM concentration, while the reduced form has a 7.5-fold stronger CEST signal at 50 ppm. The oxidized form has fast chemical exchange from the bound water, which is near 17,200 Hz, and therefore nearly too fast to generate CEST, while the chemical exchange from the bound water is a slower 11,100 Hz in the reduced form, which is more appropriate for creating a CEST signal. Another paraCEST agent has exploited a different approach by changing the oxidation state of the agent's metal ion (Fig. 13.9B). A paramagnetic Co(II) chelate with three pyrazole ligands can generate a strong 20–25% CEST amplitude at 135 ppm in its oxidized form at 8 mM concentration [27]. The reduced Co(III) chelate is diamagnetic and cannot act as a paraCEST agent. A similar approach was used to detect singlet oxygen with a derivative of Eu^{III}-complex with an anthryl moiety (Fig. 13.9C) [28].

These paraCEST agents show that the chemical shift is very sensitive to the electronic status of the metal ion or organic chelate. Using the chemical shift to monitor redox state is preferred because the chemical shift is independent of the agent's concentration, while the CEST amplitude is concentration dependent. Evidence suggests that the chemical shift of the CEST effects from the oxidized forms of both agents may have a minor dependence on pH. A second

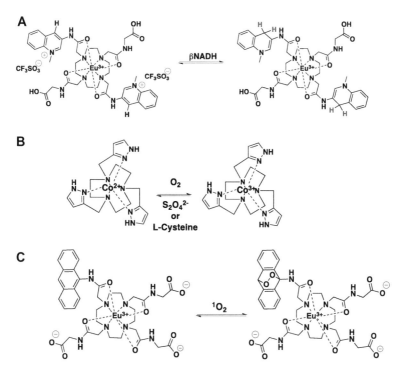

Figure 13.9 (A) An Eu^{III}-complex with two quinolinium arms that undergo a protonation/deprotonation reaction suitable for detection of redox states [26]. (B) Co^{II} is paramagnetic and, therefore, can be used as a CEST agent, while Co^{III} is diamagnetic and does not show any CEST signal [27]. (C) A derivative of Eu^{III}-complex with an anthryl moiety that reacts with singlet oxygen and changes CEST properties of the reagent [28].

redox-unresponsive, pH-dependent CEST agent may possibly be used to correct the results for this minor pH effect and improve the assessments of redox state. This improvement may facilitate in vivo studies, which have yet to be demonstrated with these two paraCEST agents.

13.7 ParaCEST Agents That Measure pH

The chemical exchange of a proton from an amide or amine to bulk water is base-catalyzed. Therefore, the CEST effect generated from

Table 13.1 ParaCEST agents that measure pH

Reference	Agent	Measurement	Chemical shift at 37°C	pH range
29	Yb-DOTAM	No ratio	−14.5, −17.7 (amides)	6.0–7.2
30	Yb-DOTAM-Gly	Amide versus H_2O of Eu-DOTAM-Gly	−16 ppm (Yb) ∼50 ppm (Eu)	6.5–8.5
31, 32	Pr-DOTAM-Gly	Amide versus H_2O	13 (amide) −70 (H_2O)	6.0–7.5
31	Nd-DOTAM-Gly	Amide versus H_2O	11 (amide) −50 (H_2O)	6.0–7.5
31	Eu-DOTAM-Gly	Amide versus H_2O	4 (amide) 50 (H_2O)	6.0–7.5
33	Yb-HPDO3A	Amide versus amide	88, 65 (amides)	5.0–7.0
34	Co-TETAM	Amide versus amide	112, 95 (amides)	6.8–7.8
35–37	Yb-DO3A-oAA	Amide versus amine	−11 (amide) 8 (amine)	6.4–7.6
38	Tm-DOTAM-Gly-Lys	Linewidth	−46 (amide)	6.0–8.0
39	Eu-DO3A-Gly$_3$-phenol	Chemical shift	49–55 ppm (H_2O)	6.0–7.6

amides and amines can be used to measure pH (Table 13.1). One of the first examples of a pH-responsive paraCEST agent generated CEST from the amide ligands of a Yb(III) macrocyclic chelate. However, the CEST amplitude also depended on the concentration of this agent, so that the CEST amplitude could not be used to accurately measure pH. To improve the pH measurement, the pH-responsive CEST amplitude from the amide protons a Yb(III) macrocyclic chelate was ratiometrically compared to the pH-unresponsive CEST amplitude from the bound water in the same organic chelate that was complexed with Eu(III). This approach was then further improved by measuring two CEST signals from the same paraCEST agent, where the pH-responsive signal was derived from saturating the amide resonance, and the pH-unresponsive signal was generated by saturating the bound water resonance. The Pr(III) macrocyclic chelate of this series was shown to generate a ratio of CEST amplitudes with the best dynamic range for measuring pH, relative to the Nd(III) and Eu(III) chelates (Fig. 13.10).

An extension of this approach used an ortho-aminoanilide ligand of a Yb(III) macrocyclic chelate, which generated two pH-dependent CEST signals from the amide and amine of the ligand. The CEST amplitude from the amine decreased at higher pH, while the amide's CEST amplitude increased with increasing pH. These differences caused the ratio of their CEST signals to have a pH dependence with a high dynamic range versus pH, relative to the paraCEST agents

ParaCEST Agents That Measure pH | 299

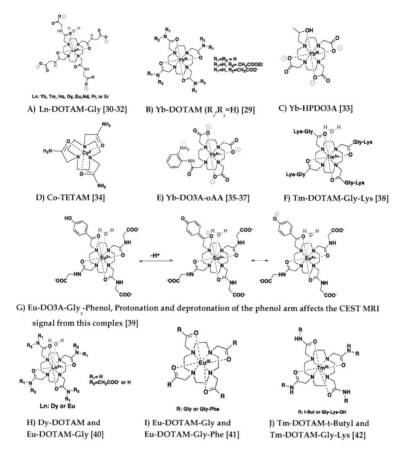

Figure 13.10 Chemical structures of paraCEST MRI contrast agents used for measuring pH or temperature.

that had a pH-unresponsive CEST signal. This higher dynamic range provided an advantage for measuring the extracellular pH during in vivo studies of mouse leg muscle and a subcutaneous tumor.

A macrocyclic metal chelate can have different conformations, and the ratio of these conformations can be dependent on pH. Furthermore, each conformation can generate CEST effects at different chemical shifts, which provides the opportunity to selectively detect each conformation. The CEST amplitudes of each conformation can be used to measure the ratio of conformations, which can then be related to pH. For example, a Yb(III) macrocyclic chelate has a twisted square antiprismatic and square antiprismatic conformations, each with R and S arm rotations. The ratio of the CEST amplitudes from the two conformations has been used to measure pH. Similarly, a Co(II) macrocyclic chelate has multiple conformations that generate four CEST signals, and the ratio of two of the CEST amplitudes is linearly correlated with pH.

As an alternative approach, the linewidth and chemical shift of a peak in a CEST spectrum can be dependent on pH. The CEST linewidth of a Tm(III) macrocyclic chelate was used to measure pH in chemical solutions and within in vivo mouse leg muscle. The chemical shift of a peak in a CEST spectrum was shown to be pH-dependent when using a Eu(III) chelate with a phenolate ligand. Importantly, the CEST peak linewidth and chemical shift are largely independent of the agent's concentration, temperature, and T1 relaxation effects, so that pH can be directly determined from a single CEST signal without requiring a ratiometric approach to assess two CEST signals.

13.8 ParaCEST Agents That Measure Temperature

The mapping of in vivo tissue temperatures has been investigated by monitoring the chemical shift of water in tissues. However, the small 0.01 ppm/°C change in chemical shift of the water signal has hampered the ability to obtain accurate temperature measurements with this MRI technique. ParaCEST agents have been shown to have CEST effects with chemical shifts that are much more sensitive to temperature (Table 13.2). These chemical shifts are a weighted average of the proton on the agent and on bulk water (a phenomenon known as MR coalescence). Following the Arrhenius equation, the chemical exchange rate of the bound water (for Dy(III) and Eu(III) chelates) or amide protons (for Tm(III) chelates) increases at higher temperatures, which shifts the weighted average of the chemical shift toward the shift value of the water. The large chemical shifts of paraCEST agents provide a good dynamic range of temperature-dependent changes in chemical shifts of their CEST effects. In addition, the large chemical shifts can avoid complications from endogenous CEST effects, facilitating the in vivo measurement

Table 13.2 ParaCEST agents that measure temperature

Reference	Agent	ppm/°C	Chemical shift (ppm)	Tested temperature range (°C)
40	Dy-DOTAM-Gly	6.9	−650 to −800	20–50
40	Eu-DOTAM-Gly	0.4	45 to 55	20–50
41	Eu-DOTAM-Gly-Phe	0.3	39 to 48	20–50
38	Tm-DOTAM-Gly-Lys	0.27	−44.6 to −46.5	35–39
42	Tm-DOTA-tBu (TSAP)	0.57	−100 to −104	35–42
42	Tm-DOTA-tBu (SAP)	0.16	−68 to −69	35–42

of temperature as shown by an in vivo study of mouse leg muscle [38]. Importantly, the chemical shift is independent of concentration and other experimental conditions, so that a ratiometric comparison with a temperature-independent CEST agent is not required to improve this temperature measurement. Because temperature is an environmental property, these studies are not limited by a low concentration of a molecular target; therefore, 10 mM of a paraCEST agent is typically used for these studies. For these many reasons, paraCEST agents are a promising approach for future in vivo temperature measurements within animal models.

13.9 Future Directions for Clinical Translation of paraCEST Agents

Of the 28 examples of responsive paraCEST agents listed in Table 13.3, only four of these agents have been successfully applied to in vivo studies of mice or mouse models of human cancers. Several improvements are needed to improve in vivo studies and to facilitate eventual clinical translation. Perhaps most importantly, the poor detection sensitivity of paraCEST agents necessitates a high 1–10 mM concentration of the agent to generate a detectable CEST signal. Yet lanthanide metal chelates are potentially toxic, which limits the administration of these lanthanide ions to 0.1 mmol/kg of body weight (which is equivalent to 0.1 mM, assuming a body density of 1 g/mL and that the contrast agent has uniform biodistribution). The use of a Co(II) chelate is an insightful remedy for this problem, because cobalt ion

Table 13.3 CEST MRI conditions used to detect responsive paraCEST agents

Agent	Application	Concentration (mM)	% CEST	Saturation power (µT)	Saturation time (s)	Reference
Tm-DOTA-DVED	Caspase-3	5.2	5	31	4	2,7
Tm-DOTA-RGGZ	Urokinase plasminogen activator	50	15	21	4	8
Tm-DO3A-GGS(Dye)GKPILF	Cathepsin D	3		OPARACHEE MRI was used to detect this agent		9
Tm-DO3A-amine	Transglutaminase	25	15	20	4	10
Yb-DOTA-αBz-βGal	β-galactosidase	20	29	25	3	11,12
Yb-DO3A-oAA-TML-ester	Esterase enzyme	25	Amide: 9; amine: 6	14.8	3	13
Eu-DOTAM-ethylester-Gly	Nucleic acid	5	33	14.1	4	14
Nd₂-DO3A-bridged benzyl	Nucleic acid	5	~15	23.5	3	15
Eu-DOTAM-methyl-bis(phenylbornate)	Glucose	10	—	24.0	2	16–18
Yb-MBDO3AM	Lactate	30	60	25	6	19
Eu-S-THP	Methyl phoshphate	5	>20	—	3	20–22
Yb-DO3A-oAA	NO and O₂	40	Amide: 7; amine: 4	4.2	4	23
Eu-DOTAM-Py	Zn²⁺	20	>25	23.5	2	24
Yb-DOTAM-imino(diacetate)	Ca²⁺	20	60	25	3	25

Eu-DOTAM-tetraamide-bis(quinolinium)	Redox	20	~14	10	5	26
Co[II]-TPT	Redox	8	22	24	3	27
Eu-DOTAM-triGly-anthracene	Singlet O_2	5	~12	9.4	4	28
Yb-DOTAM-Gly	pH	30	~65	25	4	30
Tm-DOTAM-Gly	pH	40	~40	25	4	30
Er-DOTAM-Gly	pH	40	~30	25	4	30
Ho-DOTAM-Gly	pH	30	~20	25	4	30
Dy-DOTAM-Gly	pH	30	0%	25	4	30
Pr- DOTAM-Gly	pH	30	Amide: ~42; water: ~40	Amide: 1.35; water: 17.1	4	31
Nd-DOTAM-Gly	pH	30	Amide: ~37; water: ~66	Amide: 1.35; water: 17.1	4	31
Eu-DOTAM-Gly	pH	30	Amide: ~22; water: ~80	Amide: 1.35; water: 17.1	4	31
Yb-DOTAM (Gly or methylester Gly)	pH	30	~70	—	—	29
Pr-DOTAM-Gly	pH	30	Amide: 42; water: 40	Amide: 7; water: 87.6	4	32
Nd-DOTAM-Gly	pH	30	Amide: 33; water: 70	Amide: 7; water: 87.6	4	32
Eu-DOTAM-Gly	pH	30	Amide: 25; water: ~85	Amide: 7; water: 87.6	4	32

(Contd.)

Table 13.3 (Contd.)

Agent	Application	Concentration (mM)	% CEST	Saturation power (μT)	Saturation time (s)	Reference
Eu-DO3A-Gly₃-phenol	pH	10	~30	14.1	2	39
Yb-HPDO3A	pH	20	Hydroxyl: ~40; water: ~30	24	2	33
Yb-DO3A-oAA	pH	100	Amine:18.9; amide: 11.1	20	5	35–37
		60	Amine: ~16; amide: ~5.2			
Tm-DOTAM-Gly-Lys	pH	10	~25	14	3	38
Co^II-DO3A-triamide	pH	10	~22–40	24	2	34
Dy-DOTAM	Temperature	1	—	17.3	2	40
Eu-DOTAM-Gly	Temperature	10	~25	17.3	2	40
Tm-DOTAM-Gly-Lys	Temperature	10	~25	14	3	38
Eu-DOTAM-Gly or -Gly-Phe	Temperature	10	>35	14	10	41,45
Tm-DOTAM-tButyl, Tm-DOTA-Gly-Lys-OH	Temperature	10	SAP: ~6; TSAP: ~16	20	2	42

is biologically relevant [43, 44]. In addition, macrocyclic chelates with biocompatible nickel and iron have also been shown to generate (biomarker-unresponsive) CEST effects, which could be modified to create new responsive paraCEST agents. The use of macrocyclic chelates other than tetraazacyclododecane for Co(II), Ni(II), and Fe(II) may provide additional variety for expanding the design of paraCEST agents. However, the stability of the paramagnetic form of these metal chelates can be problematic especially in biological environments, which will require further study.

To overcome the problem of needing high saturation powers, other CEST MRI protocols may possibly be used, which detect CEST agents without directly saturating the agent with a long, high-power RF pulse. For example, the OPARACHEE MRI method applies shorter RF pulses at the chemical shift of the bulk water, which results in the suppression of the bulk water signal when a paraCEST agent exchanges protons with bulk water during these RF pulses. OPARACHEE MRI requires less RF power and is ideally suited to detect the presence of a paraCEST agent. However, this method cannot selectively detect multiple CEST signals from a single agent or a combination of agents, which obviates a ratiometric analysis that is critical for detecting many types of biomarkers. As another example, FLEX MRI protocols also use shorter RF pulses to detect CEST agents by measuring the modulation of the water signal based on the chemical shift evolution of solute proton magnetization [45]. This method has been applied to detect (biomarker-unresponsive) paraCEST agents in vivo and is particularly useful for measuring the chemical exchange rate of the paraCEST agent. The development and application of FLEX MRI and other MRI protocols that are specifically optimized for individual paraCEST agents may provide a powerful approach for future molecular imaging studies.

References

1. Castelli DD, Terreno E, Longo D, and Aime S. Nanoparticle-based chemical exchange saturation transfer (CEST) agents. *NMR Biomed.*, 2013; 26(7): 839–849.

2. Yoo B and Pagel MD. A PARACEST MRI contrast agent to detect enzyme activity. *J. Am. Chem. Soc.*, 2006; 128(43): 14032–14033.
3. Caravan P, Das B, Dumas S, et al. Collagen-targeted MRI contrast agent for molecular imaging of fibrosis. *Angew. Chem. Int. Ed.*, 2007; 46(43): 8171–8173.
4. Overoye-Chan K, Koerner S, Looby RJ, et al. EP-2104R: A fibrin-specific gadolinium-based MRI contrast agent for detection of thrombus. *J. Am. Chem. Soc.*, 2008; 130(18): 6025–6039.
5. Botnar RM, Buecker A, Wiethoff AJ, et al. In vivo magnetic resonance imaging of coronary thrombosis using a fibrin-binding molecular magnetic resonance contrast agent. *Circulation*, 2004; 110(11): 1463–1466.
6. Botnar RM, Perez AS, Witte S, et al. In vivo molecular imaging of acute and subacute thrombosis using a fibrin-binding magnetic resonance imaging contrast agent. *Circulation*, 2004; 109(16): 2023–2029.
7. Yoo B, Raam MS, Rosenblum RM, and Pagel MD. Enzyme-responsive PARACEST MRI contrast agents: A new biomedical imaging approach for studies of the proteasome. *Contrast Media Mol. Imaging*, 2007; 2(4): 189–198.
8. Yoo B, Sheth VR, Howison CM, et al. Detection of in vivo enzyme activity with catalyCEST MRI. *Magn. Reson. Med.*, 2014; 71(3): 1221–1230.
9. Suchý M, Ta R, Li AX, et al. A paramagnetic chemical exchange-based MRI probe metabolized by cathepsin D: Design, synthesis and cellular uptake studies. *Org. Biomol. Chem.*, 2010; 8(11): 2560–2566.
10. Hingorani DV, Randtke EA, and Pagel MD. A catalyCEST MRI contrast agent that detects the enzyme-catalyzed creation of a covalent bond. *J. Am. Chem. Soc.*, 2013; 135(17): 6396–6398.
11. Chauvin T, Durand P, Bernier M, et al. Detection of enzymatic activity by PARACEST MRI: A general approach to target a large variety of enzymes. *Angew. Chem. Int. Ed.*, 2008; 47(23): 4370–4372.
12. Chauvin T, Torres S, Rosseto R, et al. Lanthanide(III) complexes that contain a self-immolative arm: Potential enzyme responsive contrast agents for magnetic resonance imaging. *Chem. Eur. J.*, 2012; 18(5): 1408–1418.
13. Li Y, Sheth VR, Liu G, and Pagel MD. A self-calibrating PARACEST MRI contrast agent that detects esterase enzyme activity. *Contrast Media Mol. Imaging*, 2011; 6(4): 219–228.
14. Wu YK, Carney CE, Denton M, et al. Polymeric PARACEST MRI contrast agents as potential reporters for gene therapy. *Org. Biomol. Chem.*, 2010; 8(23): 5333–5338.

15. Nwe K, Andolina CM, Hang CH, and Morrow JR. PARACEST properties of a dinuclear neodymium(III) complex bound to DNA or carbonate. *Bioconjugate Chem.*, 2009; 20(7): 1375–1382.
16. Zhang SR, Trokowski R, and Sherry AD. A paramagnetic CEST agent for imaging glucose by MRI. *J. Am. Chem. Soc.*, 2003; 125(50): 15288–15289.
17. Ren JM, Trokowski R, Zhang SR, Malloy CR, and Sherry AD. Imaging the tissue distribution of glucose in livers using a PARACEST sensor. *Magn. Reson. Med.*, 2008; 60(5): 1047–1055.
18. Trokowski R, Zhang SR, and Sherry AD. Cyclen-based phenylboronate ligands and their Eu^{3+} complexes for sensing glucose by MRI. *Bioconjugate Chem.*, 2004; 15(6): 1431–1440.
19. Aime S, Delli Castelli D, Fedeli F, and Terreno E. A paramagnetic MRI-CEST agent responsive to lactate concentration. *J. Am. Chem. Soc.*, 2002; 124(32): 9364–9365.
20. Hammell J, Buttarazzi L, Huang CH, and Morrow JR. Eu(III) complexes as anion-responsive luminescent sensors and paramagnetic chemical exchange saturation transfer agents. *Inorg. Chem.*, 2011; 50(11): 4857–4867.
21. Huang CH and Morrow JR. A PARACEST agent responsive to inner- and outer-sphere phosphate ester interactions for MRI applications. *J. Am. Chem. Soc.*, 2009; 131(12): 4206–4207.
22. Woods M, Woessner DE, Zhao PY, et al. Europium(III) macrocyclic complexes with alcohol pendant groups as chemical exchange saturation transfer agents. *J. Am. Chem. Soc.*, 2006; 128(31): 10155–10162.
23. Liu G, Li Y, and Pagel MD. Design and characterization of a new irreversible responsive PARACEST MRI contrast agent that detects nitric oxide. *Magn. Reson. Med.*, 2007; 58(6): 1249–1256.
24. Trokowski R, Ren JM, Kalman FK, and Sherry AD. Selective sensing of zinc ions with a PARACEST contrast agent. *Angew. Chem. Int. Ed.*, 2005; 44(42): 6920–6923.
25. Angelovski G, Chauvin T, Pohmann R, Logothetis NK, and Tóth É. Calcium-responsive paramagnetic CEST agents. *Bioorg. Med. Chem.* 2011; 19(3): 1097–1105.
26. Ratnakar SJ, Viswanathan S, Kovacs Z, Jindal AK, Green KN, and Sherry AD. Europium(III) DOTA-tetraamide complexes as redox-active MRI sensors. *J. Am. Chem. Soc.*, 2012; 134(13): 5798–5800.
27. Tsitovich PB, Spernyak JA, and Morrow JR. A redox-activated MRI contrast agent that switches between paramagnetic and diamagnetic states. *Angew. Chem. Int. Ed.*, 2013; 52(52): 13997–14000.

28. Song B, Wu YK, Yu MX, et al. A europium(III)-based PARACEST agent for sensing singlet oxygen by MRI. *Dalton Trans.*, 2013; 42(22): 8066–8069.
29. Zhang SR, Michaudet L, Burgess S, and Sherry AD. The amide protons of an ytterbium(III) dota tetraamide complex act as efficient antennae for transfer of magnetization to bulk water. *Angew. Chem. Int. Ed.*, 2002; 41(11): 1919–1921.
30. Aime S, Barge A, Delli Castelli D, et al. Paramagnetic lanthanide(III) complexes as pH-sensitive chemical exchange saturation transfer (CEST) contrast agents for MRI applications. *Magn. Reson. Med.*, 2002; 47(4): 639–648.
31. Aime S, Delli Castelli D, and Terreno E. Novel pH-reporter MRI contrast agents. *Angew. Chem. Int. Ed.*, 2002; 41(22): 4334–4336.
32. Terreno E, Delli Castelli D, Cravotto G, Milone L, and Aime S. Ln(III)-DOTAMGly complexes: A versatile series to assess the determinants of the efficacy of paramagnetic chemical exchange saturation transfer agents for magnetic resonance imaging applications. *Invest. Radiol.*, 2004; 39(4): 235–243.
33. Delli Castelli D, Terreno E, and Aime S. Yb-III-HPDO3A: A dual pH- and temperature-responsive CEST agent. *Angew. Chem. Int. Ed.*, 2011; 50(8): 1798–1800.
34. Dorazio SJ, Olatunde AO, Spernyak JA, and Morrow JR. CoCEST: Cobalt(II) amide-appended paraCEST MRI contrast agents. *Chem. Commun.*, 2013; 49(85): 10025–10027.
35. Liu G, Li Y, Sheth VR, and Pagel MD. Imaging in vivo extracellular pH with a single paramagnetic chemical exchange saturation transfer magnetic resonance imaging contrast agent. *Mol. Imaging*, 2012; 11(1): 47–57.
36. Sheth VR, Li Y, Chen LQ, Howison CM, Flask CA, and Pagel MD. Measuring in vivo tumor pHe with CEST-FISP MRI. *Magn. Reson. Med.*, 2012; 67(3): 760–768.
37. Sheth VR, Liu G, Li Y, and Pagel MD. Improved pH measurements with a single PARACEST MRI contrast agent. *Contrast Media Mol. Imaging*, 2012; 7(1): 26–34.
38. McVicar N, Li AX, Suchý M, Hudson RHE, Menon RS, and Bartha R. Simultaneous in vivo pH and temperature mapping using a PARACEST-MRI contrast agent. *Magn. Reson. Med.*, 2013; 70(4): 1016–1025.
39. Wu YK, Soesbe TC, Kiefer GE, Zhao PY, and Sherry AD. A responsive europium(III) chelate that provides a direct readout of pH by MRI. *J. Am. Chem. Soc.*, 2010; 132(40): 14002–14003.

40. Zhang SR, Malloy CR, and Sherry AD. MRI thermometry based on PARACEST agents. *J. Am. Chem. Soc.*, 2005; 127(50): 17572–17573.
41. Li AX, Wojciechowski F, Suchý M, et al. A sensitive PARACEST contrast agent for temperature MRI: Eu^{3+}-DOTAM-glycine (Gly)-phenylalanine (Phe). *Magn. Reson. Med.*, 2008; 59(2): 374–381.
42. Stevens TK, Milne M, Elmehriki AAH, Suchý M, Bartha R, and Hudson RHE. A DOTAM-based paraCEST agent favoring TSAP geometry for enhanced amide proton chemical shift dispersion and temperature sensitivity. *Contrast Media Mol. Imaging*, 2013; 8(3): 289–292.
43. Lukaski HC. Vitamin and mineral status: Effects on physical performance. *Nutrition* 2004; 20(7): 632–644.
44. Lexa D and Saveant JM. The electrochemistry of vitamin B12. *Acc. Che. Res.*, 1983; 16(7): 235–243.
45. Coman D, Kiefer GE, Rothman DL, Sherry AD, and Hyder F. A lanthanide complex with dual biosensing properties: CEST (chemical exchange saturation transfer) and BIRDS (biosensor imaging of redundant deviation in shifts) with europium DOTA-tetraglycinate. *NMR Biomed.*, 2011; 24(10): 1216–1225.

Chapter 14

Saturating Compartmentalized Water Protons: Liposome- and Cell-Based CEST Agents

Daniela Delli Castelli, Giuseppe Farrauto, Enzo Terreno, and Silvio Aime

Molecular and Preclinical Imaging Centers, Department of Molecular Biotechnology and Health Sciences, University of Torino, Torino, Italy
daniela.dellicastelli@unito.it, silvio.aime@unito.it

14.1 Introduction

Since the early days of chemical exchange saturation transfer (CEST) contrast agents, it was clear that sensitivity would be the Achilles' heel of these probes. The first series of compounds tested as CEST agents, by Balaban et al., were small diamagnetic molecules whose detection threshold was in the mM range. Obviously, the low sensitivity in generating contrast reduced the clinical potential of this novel class of diamagnetic agents. This is particularly true in the era of molecular imaging whose applications require the development of imaging reporters able to visualize very diluted targets (in the nano/picomolar range). Therefore, in the last 15

Chemical Exchange Saturation Transfer Imaging: Advances and Applications
Edited by Michael T. McMahon, Assaf A. Gilad, Jeff W. M. Bulte, and Peter C. M. van Zijl
Copyright © 2017 Pan Stanford Publishing Pte. Ltd.
ISBN 978-981-4745-70-3 (Hardcover), 978-1-315-36442-1 (eBook)
www.panstanford.com

years, much work in CEST area has addressed strategies to enhance the detection sensitivity of these agents. The saturation transfer efficiency depends on many parameters. Among them, the most important and easiest to control are the number of equivalent mobile protons to be irradiated and their exchange rate with bulk water (k_{ex}). Thus, the sensitivity issue has been primarily faced with two approaches: (1) increasing k_{ex}, and (2) increasing the number of mobile protons per single molecule. As, for any CEST agent, the *condicio sine qua non* is that the chemical shift separation between the exchanging protons has to be larger than their exchange rate ($\Delta\omega > k_{ex}$), it is evident that the task of increasing k_{ex} endlessly is not desirable. A strategy that has been pursued to improve the sensitivity via k_{ex}, by using paramagnetic molecules (paraCEST see Chapter 11) in virtue of their ability to induce large $\Delta\omega$ values, has led to important achievements, although the sensitivity enhancements of these molecules are still far below the detection threshold imposed by molecular imaging applications. Increasing the number of mobile protons per single molecule has been pursued following different approaches: (1) use of polyaminoacids or RNA-like polymers [1, 2]; (2) formation of supramolecular adducts between paramagnetic shift reagents (SRs) and diamagnetic substrates rich in mobile protons [3]; (3) design of macromolecular/nanoparticulate scaffolds (polymers, dendrimers, micelles, perfluorocarbon nanoemulsions, silica particles, virus capsides) covalently conjugated with a large number of paraCEST agents [4–9]; and (4) design of systems containing water protons compartmentalized in phospholipid-based vesicles (lipoCEST and cellCEST) [10, 11]. The latter approach allowed to reach the highest sensitivity reported so far for CEST agents and will be the topic of this chapter.

The rational of this approach was to use vesicles entrapping a huge number of water protons, whose exchange with the bulk solvent and resonance frequency can be modulated to fulfill the $\Delta\omega > k_{ex}$ condition. The number of water molecules that can be entrapped in the inner core of liposomes ranges from millions to billions depending on the size of the particle, and even many more protons can be saturated using cells. The necessary removal of the isochronicity between intravesicular and bulk protons can be successfully accomplished by entrapping a paramagnetic SR inside the vesicle.

LipoCEST/cellCEST agents have introduced a new concept in the field of CEST agents, because it is the whole (intact) vesicle to act as contrast agent itself, and this feature opened new exciting applications for CEST probes, especially in the emerging field of imaging drug delivery and release.

14.2 Basic Features of lipoCEST/cellCEST Agents

As already hinted in the introduction, unlike the other subclasses of CEST agents, lipoCEST/cellCEST systems exploit the solvent water protons in the inner compartment of the vesicle as a pool of exchangeable spins to be irradiated. The number of involved spins is extremely large ($>10^6$ in a typical lipoCEST system). As the resonance frequency of the water protons inside (intravesicular) and outside (bulk) the vesicle is almost the same, there is the need to induce a shift difference that exceeds the exchange rate of the intravesicular protons ($\Delta\omega > k_{ex}$). This goal can be successfully accomplished by entrapping a paramagnetic SR in the intravesicular compartment (Fig. 14.1).

In the following subsection, the basics of the mechanism through which a paramagnetic species affects the resonance frequency of the intravesicular water protons will be presented. For the sake of simplicity, the discussion will be focused on lipoCEST agents, but the same rules also apply to cellCEST systems.

14.2.1 Chemical Shift of Intravesicular Water Protons in Presence of Paramagnetic SR

The chemical shift separation (Δ) between intraliposomal (IL) and bulk water protons when a paramagnetic SR is confined in the inner cavity of the vesicles is dependent on two terms:

$$\Delta^{IL} = \Delta^{IL}_{HYP} + \Delta^{IL}_{BMS} \qquad (14.1)$$

where HYP and BMS refer to hyperfine and bulk magnetic susceptibility contributions, respectively. The hyperfine effect derives

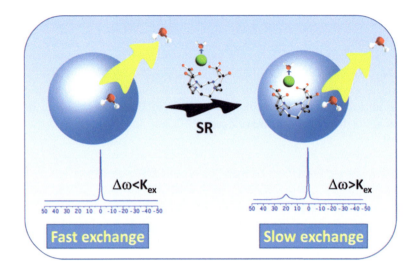

Figure 14.1 Schematic representation of a lipoCEST/cellCEST agent.

from the chemical interaction between the paramagnetic center and the intraliposomal water protons. It can operate through bonds (contact shift) or space (pseudo-contact or dipolar shift). Typically, SRs are lanthanide complexes with at least one fast exchanging water molecule ($q = 1$) coordinated to the metal. In this case, the HYP term is dependent on the shift of the water protons at the metal site (Δ_M^{IL}), weighted by their metal-bound molar fraction:

$$\Delta_{HYP}^{IL} = \frac{q[SR]^{IL}}{55.6} \Delta_M^{IL} \qquad (14.2)$$

where $[SR]^{IL}$ is the molar concentration of the metal complex in the inner core of the liposome. Δ_M^{IL} is proportional to magnetic ($\Delta\chi$, magnetic anisotropy) and structural (G) properties of the SR:

$$\Delta_M^{IL} \propto \Delta\chi, G \quad \Delta\chi = C_J A_2^0 \langle r^2 \rangle ; G \propto \frac{\cos^2\theta}{r^3} \qquad (14.3)$$

C_J is Bleaney's constant, which characterizes each lanthanide and can have positive (Eu, Er, Tm, and Yb), negative (Ce, Pr, Nd, Sm, Tb, Dy, and Ho), or zero (Gd) values. The term $A_2^0 \langle r^2 \rangle$ refers to the crystal field parameters of the complex and can assume positive

Basic Features of lipoCEST/cellCEST Agents | 315

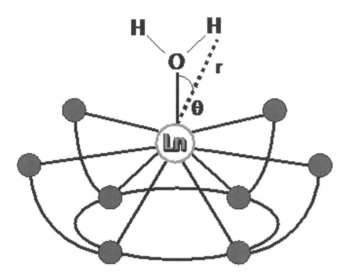

Figure 14.2 Schematic view of a lanthanide(III) complex (antiprismatic structure) highlighting the distance r between the paramagnetic metal center and the bound water protons and the angle θ formed between the principal magnetic axes of the complex (here corresponding to the metal-O_w bond) and the r vector.

or negative values. The role played by the structure of the SR is expressed by (i) the angle (θ) between the principal magnetic axes of the complex and the vector connecting the paramagnetic metal with the coordinated water protons and (ii) the distance (r) between the metal and the water protons (Fig. 14.2).

Hence, for a given ligand able to form isostructural complexes along the lanthanide series, the magnitude and sign of Δ_M^{IL} are controlled by C_J values, and the largest positive and negative values are predicted for Tm and Dy, respectively. On the other hand, for a specific lanthanide, Δ_M^{IL} is defined by the structure of the complex. In particular, from the geometric point of view, the maximum attainable shift occurs when the coordinated water molecule aligns with the principal magnetic axes of the SR. This requirement is fulfilled by lanthanide complexes of macrocyclic polyaminocarboxylic ligands with C4 symmetry such as DOTA and DOTMA (Chart 14.1), whose Δ_M^{IL} values are larger than less

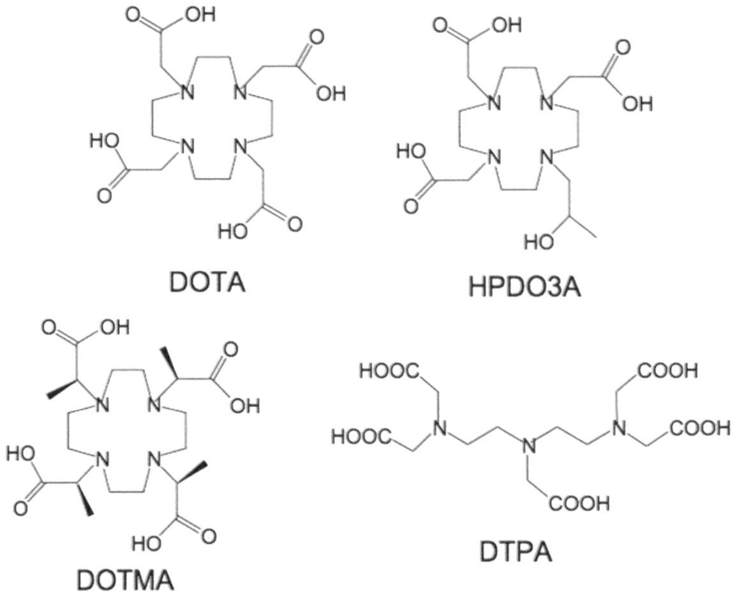

Chart 14.1

symmetric macrocyclic (e.g., HPDO3A) or linear (e.g., DTPA) ligands (Fig. 14.3) [12].

The contribution from bulk magnetic susceptibility, Δ_{BMS}^{IL}, does not require a chemical interaction between the water protons and the paramagnetic metal, as it depends on (i) the shape and orientation with respect to the external magnetic field of the paramagnetic vesicle, and (ii) the effective magnetic moment (μ_{eff}) of the metal [13, 14]. In a spherical compartment, as in ordinary liposomes, BMS effectively averages to zero and, therefore, δ^{IL} is determined by the hyperfine contribution only. Equations 14.2 and 14.3 indicate that Δ_{HYP}^{IL} can be increased by changing the characteristics of the SR (mainly q and G) as well as its intravesicular concentration $[SR]^{IL}$.

As far as in vivo applications are concerned, the maximum amount of SR that can be encapsulated inside a liposome is controlled by osmolarity issues. In particular, the intraliposomal solution should be isotonic (approximately 300 mOsM) to avoid osmotic stress for the vesicles. Hence, to increase the Δ_{HYP}^{IL} value,

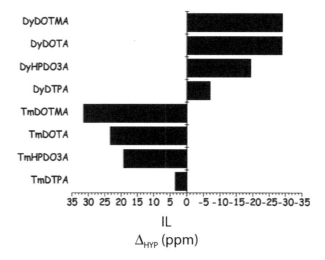

Figure 14.3 Lanthanide-induced chemical shift of water protons calculated for 1 M solutions of different SRs (25°C). Reprinted with permission from Ref. [12], Copyright 2013, John Wiley and Sons.

neutral SRs are preferred, such as Ln-HPDO3A complexes, as they can be loaded up to an intraliposomal concentration of 0.3 M. Conversely, mono-ionic SRs (e.g., Ln-DOTA or Ln-DOTMA) may reach a theoretical maximum concentration of 0.15 M. As the shifting efficiency of such ionic chelates is less than twice the value of Δ_M^{IL} of Ln-HPDO3A (see Fig. 14.3), the latter complexes have been most used for the formulation of lipoCEST/cellCEST agents.

The concentration limit imposed by the osmotic rules restricts the maximum attainable Δ^{IL} value to approximately 4 ppm (downfield or upfield according to the sign of Bleaney's constant). Unfortunately, mobile protons from endogenous molecules (e.g., amide protons) share the same chemical shift range; therefore, the in vivo detection of lipoCEST agents might be hampered by the background contrast. To face this issue, routes to further increase Δ^{IL} values have been explored. A possible approach has been recently proposed by Chahida et al., by loading liposomes with a neutral $q = 2$ complex (Tm-DO3A), thus attaining a Δ^{IL} value around 10 ppm [15].

However, a very valuable way to enlarge the shift separation between intraliposomal and bulk protons has been found by controlling the Δ^{IL}_{BMS} contribution (Eq. 14.1). Such a contribution is inversely correlated to the temperature (T) and proportional to $[SR]^{IL}$, to the effective magnetic moment of the paramagnetic metal (μ_{eff}), and to a factor (s), whose value and sign depend on the shape and orientation of the particles within B_0 [16]:

$$\Delta^{IL}_{BMS} \propto \frac{[SR]^{IL}\mu_{eff}^2 s}{T} \qquad (14.4)$$

Hence, the exploitation of BMS has meant to deal with non-spherical liposomes. Luckily, liposomes are very soft materials and their phospholipid-based membrane is permeable to water, but not to other hydrophilic solutes. Thus, the application of osmotic gradients can induce a net flux of solvent from bulk to the inner cavity (liposomes in hypo-osmotic medium) or vice versa (liposomes in hyper-osmotic medium). Whereas in the former case particles swell maintaining their shape, in the latter condition, they shrink with a consequent loss of sphericity. It is well established that phospholipid-based bilayers containing systems spontaneously orient in a magnetic field [17]. The driving force of this process is the interaction between B_0 and the magnetic susceptibility anisotropy ($\Delta\chi$) of the lipid bilayer [18]. In particular, it is the sign of $\Delta\chi$ that defines the orientation (parallel or perpendicular) of the particles in the magnetic field. Systems with $\Delta\chi$ values <0 have the principal axes of magnetic anisotropy tensor perpendicular to B_0, whereas particles with $\Delta\chi$ values >0 adopt a parallel orientation (Fig. 14.4).

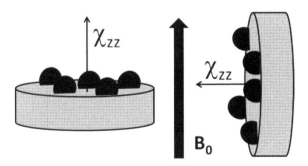

Figure 14.4 Orientation of aspherical particles within a magnetic field depends on the angle between the principal axis of the magnetic susceptibility tensor (χ_{ZZ}) and B_0. Adapted from Ref. [18].

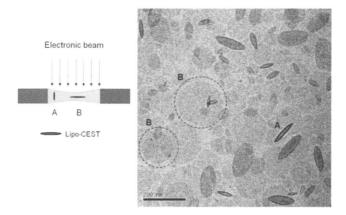

Figure 14.5 Cryo-TEM image of aspherical lipoCEST agents. Liposomes with the short axis positioned in plane (A) or perpendicular (B) with the electron beam (which is perpendicular to the image plane), as well as vesicles with all different orientations were observed. Reprinted with permission from Ref. [12], Copyright 2013, John Wiley and Sons.

As far the shift of the intraliposomal protons is concerned, the former orientation induces a positive BMS shift, whereas the latter one is associated with a negative Δ_{BMS}^{IL} value. Phospholipid-based bilayers have $\Delta\chi$ values <0; therefore, Δ^{IL} can be increased by the liposome encapsulation of an SR with a positive C_J value such as Tm.

Practically, the preparation of non-spherical lipoCEST can be accomplished by encapsulating in a liposome a hypotonic solution of the SR (e.g., 40 mOsM instead of 300 mOsM as typically done for spherical lipoCEST) [19]. When the vesicles are suspended in an isotonic medium, they leak water to reach the osmotic equilibrium and assume a pseudo-lenticular shape (Fig. 14.5) [20].

Interestingly, the BMS contribution exceeded the hyperfine one, thus allowing up to a three-fold increase in Δ^{IL} [12]. The BMS effect is directly correlated to the concentration of the SR entrapped in the liposomes; thus, the maximum payload of SR is still limited by osmotic rules. Hence, different strategies were designed to further increase $[SR]^{IL}$, without affecting the osmolarity of the intraliposomal cavity, e.g., through the entrapment of poly-metallic neutral complexes or through the incorporation of amphiphilic SRs into the liposome membrane [21, 22].

The incorporation of an SR in the membrane of shrunken lipoCEST agent offers the intriguing possibility to change the orientation of the nanovesicles in the field, and consequently the sign of Δ_{BMS}^{IL}. To do that, it is mandatory to incorporate in the membrane some amount (10–20% in moles) of a paramagnetic complex with a positive $\Delta\chi$ value. As highlighted in Fig. 14.6, the sign of $\Delta\chi$ is not only defined by the C_J value of the lanthanide (see Eq. 14.3), but also results from the crystal field parameters of the complex (Eq. 14.3).

The data reported in Fig. 14.6 suggest that the sign of $\Delta\chi$ seems to be correlated to the way through which the aliphatic chains are linked to the ligand.

A very elegant demonstration of the correlation between Δ^{IL} and the field orientation of non-spherical lipoCEST has been published by Burdinski et al. [23]. They coated the inner surface of a glass capillary (diameter 100 µm) with a monolayer of β-cyclodextrin. Then, two osmotically shrunken lipoCEST formulations were prepared: $Tm^{inner}/Dy^{bilayer}$ (Tm-HPDO3A in the inner core and a Dy-amphiphilic complex in the bilayer), and $Dy^{inner}/Tm^{bilayer}$ (Dy-HPDO3A in the inner core and a Tm-amphiphilic complex in the bilayer). The formulations were designed to have lipoCEST with positive or negative $\Delta\chi$ values. Furthermore, the nanovesicles were embedded with a phospholipid conjugated with adamantane, which is known to be a strong β-cyclodextrin binder. The experiments were carried out by filling up the capillary with the corresponding lipoCEST agent ($Tm^{inner}/Dy^{bilayer}$ or $Dy^{inner}/Tm^{bilayer}$). Then, Z-spectra were recorded to determine Δ^{IL} either before or after the removal of the liposomes not bound to the capillary wall. The experiment was performed twice, placing the capillary aligned or perpendicular to B_0 (Fig. 14.7).

The shift of the intraliposomal protons was positive for the unbound vesicles aligned with the field ($Tm^{inner}/Dy^{bilayer}$ formulation, $\Delta\chi < 0$), and negative for the free liposomes incorporating the Tm-complex ($Dy^{inner}/Tm^{bilayer}$ formulation, $\Delta\chi > 0$). Once the bulk liposomes were flushed away, the bound nanoparticles in the capillary placed parallel to B_0 were forced to align with the field, and δ^{IL} was positive for both the samples. Conversely, when the capillary is turned by 90°, there was no preferential orientation for the bound

Basic Features of lipoCEST/cellCEST Agents | 321

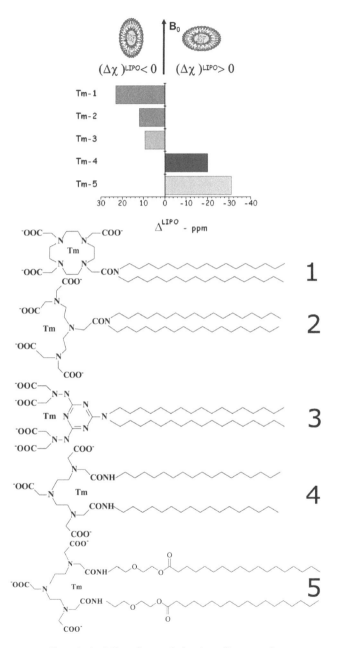

Figure 14.6 Chemical shift values of the intraliposomal water protons (25°C) of aspherical lipoCEST agents containing Tm-HPDO3A in the inner cavity and the reported amphiphilic complexes in the vesicle bilayer.

322 | Saturating Compartmentalized Water Protons

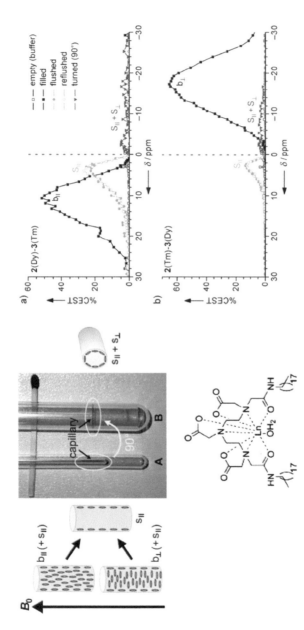

Figure 14.7 Top left: Parallel and perpendicular alignment of aspherical lipoCEST agents with respect to B_0 field in bulk solution (b) and at the capillary surface (s). The photo shows the capillaries in NMR tubes with parallel (A) or perpendicular (B) orientations. Capillaries were loaded with two lipoCEST formulations: Tm-2/Dy-HPDO3A (2(Tm)-3(Dy)) or Dy-2/Tm-HPDO3A (2(Dy)-3(Tm)). Ligand 2 is reported bottom right. Right: CEST spectra at 7 T. Capillaries were filled, flushed with buffer, and turned through 90°. Reprinted with permission from Ref. [23], Copyright 2010, John Wiley and Sons.

liposomes, and the shift values spread out over a broad frequency range. The exploitation of the BMS effect significantly extended the window of the accessible irradiation frequencies from the original 8 ppm (+4/−4 ppm) observed for the first spherical lipoCEST agents to approximately 80 ppm, thus improving in vivo CEST contrast detection and multiplex imaging.

Finally, it may be worth noting that all the theoretical aspects presented in this section also apply to any kind of nanovesicular CEST system as it has been demonstrated in the case of polymeric nanovesicles [24].

14.2.2 CEST Contrast in lipoCEST/cellCEST: Effect of Exchange Rate and Size

As discussed in the introduction, the uniqueness of lipoCEST/cellCEST agents relies on the huge number of saturable mobile protons that belong to the inner compartment of the vesicle. Therefore, the CEST contrast will be dependent on such number (controlled by the vesicle size) as well as the exchange rate across the membrane. For a spherical vesicle, the exchange rate (k_{ex}) depends on the water permeability of the membrane (P_W) and the surface/volume ratio of the vesicles (S/V) according to the following equation:

$$k_{ex} = P_W \frac{S}{V} = P_W \frac{3}{r} \tag{14.5}$$

where r is the radius of the vesicle. Hence, a size reduction accelerates the exchange with a consequent enhancement of CEST contrast if the slow-to-intermediate exchange regime is maintained. However, this benefit is compromised by the size-associated diminution in the number of irradiated mobile protons [25]. It has been concluded that a size reduction improves the CEST effect when it is normalized to the molar fraction of intraliposomal water. Conversely, when the CEST performance is evaluated based on the concentration of vesicles, the contribution of the number of the intraliposomal protons predominates so that a lower concentration of a big lipoCEST agent works better than the same amount of smaller-sized vesicles [20].

The permeability of a phospholipid membrane is strongly affected by the bilayer composition. Unsaturated phospholipids generate imperfection in the membrane package, thus giving rise to a more permeable membrane. For liposomes, such membranes suffer from poor in vivo stability and the vesicles are easily recognized and de-assembled by the reticulo-endothelial system. Cholesterol is a common component of membranes, and it is commonly used in liposome formulation to improve vesicles stability. For CEST purposes, it is noteworthy that the presence of cholesterol reduces water permeability [26].

14.2.3 Liposomes Loaded with CEST Agents

In addition to acting as a reservoir of a huge number of exchangeable protons, vesicles like liposomes or even cells can be used for their ability to transport conventional CEST agents: diamagnetic, paramagnetic, or hyperpolarized. In this case, the pool of saturated spins belongs to the CEST agent itself (i.e., there is no necessity to modify the chemical shift of the intraliposomal protons), but the mechanism for contrast generation may be still mediated by the exchange between intraliposomal and bulk pool (k_{ex}). In principle, if the exchange rate of the protons of the CEST molecule (k_{CEST}) is much lower than k_{ex}, then the overall CEST effect is not modulated by the liposome membrane and corresponds to the value that can be measured for the free agent (i.e., not loaded in the vesicles) at the same concentration. On the other hand, if $k_{CEST} \geq k_{ex}$, then the CEST effect will be quenched. Since the k_{CEST} values for diaCEST agents are often lower than paraCEST probes, it is expected that the quenching of the CEST contrast is favored for the liposomes loaded with paramagnetic agents. However, even if reduced, the CEST contrast from liposomes loaded with paraCEST agents has been successfully exploited for the design of an improved pH responsive agent [27]. A particular option is the formulation of liposomes with low water permeability loaded with a paramagnetic agent able to act as CEST and SR agent at the same time [28]. In this case, the lipoCEST contrast reports on the vesicle integrity, whereas the paraCEST contrast can be detected only following the release of the probe. An example of this approach will be presented in the next section. It has also been demonstrated that loading liposomes with different diaCEST molecules preserves the "multicolour" characteristics of the free agents, thus allowing the in vivo visualization of the delivery and accumulation of the nanocarrier [29, 30].

14.3 Applications

14.3.1 *LipoCEST Agents*

The advent of lipoCEST agents offered the valuable opportunity to integrate the classical application fields of CEST agents (multiplex imaging, development of responsive agents) with the longstanding, well-established, and successful use of liposomes as drug-delivery carriers [31]. Furthermore, lipoCEST can also be used as multimodal MRI agents, being detectable as T_2 or even T_1 agents. The former contrast mode is derived from the T_2 magnetic susceptibility effects that accompany the entrapment of a large amount of paramagnetic centers in the inner core of the liposomes [32, 33]. Though not yet exploited in vivo, lipoCEST agents can also generate T_1 contrast. In fact, in spite of the impossibility for Gd to induce hyperfine chemical shift effects ($C_J^{Gd} = 0$), the encapsulation of Gd(III)-complexes in non-spherical liposomes creates a BMS effect. Thus, osmotically shrunken liposomes loaded with Gd may act as trimodal (T_1, T_2, and CEST) MRI agents with good potential in the field of imaging drug delivery and release [34].

In another study, the multimodal properties of lipoCEST were exploited to investigate the trafficking of liposomes in a tumor environment [35]. In that case, a spherical Tm-HPDO3A-loaded lipoCEST preparation was injected in mice grafted with a syngeneic B16 melanoma model, and both T_2 and CEST contrast were monitored over time. In addition, liposomes with the same formulation, but encapsulating Gd-HPDO3A, were injected and followed by T_1 and T_2 contrast. The study outlined the great potential of the multicontrast properties of these agents, which allowed the attainment of relevant information about the dynamics underlying the cellular uptake of the liposomes and the intracellular release of their content.

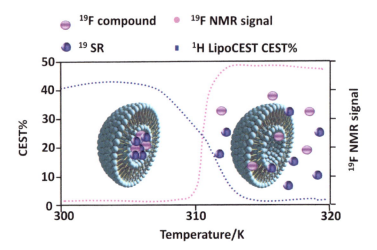

Figure 14.8 CEST effect and ^{19}F-NMR signal intensity of a thermosensitive lipoCEST agent containing Tm-HPDO3A and NH$_4$PF$_6$ in the inner cavity of the nanovesicles. Adapted from Ref. [36].

Besides multicontrast proton MRI, lipoCEST agents have also been considered for dual ^{1}H/^{19}F-MRI applications. Langereis et al. prepared a spherical thermosensitive lipoCEST formulation where the SR (Tm-HPDO3A) was co-encapsulated with the fluorine-rich PF6$^-$ ion [36]. The integrity of the liposomes could be monitored by CEST contrast (^{19}F-MRI was undetectable for the strong signal broadening caused by the SR), where the heat-triggered release of the encapsulated material was revealed by the appearance of the ^{19}F-MRI signal (CEST contrast disappeared as a consequence of the SR release) (Fig. 14.8).

Our group has reported another example of lipoCEST-based multicontrast MRI agent potentially acting as a responsive system [37]. The agent consisted of a lipoCEST agent suitably functionalized to expose Gd-complexes only on the outer side of the bilayer. The linker that connected the Gd-units to the liposome was specifically designed to contain a site prone to be cleaved in response to stimuli such as pH, redox potential, or acting as an enzymatic substrate (Fig. 14.9). In the absence of the stimulus, the T_1 shortening of the bulk protons induced by the Gd-units nullifies the CEST contrast. Conversely, when the stimulus is active, the Gd-complexes

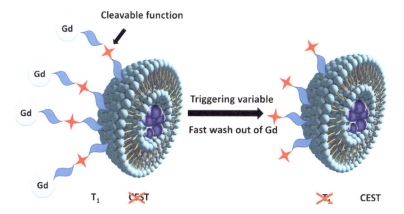

Figure 14.9 Schematic representation of a dual T_1/CEST liposomal ^{1}H-MRI agent. The system can act as T_1 agent only as long as the Gd-complexes are bound to the lipoCEST. When the T_1 units are detached from the liposomes under the action of a specific variable and then rapidly washed out, the CEST contrast is activated.

detach from the nanoparticles and are quickly washed out from the tissue in which they accumulated. The stimulus-dependent removal of the T_1 agent switches on the CEST contrast. This mechanism was successfully demonstrated in vitro using a disulfide bond as redox-sensitive site. However, as CEST agents are very suitable for multiplex imaging procedures, several studies relied on the detection of different lipoCEST agents in the same MRI experiment.

A recent example dealt with the use of a mixture of liposome-based CEST agents to visualize the release of the liposomal content under the action of specific stimulations [28]. In particular, two formulations were tested, whose release was stimulated by pH or ultrasound. The former system was a conventional spherical lipoCEST agent loaded with Tm-HPDO3A (intraliposomal protons at 3 ppm from bulk) that could be detected (even in co-presence of the sono-sensitive agent) as long as the pH was >6. Below this limit, liposomes start degrading and the contrast at 3 ppm switched off. However, the released SR can also act as a paraCEST agent, offering the resonance of the hydroxyl proton at 70 ppm. The CEST effect at 70 ppm cannot be detected when the SR is entrapped because

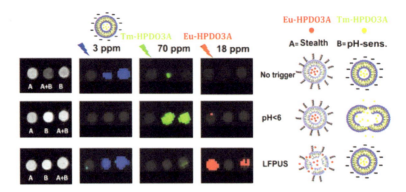

Figure 14.10 T_{2w} (left column) and CEST maps (7 T, 20°C) of a phantom consisting of three tubes filled with sono-sensitive stealth liposomes encapsulating Eu-HPDO3A (A), pH-sensitive liposomes encapsulating Tm-HPDO3A (B), and a mixed sample (A+B). The rows indicate the treatment at which the samples were exposed, whereas the columns represent the different saturation frequencies applied. Reprinted from Ref. [28], Copyright 2013, with permission of Springer.

of the limiting effect of the liposome membrane. Hence, the pH-stimulated release of the liposomal content was associated with the disappearance of the CEST contrast at 3 ppm and the concomitant detection of the contrast at 70 ppm (Fig. 14.10). The sono-sensitive formulation was not a real lipoCEST agent, but rather a vesicle-based CEST agent (see Section 14.2.3) because the entrapped complex (Eu-HPDO3A) is a very weak SR for water protons due to the small C_1 value of Eu. Therefore, this liposome was CEST-silent in the absence of the US stimulation (no lipoCEST effect and quenched contrast from the entrapped paraCEST agent). But the US-triggered release of the complex resulted in a CEST contrast detectable at 18 ppm, resonance corresponding to the hydroxyl proton of the released chelate. Also the extended frequency window of non-spherical lipoCEST agents was exploited for multiplex imaging. Our group reported the first example demonstrating the feasibility of detecting the individual CEST contrast from a mixture of two lipoCEST agents injected ex vivo in bovine muscle [38]. The two working frequencies were 3 ppm (spherical formulation) and 18 ppm (osmotically shrunken formulation), and the results are shown in Fig. 14.11. The first in vivo proof of concept of multiplex imaging

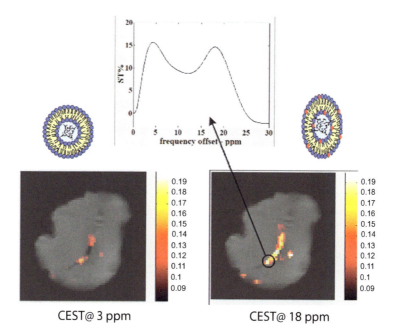

Figure 14.11 CEST maps (7 T and 39°C) at 3 ppm (left) and 18 ppm (right) superimposed on a proton density MR image of a bovine muscle injected with 100 μL of an aqueous suspension containing both a spherical and an aspherical lipoCEST agent. The color coding is the same for both agents. Top: CEST profile of a selected region that highlights the simultaneous presence of the two probes. Reprinted with permission from Ref. [38], Copyright 2008, John Wiley and Sons.

with lipoCEST agents, where different pairs of formulations were subcutaneously or intramuscularly injected in healthy mice and individually detected followed 1 year later [19].

Finally, the excellent sensitivity displayed by lipoCEST agents prompted the assessment of their potential in targeting experiments. The most relevant report was published by Flament et al., who formulated a spherical lipoCEST agent conjugated with RGD, a tripeptide targeting the α_v-β_3 integrin overexpressed in tumor angiogenesis [39]. After intravenous injection of the agent in mice bearing U87MG glioblastoma, CEST contrast was detected in the tumor in the first two hours (Fig. 14.12). Though lower than 2%, the contrast measured for the targeted agent displayed a much slower

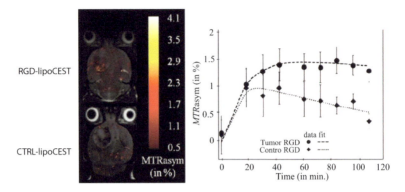

Figure 14.12 Left: CEST maps of mice bearing U87MG-induced intracerebral tumor 60 min after intravenous injection of RGD- or CTRL-lipoCEST. Right: temporal evolution of the CEST contrast in tumor and controlateral brain regions after injection of RGD-lipoCEST agent. Reprinted with permission from Ref. [39], Copyright 2012, John Wiley and Sons.

tumor washout with respect to the contrast dynamics observed after the injection of an untargeted lipoCEST used as control. The longer persistence of the contrast for the targeted agent is an indication of the effective binding to the endothelial receptor, which was further confirmed by immunofluorescence.

14.3.2 CellCEST Agents

Liposomes have been commonly used as cellular models. Both systems can be considered vesicles limited by the water permeability of phospholipid-based bilayers. Interestingly, cells are not spherical objects and, in particular, red blood cells (RBCs) display the well-known highly asymmetrical bi-concave shape. This property makes them well suited as containers for the design of CEST agents potentially more sensitive than lipoCEST based on the increased number of intravesicular protons. In fact, it can be estimated that the sensitivity of this type of CEST probes could reach the sub-picomolar range (in terms of cells concentration).

In order to develop cellCEST agents, several paramagnetic Ln-HPDO3A complexes (Ln = Eu, Gd, Dy, Tm, and Yb) have been loaded inside RBCs to act as SRs for the intracellular spins. Such chelates

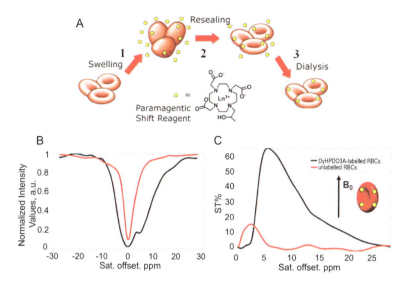

Figure 14.13 (A) Scheme of RBCs labeling by hypotonic swelling procedure. RBCs are placed in a hypotonic solution in the presence of Ln-HPDO3A allowing internalization of the SR (step 1). Then the external medium is brought to an isotonic osmolarity and the morphology of the cells is restored (i.e., resealing of the RBC, step 2). Finally, the lanthanide-loaded cells are extensively washed with phosphate buffered saline (PBS) to eliminate SRs that have not been internalized (step 3). (B) Z-spectra and (C) CEST spectra of unlabeled (*red*) and Dy-HPDO3A-labeled (*black*) RBCs. Reprinted with permission from Ref. [11], Copyright 2014, American Chemical Society.

have high thermodynamic and kinetic stabilities and are very well tolerated by cells even at high intracellular concentrations [40]. A safe and efficient procedure to internalize Ln-HPDO3A complexes inside RBCs is through hypotonic swelling (Fig. 14.13A) [41]. This labeling procedure allowed the internalization of a large amount of Ln-complexes inside RBCs (approximately 3×10^8 Ln-complex/cell, corresponding to an intracellular concentration of 4–5 mM) without any effect on their shape, size, morphology, and physiology [11]. The Z-spectra and CEST spectra of a suspension of Ln-loaded-RBCs suspended in phosphate buffered saline (PBS) (Fig. 14.13B,C) were very different from those of unlabeled RBCs. In particular, in Dy-HPDO3A-loaded RBCs, a clear asymmetry centered at 6.5 ppm from bulk water with a corresponding CEST percentage value

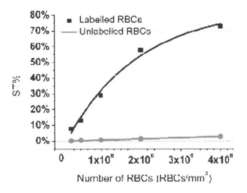

Figure 14.14 Correlation between number of RBCs/μL and CEST effect for Dy-HPDO3A-labeled (*black*) and unlabeled (*red*) cells. Reprinted with permission from Ref. [11], Copyright 2014, American Chemical Society.

of 65% was observed. In the case of unlabeled RBCs, a right-hand asymmetry of the Z-spectrum at 2–3 ppm downfield of the water resonance peak was detected. This effect was caused by the presence of a pool of exchanging protons that are naturally present in erythrocytes. Due to the relatively low maximum achievable intracellular concentration of SR, the chemical shift of the bulk-exchanging intracellular protons is exclusively due to the BMS contribution. As RBCs are naturally oriented with their long axes parallel to B_0, a positive Δ value is measured. Furthermore, as Δ_{BMS} is proportional to μ_{eff}, Dy is the lanthanide of choice to maximize the shift.

The detection of CEST contrast by Dy-HPDO3A-loaded RBCs was very sensitive (Fig. 14.14), with a threshold of approximately 2.5×10^5 cells/μL (in a mouse, this value corresponds to approximately 5% of physiologically circulating RBCs). The unlabeled RBCs did not show a significant effect (ST% < 5%) at the specific frequency of the intracellular shifted water neither at the higher number of cells used.

Since RBCs are fully retained in the vascular space, without escaping into the extravascular compartment, Dy-RBCs have been proposed as MRI reporters of tumor vascular volume. Figure 14.15 reports a representative MR image of a mouse bearing a transplantable breast cancer tumor. The CEST-spectrum of the tumor

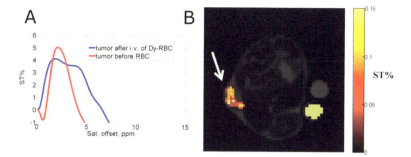

Figure 14.15 (A) CEST spectra of tumor region pre and post Dy-RBCs injection; (B) ST map at 4 ppm for vascular volume map. Reprinted with permission from Ref. [11], Copyright 2014, American Chemical Society.

region after the administration of Dy-loaded RBCs clearly shows the presence of CEST contrast at 4 ppm. The CEST percent is not homogeneous in the tumor region, which indicates heterogeneity in the blood vessels' distribution in the tumor mass.

A further evolution of the concepts around cellCEST agents led us to design lipoCEST/cellCEST aggregates that may be exploited as innovative theranostic agents [42]. It has been shown that spherical lipoCEST agents encapsulated with Dy-HPDO3A in their inner aqueous cavity and endowed with a residual positive charge on their surface can be electrostatically anchored on the membrane of RBCs. The adhesion of the liposomes on the cell surface triggered a shift of the intracellular water pool by inducing BMS effects from the external side of the RBCs (Fig. 14.16). It followed that these Dy-lipoCEST/RBCs aggregates yielded two CEST pools represented by intraliposomal and cytoplasmatic protons, respectively. The intraliposomal pool displayed a negative shift due to the hyperfine contribution given by Dy-HPDO3A, whereas the intracellular signal showed a positive shift originated by the BMS effect induced by the lipoCEST agents anchored on the external surface of the RBCs. An in-depth study highlighted that the shift of the intracellular pool can be properly modulated by acting on the liposomes size, the $[SR]^{IL}$ value, the magnetic properties of the encapsulated SR, and the number of vesicles anchored on RBCs surface. Dy-lipoCEST/RBC aggregates (positive shift) can be used to evaluate the vascular volume of the

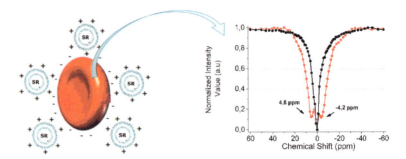

Figure 14.16 Left: scheme of binding of cationic paramagnetic liposomes to RBCs to form Dy-lipoCEST/RBCs aggregates. Right: Z-spectra of control RBCs (black) and Dy-lipoCEST/RBCs aggregates (red). Reprinted with permission from Ref. [42], Copyright 2014, American Chemical Society.

tumor region since they are fully confined in the intravascular space. Simultaneously, the quantification of Dy-lipoCEST (negative shift) contrast can report on the release of the liposomes in the tumor region. After the intravenous administration of Dy-lipoCEST/RBCs aggregates, both the positive and negative CEST pools can be visualized (Fig. 14.17).

The CEST contrast generated by RBCs in the entire tumor region rapidly decreased 1 h after the administration of the aggregates. Conversely, the contrast from Dy-lipoCEST did not change so much and it was still detected 1 h post injection. These data can provide valuable information about the vascular volume, as well as on the amount of lipoCEST agents that detached from RBCs and likely extravasated in the tumor extracellular matrix. Therefore, lipoCEST/cellCEST aggregates may represent an interesting route to improve the circulation lifetime of the liposomes, whereas the CEST contrast modality may allow the assessment of the release and accumulation of the liposomes at the target site.

The use of cells as CEST reporters opens a number of interesting directions. First, one may envisage a role in cell tracking and homing. The shift of the intracellular water decrease upon the proliferation of cells and this parameter can be exploited to assess the number of cellular divisions that have occurred. The use of other paramagnetic metal complexes than Ln-HPDO3A may eventually provide systems with enhanced ability to shift further the water resonance. Next, as

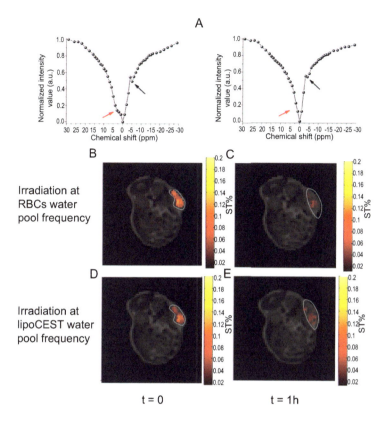

Figure 14.17 (A) Z-spectrum of in vivo liposome/RBCs aggregates, at $t = 0$ (left) and $t = 1$ h (right), reporting positive (3.2 ppm) and negative (−4.2 ppm) signals belonging to RBCs and lipoCEST, respectively. CEST percent map of tumor region with irradiation RF offset at 3.2 ppm at $t = 0$ (B) and $t = 1$ h (C). CEST percent map with irradiation RF offset at −4.2 ppm signal at $t = 0$ (D) and $t = 1$ h (E). The ROI has been circled with a white line. Reprinted with permission from Ref. [42], Copyright 2014, American Chemical Society.

the CEST effect is sensitive to the exchange rate of water protons across the cellular membrane, cellCEST may allow getting more insights into the determinants of membrane permeability and its relationship with the upsurge of pathological states. Finally, the results obtained from lipoCEST/RBC aggregates may prompt new routes to visualize molecular targets on the cellular membrane once the lipoCEST agent is functionalized with the proper targeting vector.

14.3.3 Liposomes Loaded with CEST Agents

Liposomes loaded with conventional CEST agents have been demonstrated to have a great potential for in vivo translation in several application fields. The ability of liposomes to accumulate in tumors via the well-known enhanced permeability and retention (EPR) effect has been exploited by Chan et al., who demonstrated the feasibility to monitor the tumor accumulation of liposomes co-loaded with the anticancer drug doxorubicin and barbituric acid (BA) acting as biocompatible diaCEST molecule [43]. Interestingly, the approach was validated using a molecule (TNF-α) in clinical trials as permeability enhancer of tumor endothelium. Figure 14.18 illustrates the enhancement of the CEST contrast detected in the tumor associated with the increased permeability.

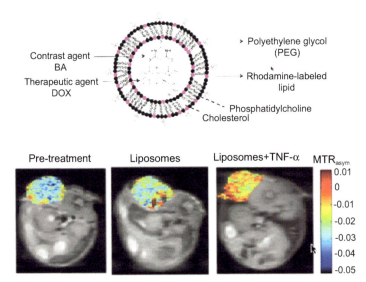

Figure 14.18 Left: schematic representation of the liposomes co-loaded with doxorubicin and barbituric acid (BA). The fluorescent rhodamine-labeled phospholipid was added for in vitro fluorescence microscopy. Right: CEST maps at 5 ppm for representative animals before the treatment (left), after the systemic injection of liposomes to a mouse bearing a colorectal adenocarcinoma xenograft before (middle) and after (right) the treatment with TNF-α. Reprinted from Ref. [43], Copyright 2014, with permission from Elsevier.

Later, the same team demonstrated that liposomes loaded with BA can be successfully used to monitor by CEST imaging the mucosal delivery of hydrophilic drugs [44]. The authors developed an optimized formulation that coupled good encapsulation efficiency with a uniform coverage of the vaginal mucosa.

Liposomes loaded with the pH-sensitive diaCEST molecule arginine have been proposed as MRI nanosensors to monitor cell viability in vivo [45]. The vesicles were embedded in alginate-

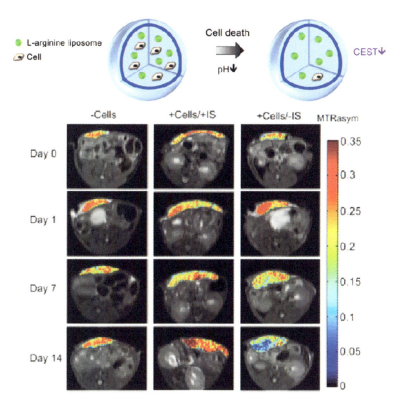

Figure 14.19 MRI-CEST images (CEST@2 ppm) of Balb/c mice subcutaneously transplanted with 2500 alginate microcapsules embedded with arginine-loaded liposomes without cells (−Cells), with Luc-transfected hepatocytes while receiving immunosuppression (+Cells/+IS), and with capsules containing cells but without immunosuppression (+Cells/−IS). Reprinted by permission from Macmillan Publishers Ltd: *Nature Materials* [45], Copyright 2013.

based microcapsules along with hepatocytes suitably transfected to express luciferase (Fig. 14.19). The CEST contrast, upon saturation of the mobile protons of arginine, decreases at acidic pH, a condition that accompanies cell death. Cell-loaded microcapsules were injected subcutaneously into two groups of mice and CEST contrast was followed for 2 weeks. One group of animals was immunosuppressed to prolong cell lifetime, whereas the other did not receive any treatment. The series of MRI-CEST images reported in Fig. 14.19 highlight the good performance and sensitivity of the approach. In addition, this method was able to discriminate between apoptotic and dead cells, as demonstrated by in vitro experiments. Importantly, alginate-encapsulated cells are currently being tested in clinical trials, thus envisaging a possible clinical translation of this technology.

Finally, in addition to diamagnetic and paramagnetic CEST agents, also hyperCEST systems have been loaded in liposomes [46]. Schnurr et al. incorporated the lipophilic cryptophane-A

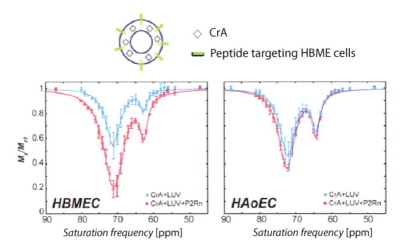

Figure 14.20 Z-spectra (9.4 T, 37°C, 45–90 ppm range) of HBMECs (left) and HAoECs (right) pellets incubated for 4 h at 37°C with CrA+LUV (cyan) and CrA+LUV+P2Rn (magenta). The resonance at 62 ppm is related to ^{129}Xe temporarily encapsulated in CrA that was released after washing the cells and is dissolved in solution. The resonance at 71 ppm is related to Xe temporarily encapsulated in CrA associated with cells. Reprinted with permission from Ref. [46], Copyright 2014, John Wiley and Sons.

(CrA) cage for ^{129}Xe in a liposome bilayer (Fig. 14.20) with the aim of designing a hyperCEST probe for targeting brain vascular endothelium. Targeting was accomplished exposing on the liposome surface an arginine-rich peptide (P2Rn). A good difference in CEST performance was observed between the target (HMBECs, human brain microvascular endothelial cells) and control (HAoECs, human aortic endothelial cells) cells after in vitro incubation with the liposomal hyperCEST agent. The use of liposomes was pursued to circumvent the direct targeting of CrA that would have required cumbersome chemical modifications and thus lends flexibility in the development of future CrA-based contrast agents for the design of theranostic Xe-based MRI applications.

References

1. Goffeney N, Bulte JWM, Duyn J, Bryant Jr. LH, and van Zijl PCM. Sensitive NMR detection of cationic-polymer-based gene delivery systems using saturation transfer via proton exchange. *J. Am. Chem. Soc.*, 2001; 123: 8628–8629.
2. Snoussi K, Bulte JWM, Gueron M, and van Zijl PCM. Sensitive CEST agents based on nucleic acid imino proton exchange: Detection of poly(rU) and of a dendrimer-poly(rU) model for nucleic acid delivery and pharmacology. *Magn. Reson. Med.*, 2003; 49: 998–1005.
3. Aime S, Delli Castelli D, and Terreno E. Supramolecular adducts between poly-L-arginine and [TmIIIdotp]: A route to sensitivity-enhanced magnetic resonance imaging-chemical exchange saturation transfer agents. *Angew. Chemie Int. Ed.*, 2003; 42: 4527–4529.
4. Meser Ali M, Yoo B, and Pagel MD. Tracking the relative in vivo pharmacokinetics of nanoparticles with PARACEST MRI. *Mol. Pharm.*, 2009; 6: 1409.
5. Winter PM, Cai K, Chen J, Adair CR, Kiefer GE, Athey PS, Gaffney PJ, Buff CE, Robertson JD, Caruthers SD, Wickline SA, and Lanza GM. Targeted PARACEST nanoparticle contrast agent for the detection of fibrin. *Magn. Reson. Med.*, 2006; 56: 1384.
6. Wu Y, Zhao P, Kiefer GE, and Sherry AD. Multifunctional polymeric scaffolds for enhancement of paraCEST contrast sensitivity and performance: Effects of random copolymer variations. *Macromolecules*, 2010; 43: 6616–6624.

7. Evbuomwan ON, Merritt EM, and Kiefer GE. Nanoparticle-based paraCEST agents: The quenching effect of silica nanoparticles on the CEST signal from surface-conjugated chelates. *Contrast Media Mol. Imaging*, 2012; 7: 19–25.
8. Vasalatiy O, Gerard D, Zhao P, Sun X, and Sherry AD. Labeling of adenovirus particles with paraCEST agents. *Bioconjug. Chem.*, 2008; 19: 598–606.
9. Evbuomwan OM, Kiefer G, and Sherry AD. Amphiphilic EuDOTAtetraamide complexes form micelles with enhanced CEST sensitivity. *Eur. J. Inorg. Chem.*, 2012; 12: 2126–2134.
10. Aime S, Delli Castelli D, and Terreno E. Highly sensitive MRI chemical exchange saturation transfer agents using liposomes. *Angew. Chem. Int. Ed.*, 2005; 44: 5513–5515.
11. Ferrauto G, Delli Castelli D, Di Gregorio E, Langereis S, Burdinski D, Grüll H, Terreno E, and Aime S. Lanthanide-loaded erythrocytes as highly sensitive chemical exchange saturation transfer MRI contrast agents. *J. Am. Chem. Soc.*, 2014; 136: 638–641.
12. Delli Castelli D, Terreno E, Longo D, and Aime S. Nanoparticle-based chemical exchange saturation transfer (CEST) agents. *NMR Biomed.*, 2013; 26: 839–849.
13. Chu SC-K, Xu Y, Balschi JA, and Springer Jr CS. *Magn. Reson. Med.*, 1990; 13: 239–262.
14. Kuchel PW, Chapman BE, Bubb WA, Hansen PE, Durrant CJ, and Hertzberg MP. *Concepts Magn. Reson. Part A*, 2003; 18: 56–71.
15. Chahida B, Vander Elst L, Flament J, Boumezbeur F, Medina C, Port M, Muller RN, and Lesieur S. Entrapment of a neutral Tm(III)-based complex with two inner-sphere coordinated water molecules into PEG-stabilized vesicles: Towards an alternative strategy to develop high-performance lipoCEST contrast agents for MR imaging. *Contrast Media Mol. Imaging*, 2014; 9: 391–399.
16. Corsi DM, Platas-Iglesias C, van Bekkum H, and Peters JA. Determination of paramagnetic lanthanide(III) concentrations from bulk magnetic susceptibility shifts in NMR spectra. *Magn. Res. Chem.*, 2001; 39: 723–726.
17. Ottiger M and Bax A. Characterization of magnetically oriented phospholipid micelles for measurement of dipolar couplings in macromolecules. *J. Biomol. NMR*, 1998; 12: 361–372.
18. Prosser RS and Shiyanovskaya IV. Lanthanide ion assisted magnetic alignment of model membranes and macromolecules. *Concepts Magn. Reson.*, 2001; 13: 19–31.

19. Aime S, Delli Castelli D, and Terreno E. Lanthanide-loaded paramagnetic liposomes as switchable magnetically oriented nanovesicles. In Nejat Düzgüneş, ed.: *Methods in Enzymology*, Vol. 464, Burlington: Academic Press, 2009, pp. 193–210.
20. Terreno E, Delli Castelli D, Violante E, Sanders HM, Sommerdijk NA, and Aime S. Osmotically shrunken lipoCEST agents: An innovative class of magnetic resonance imaging contrast media based on chemical exchange saturation transfer. *Chem. Eur. J.*, 2009; 15: 1440–1448.
21. Terreno E, Barge A, Beltrami L, Cravotto G, Delli Castelli D, Fedeli F, Jebasingh B, and Aime S. Highly shifted LIPOCEST agents based on the encapsulation of neutral polynuclear paramagnetic shift reagents. *Chem. Commun.*, 2008; 600–602.
22. Terreno E, Cabella C, Carrera C, Delli Castelli D, Mazzon R, Rollet S, Stancanello J, Visigalli M, and Aime S. From spherical to osmotically shrunken paramagnetic liposomes: An improved generation of LIPOCEST MRI agents with highly shifted water protons. *Angew. Chem. Int. Ed.*, 2007; 46: 966–968.
23. Burdinski D, Pikkemaat JA, Emrullahoglu M, Costantini F, Verboom W, Langereis S, Grüll H, and Huskens J. Targeted lipoCEST contrast agents for magnetic resonance imaging: Alignment of aspherical liposomes on a capillary surface. *Angew. Chem. Int. Ed.*, 2010; 49: 2227–2229.
24. Grull H, Langereis S, Messager L, Delli Castelli D, Sanino A, Torres E, Terreno E, and Aime S. Block copolymer vesicles containing paramagnetic lanthanide complexes: A novel class of T1- and CEST MRI contrast agents. *Soft Matter*, 2010; 6: 4847–4850.
25. Zhao JM, Har-el Y, McMahon MT, Zhou J, Sherry AD, Sgouros G, Bulte JWM, and van Zijl PCM. Size-induced enhancement of chemical exchange saturation transfer (CEST) contrast in liposomes. *J. Am. Chem. Soc.*, 2008; 130: 5178–5184.
26. Terreno E, Sanino A, Carrera C, Delli Castelli D, Giovenzana GB, Lombardi A, Mazzon R, Milone L, Visigalli M, and Aime S. Determination of water permeability of paramagnetic liposomesof interest in MRI field. *J. Inorg. Biochem.*, 2008; 1112–1118.
27. Opina AC, Ghaghada KB, Zhao P, Kiefer G, Annapragada A, and Sherry AD. TmDOTA-tetraglycinate encapsulated liposomes as pH-sensitive lipoCEST agents. *PLoS One*, 2011; 6: e27370.
28. Delli Castelli D, Boffa C, Giustetto P, Terreno E, and Aime S. Design and testing of paramagnetic liposome-based CEST agents for MRI visualization of payload release on pH-induced and ultrasound stimulation. *J. Biol. Inorg. Chem.*, 2014; 19: 207–214.

29. Liu G, Moake M, Har-el Y, Long CM, Chan KWY, Cardona A, Jamil M, Walczak P, Gilad AA, Sgouros G, van Zijl PCM, Bulte JWM, and McMahon MT. In vivo multi-color molecular MR imaging using DIACEST liposomes. *Magn. Reson. Med.*, 2012; 67: 1106–1113.
30. Chan KWY, Bulte JWM, and McMahon MT. Diamagnetic chemical exchange saturation transfer (diaCEST) liposomes: Physicochemical properties and imaging applications. *Wiley Interdiscip. Rev. Nanomed. Nanobiotechnol.*, 2014; 6: 111–124.
31. Heneweer C, Gendy SE, and Peñate-Medina O. Liposomes and inorganic nanoparticles for drug delivery and cancer imaging. *Ther. Deliv.*, 2012; 3: 645–656.
32. Terreno E, Delli Castelli D, Viale A, and Aime S. Challenges for molecular magnetic resonance imaging. *Chem. Rev.*, 2010; 110: 3019–3042.
33. Mulas G, Ferrauto G, Dastrù W, Anedda R, Aime S, and Terreno E. Insights on the relaxation of liposomes encapsulating paramagnetic Ln-based complexes. *Magn. Reson. Med.*, 2014; 74(2): 468–473.
34. Aime S, Delli Castelli D, Lawson D, and Terreno E. Gd-loaded liposomes as T1, susceptibility, and CEST agents, all in one. *J. Am. Chem. Soc.*, 2007; 129: 2430–2431.
35. Delli Castelli D, Dastrù W, Terreno E, Cittadino E, Mainini F, Torres E, Spadaro M, and Aime S. In vivo MRI multicontrast kinetic analysis of the uptake and intracellular trafficking of paramagnetically labeled liposomes. *J. Control. Release*, 2010; 144: 271–279.
36. Langereis S, Keupp J, van Velthoven JL, de Roos IH, Burdinski D, Pikkemaat JA, and Grüll H. A temperature-sensitive liposomal ^{1}H CEST and ^{19}F contrast agent for MR image-guided drug delivery. *J. Am. Chem. Soc.*, 2009; 131: 1380–1381.
37. Terreno E, Boffa C, Menchise V, Fedeli F, Carrera C, Delli Castelli D, Digilio G, and Aime S. Gadolinium-doped lipoCEST agents: A potential novel class of dual [1]H-MRI probes. *Chem. Commun.*, 2011; 47: 4667–4669.
38. Terreno E, Delli Castelli D, Milone L, Rollet S, Stancanello J, Violante E, and Aime S. First ex-vivo MRI co-localization of two LIPOCEST agents. *Contrast Media Mol. Imaging*, 2008; 3: 38–43.
39. Flament J, Geffroy F, Medina C, Robic C, Mayer JF, Mériaux S, Valette J, Robert P, Port M, Le Bihan D, Lethimonnier F, and Boumezbeur F. In vivo CEST MR imaging of U87 mice brain tumor angiogenesis using targeted lipoCEST contrast agent at 7 T. *Magn. Reson. Med.*, 2013; 69: 179–187.

40. Di Gregorio E, Ferrauto G, Gianolio E, and Aime S. Gd loading by hypotonic swelling: An efficient and safe route for cellular labeling. *Contrast Media Mol. Imaging,* 2013; 8: 475–486.
41. Rossi L, Serafini S, Pierigé F, Antonelli A, Cerasi A, Fraternale A, Chiarantini L, and Magnani M. Erythrocyte-based drug delivery. *Expert Opin. Drug Deliv.*, 2005; 2: 311–322.
42. Ferrauto G, Di Gregorio E, Baroni S, and Aime S. Frequency-encoded MRI-CEST agents based on paramagnetic liposomes/RBC aggregates. *Nano Lett.*, 2014; 14: 6857–6862.
43. Chan KWY, Yu T, Qiao Y, Liu Q, Yang M, Patel H, Liu G, Kinzler KW, Vogelstein B, Bulte JWM, van Zijl PCM, Hanes J, Zhou S, and McMahon MT. A diaCEST MRI approach for monitoring liposomal accumulation in tumors. *J. Control. Release*, 2014; 180: 51–59.
44. Yu T, Chan KWY, Anonuevo A, Song X, Schuster BS, Chattopadhyay S, Xu Q, Oskolkov N, Patel H, Ensign LM, van Zjil PCM, McMahon MT, and Hanes J. Liposome-based mucus-penetrating particles (MPP) for mucosal theranostics: Demonstration of diamagnetic chemical exchange saturation transfer (diaCEST) magnetic resonance imaging (MRI). *Nanomed. Nanotech. Biol. Med.*, 2015; 11: 401–405.
45. Chan KWY, Liu G, Song X, Kim H, Yu T, Arifin DR, Gilad AA, Hanes J, Walczak P, van Zijl PCM, Bulte JWM, and McMahon MT. MRI-detectable pH nanosensors incorporated into hydrogels for in vivo sensing of transplanted cell viability. *Nat. Mater.*, 2013; 12: 268–275.
46. Schnurr M, Sydow K, Rose HM, Dathe M, and Schröder L. Brain endothelial cell targeting via a peptide-functionalized liposomal carrier for xenon hyper-CEST MRI. *Adv. Healthcare Mater.*, 2015; 4: 40–45.

Section IV

Emerging Clinical Applications of CEST Imaging

Chapter 15

Principles and Applications of Amide Proton Transfer Imaging

Jinyuan Zhou,[a,b] Yi Zhang,[a] Shanshan Jiang,[a] Dong-Hoon Lee,[a] Xuna Zhao,[a] and Hye-Young Heo[a]

[a]*Department of Radiology, Johns Hopkins University School of Medicine, 600 N Wolfe Street, Park 336, Baltimore, MD 21287, USA*
[b]*F. M. Kirby Research Center, Kennedy Krieger Institute, 707 N Broadway, Baltimore, MD 21205, USA*
jzhou@mri.jhu.edu

15.1 Introduction

Magnetic resonance imaging (MRI) is a versatile technology that employs water content and water relaxation properties to image the anatomy and physiology of the body. Currently, MRI has been widely used in hospitals for the localization, diagnosis, and characterization of cancer [1, 2], stroke [3], and other diseases, and for the assessment of treatment effects. In addition to the conventional MRI sequences, such as T_1-weighted (T_1w) and T_2-weighted (T_2w), several advanced, functional MRI techniques [4, 5], such as perfusion imaging, diffusion imaging, and proton MR spectroscopic imaging, have been emerging in clinical research protocols. As MRI is applied further at the molecular level, more

---

Chemical Exchange Saturation Transfer Imaging: Advances and Applications
Edited by Michael T. McMahon, Assaf A. Gilad, Jeff W. M. Bulte, and Peter C. M. van Zijl
Copyright © 2017 Pan Stanford Publishing Pte. Ltd.
ISBN 978-981-4745-70-3 (Hardcover), 978-1-315-36442-1 (eBook)
www.panstanford.com

possibilities for disease diagnosis and treatment assessment will become evident [6, 7]. Currently, most molecular and cellular MRI studies rely on the administration of paramagnetic or superparamagnetic metal-based substrates that are potentially harmful. Ideally, molecular imaging would exploit endogenous molecules that can be probed non-invasively using existing hardware. Pioneered by Balaban et al. [8–10], chemical exchange saturation transfer (CEST) imaging is a new molecular MRI method that allows detection of low-concentration, endogenous or exogenous chemicals with exchangeable protons through the water signal [11–13]. In the past several years, many new CEST contrast agents have been designed—diamagnetic [14], paramagnetic (paraCEST) [15, 16], and even hyperpolarized (hyperCEST) [17]—and the validation of new types of applications is progressing rapidly on many fronts.

According to their mobility, proteins in biological tissue can be roughly distinguished into two types: bound proteins (such as nuclear proteins and membrane proteins) that possess solid-like properties and have a very short proton transverse relaxation time T_2 (~10 µs); and mobile proteins (such as cytosolic proteins, many endoplasmic reticulum proteins, and secreted proteins) that rotate rapidly and have a relatively long proton T_2 (tens of ms). Bound semi-solid proteins can be assessed by conventional magnetization transfer (MT) imaging [18, 19], but mobile cytosolic proteins have relatively rarely been examined. Behar et al. [20, 21] and Kauppinen et al. [22] first detected and identified several macromolecular peaks in brain proton MR spectra in the low-frequency range (0–4 ppm) with respect to water. These signals were attributed to the aliphatic protons from mobile proteins, polypeptides, and lipids [20–23]. In addition, using specially designed pulse sequences, a composite resonance around 8.3 ± 0.5 ppm can be observed in the proton spectra of cancer cells and animal brain in situ [24–26]. Based on the CEST principle [8–10], using the exchangeable amide protons in the protein backbone, Zhou et al. demonstrated that these proteins can be detected in vivo using the water signal, thus enabling imaging on clinical MRI scanners. This technique was called amide proton transfer (APT) imaging [26]. Together with other CEST types, the APT approach has opened a new venue of molecular MRI research, as summarized in numerous excellent review articles [11–13, 27–

31]. In this chapter, we review the basic principles and applications of this novel APT imaging technique.

15.2 APT Imaging Principle and Theory

Amide protons (NH) in the backbone of solute proteins resonate around 8.25 ppm in the proton MR spectrum [32], approximately 3.5 ppm downfield of the water signal, and interact with water protons through chemical exchange (Fig. 15.1a). The exchange rate (k) of these protons is pH dependent and typically decreases tenfold per unit of pH drop [33]. The endogenous mobile proteins and peptides in biological tissue consists of a new class of CEST agents [34, 35] for which an entirely new protein-based MRI technique, dubbed APT imaging [26], has been developed. In principle, APT imaging can be acquired using a pulse sequence that is similar to standard MT experiments [18, 19]. However, while the APT effect originates from multiple types of amide protons, their composite resonance is visible as a relatively narrow signal at a specific frequency offset, namely, 3.5 ppm downfield from the water resonance, and the applied radiofrequency (RF) saturation power is always relatively low (a few µT). Technically, the APT effect is measured as a reduction in bulk water intensity due to the chemical exchange of water protons with magnetically labeled backbone amide protons of endogenous mobile proteins (Fig. 15.1b). Thus, specific molecular information is obtained indirectly through the bulk water signal usually used in imaging. Such labeling is accomplished using selective RF irradiation at the MR frequency of the backbone amide protons, roughly 3.5 ppm downfield of the water resonance, causing saturation (or signal destruction) that is transferred to water protons. Although the effect of mM-concentration amide protons on the water resonance (~110 M protons) may not be detectable in a single transfer, the repeated saturation and exchange in which unsaturated water protons are replaced by saturated solute protons allow sensitivity enhancements of 100–1000 fold.

When performing such APT experiments in vivo, the direct water saturation (DS) and conventional MT effects will interfere with the measurements. The sum of all saturation effects is generally called

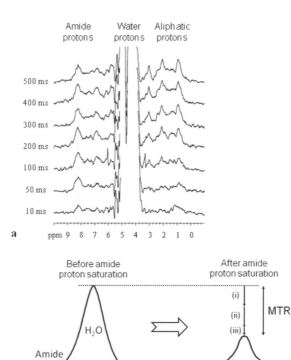

Figure 15.1 (a) In vivo water-exchange (WEX) spectra for the rat brain at 4.7 T as a function of mixing time. Note the early appearance of amide signals around 8.3 ppm and the subsequent slower label transfer to the aliphatic protons downfield from the water resonance. (Reprinted with permission from Ref. [26], Copyright 2003, Nature Publishing Group.) (b) APT signal enhancement principle. The two exchangeable proton pools are the dilute amide protons (at 3.5 ppm downfield of the water resonance), and the large bulk water protons. The RF irradiation selectively saturates exchangeable amide protons, which subsequently exchange with unsaturated water protons. Once on the solute, these protons become saturated and the process repeats itself. If the saturation time is long enough (on the order of seconds), the bulk water signal decreases significantly due to progressive saturation transfer, enabling the detection of low concentrations. Note that the APT effect always coexists with the large conventional MT and DS effects, and the asymmetry analysis regarding the water resonance is often used to selectively assess the APT signal.

the MT ratio, $MTR = 1 - S_{sat}/S_0$, where S_{sat} and S_0 are the signal intensities with and without RF irradiation. The APT signal is usually measured through the so-called MTR asymmetry analysis, at an offset of 3.5 ppm [26]:

$$MTR_{asym}(3.5) = MTR(+3.5) - MTR(-3.5)$$
$$= [S_{sat}(-3.5) - S_{sat}(+3.5)]/S_0$$
$$= APTR + MTR'_{asym}(3.5) \quad (15.1)$$

where APTR is the proton transfer ratio associated with amide protons, and $MTR'_{asym}(3.5)$ consists of the possible inherent asymmetry of the conventional MT effect and the possible nuclear Overhauser effects (NOEs) of aliphatic protons of mobile macromolecules [36]. Because of the presence of $MTR'_{asym}(3.5)$, APT images defined by $MTR_{asym}(3.5)$ should, in principle, be called APT-weighted (APTw) images [37]. Currently, an actively explored issue is to develop a more appropriate theory or approach with the ability to remove various interfering effects [38, 39]. Several alternative post-processing approaches (non-MTR asymmetry) [40–48] or image acquisition schemes [49–54] have been proposed to isolate APT signals from other saturation effects. Although promising, most of these attempts must be further evaluated in a clinical setting. On the other hand, it is important to keep in mind that the NOE is actually a positive confounding factor that enhances the APTw image contrast in the tumor. Moreover, the APTw hyperintensity in the tumor is often dominated by the APT effect, and the MTR asymmetry at 3.5 ppm is a reliable and valid metric for APT imaging of gliomas at 3 T [47, 48, 55].

Based on a two-pool exchange model under slow exchange conditions, the APTR contribution to the MTR_{asym} is given by [56]:

$$APTR = \{[NH]/(2 \cdot [H_2O])\} \cdot \alpha \cdot k_{sw} \cdot T_{1w}(1 - e^{-t_{sat}/T_{1w}}), \quad (15.2)$$

where k_{sw} is the average solute-to-water exchange rate over all exchangeable protons participating in the effect, [NH] is the total mobile amide proton concentration (~72 mM in brain tissue [26]), [H_2O] is the water proton concentration (55.6 M), T_{1w} is the longitudinal relaxation time of water, and t_{sat} is the RF saturation time. When interpreting these data, it is important to keep the CEST mechanism in mind. According to Eq. (15.2), the APT-MRI signal depends primarily on the mobile amide proton concentration and amide proton exchange rates (which are dependent on tissue pH). In the case of ischemic stroke, there is a drop in pH (~0.5 pH units [57]) in the lesion at the acute stage. The observed APT-MRI hypointensity would be dominated by the pH reduction because the changes in the protein content are minimal. In the case of brain tumors, on the other hand, because only a small intracellular pH increase is detected in the tumor [58], with respect to the normal brain tissue, the measured APT effect in the tumor would be attributed primarily to changes in the protein content [59, 60]. Stroke and tumor are two examples of important APT-MRI applications, which are discussed in more detail below.

15.3 APT Imaging of Stroke

Zhou et al. [26] first measured pH-sensitive APT effects during several physiological alterations in the rat brain on a 4.7 T MRI scanner, in 2003. They found reduced $MTR_{asym}(3.5)$ in the postmortem brain, which could be attributed to a pH reduction that causes a slower exchange. These APT effects were small (a few percent on the water signal) but corresponded to a detection sensitivity of molar concentration. By using ^{31}P spectroscopy for pH assessment (pH 7.11 in vivo versus 6.66 postmortem), they further calibrated the change in $APTR$ in a global ischemic model. Using the assumption of base-catalyzed exchange in the physiological pH range (known to be the case for most amide protons [33]), the $APTR$-pH relationship in brain was determined to be:

$$APTR = 5.73 \times 10^{pH-9.4}. \tag{15.3}$$

This equation indicates that the sensitivity for measuring pH changes in the physiological range (pH 6.5–7.5) is quite good, resulting in an $APTR$ reduction of 65% between pH 7.11 and 6.66. In the case of stroke, the $MTR_{asym}(3.5)$ images calculated from Eq. (15.1) are called the pH-weighted (pHw) images.

After demonstrating the existence of APT effects in a global ischemic model, Zhou et al. subsequently used APT to study a permanent middle cerebral artery occlusion (MCAO) model in the rat brain [26]. Using Eq. (15.3), an absolute pH image could be generated, correctly outlining the ischemic area in the caudate nucleus (Fig. 15.2a), a region commonly affected by infarction following MCAO. Ischemia was confirmed by histology (acquired 8 h later). During the acute stage, no infarct was visible on the T_2w image. The average ischemic pH was 6.52 ± 0.32 ($n = 7$).

Following this initial study, Sun et al. investigated the possibility of using the APT approach to detect a separate pH-based acidosis penumbra [61–63]. They imaged adult rats with permanent MCAO using multi-parametric MRI over the first 3.5 h post-occlusion (when the animals remained in the magnet). The endpoint used was the stroke area defined by T_2 hyperintensity at 24 h. The experiment was designed to minimize the occlusion. In all images, the large MCA area showed hypoperfusion on cerebral blood flow (CBF) images, but no T_1 and T_2 changes were found during the first 3.5 h of imaging. Notably, several animals showed negligible apparent diffusion coefficient (ADC) effects in the hyperacute period, despite the presence of perfusion and pH effects, confirming that pH changes occur before ADC changes, as expected from the ischemic flow thresholds. These data suggest that the hypoperfused area, which showed a decrease in pH without an ADC abnormality, corresponds to the ischemic acidosis penumbra, while the hypoperfused region at normal pH corresponds to benign oligemia [61].

APT-based pH imaging has been applied to permanent brain ischemia in rats by a few other investigators [64–68]. In addition, the initial clinical data (Fig. 15.2c) suggest that APT imaging is feasible for stroke patients and may provide extra information, by virtue of pH changes, about the potential of progression to infarct versus spontaneous recovery in ischemia [69–72]. It is expected that APT imaging can provide an additional marker that can be used for the diagnosis and prognosis of stroke patients [27]. It is also expected that the pH/diffusion mismatch is better than the current standard of care criteria (perfusion/diffusion mismatch) at identifying the ischemic penumbra. This may help avoid unnecessary treatment, and perhaps be used as a reference to recommend stroke patients for treatment at longer times post-ictus, even later than the current therapeutic window.

354 | Principles and Applications of Amide Proton Transfer Imaging

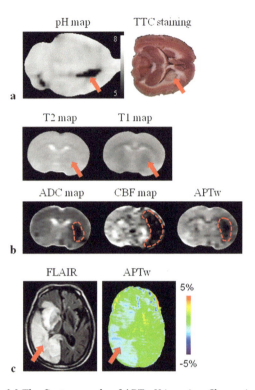

Figure 15.2 (a) The first example of APT-pH imaging. Shown is an absolute pH map of the rat brain for an MCAO model (2 h after occlusion), calculated by Eq. (15.3). APT imaging was acquired on a 4.7 T clinical MRI scanner. A train of 400 Gaussian pulses (length 6.6 ms, flip angle 180°, delay 3.4 ms, total duration 4 s, average saturation power 2 μT) was used (TR 10 s, TE 50 ms, 40 scans). The caudate nucleus (arrow) shows the largest change in pH. (Reprinted with permission from Ref. [26], Copyright 2003, Nature Publishing Group.) (b) Another example of APT imaging of rat brain ischemia (1 h after occlusion) at 4.7 T. Weak continuous wave saturation was used for APT imaging (irradiation power 1.5 μT, irradiation time 4 s, TR 10 s, TE 30 ms, 16 scans, scan time 8 min). The display windows are T_2 (0 to 100 ms), T_1 (0.5 to 2.5 s), ADC (0 to 2×10^{-9} m^2/s), CBF (0–200 ml/100g/min), and APT (−12% to 8% of the bulk water signal intensity). No signal abnormality was seen on the T_2 and T_1 maps. ADC and CBF maps show a significant mismatch, and the difference in areas is a classical penumbra. The mismatch between the ADC map and pHw image identifies a unique penumbra that is smaller than that identified by the CBF map and the ADC map. (c) FLAIR and APTw MR images of a patient with a stroke at 5 days post-onset. MRI was acquired on a 3 T clinical MRI scanner. The APT imaging parameters were saturation time 0.5 s; saturation power 2 μT; and TR 3 s. The lesion (arrow) is hypointense on the APTw images. (Reprinted with permission from Ref. [69], Copyright 2011, John Wiley and Sons.).

15.4 Differentiation between Ischemia and Hemorrhage

There are two major types of stroke, ischemic and hemorrhagic, and accurate diagnosis of hemorrhagic versus ischemic stroke is crucial for all stroke patients [73, 74]. Head CT is currently the primary neuroimaging modality for the diagnosis of acute stroke in the emergency room [75]. The main diagnostic advantage of CT in the hyperacute phase (<12 h) of stroke is its ability to detect or rule out hemorrhage. However, CT has a limited sensitivity for the identification of early cerebral ischemia in the hyperacute stage. MRI is increasingly being used for the diagnosis of acute strokes. Several studies have shown that gradient-echo MRI sequences are as accurate as CT for the detection of acute hemorrhage, and far superior to CT in the chronic stage [76–78]. However, like CT, these gradient-echo MRI sequences are not sensitive for the detection of early ischemia. Currently, for safety reasons, both CT and MRI are often obtained in stroke centers for the initial evaluation of patients. Ideally, patients would undergo only one imaging modality, MRI or CT, in order to expedite treatment and minimize cost. If MRI is used rather than CT, MRI must be able to reliably identify hemorrhagic strokes; moreover, a single MRI scan that can simultaneously detect hemorrhagic strokes and ischemic strokes at the hyperacute stage is very desirable.

In a recent study Wang et al. [79] explored the capabilities of APT imaging in differentiating between intracerebral hemorrhage (ICH) and ischemia at the hyperacute stage in rat models. The rat ICH model was induced by injecting bacterial collagenase VII-S (0.75 U in 1 µl saline) into the caudate nucleus, and the permanent ischemic stroke model was induced by inserting a 4-0 nylon suture into the lumen of the internal carotid artery to occlude the origin of the middle cerebral artery. MRI data were acquired on a 4.7 T Biospec animal imager. Figures 15.3a,b compare multi-parametric MRI features of hyperacute ICH and ischemia using rat models. As expected, both T_1 and T_2 were not sensitive enough to detect early ischemia, and ADC and CBF did not seem to be specific for ICH and ischemia. Notably, APTw showed quite a different APT-MRI appearance in ICH (hyperintense compared to the contralateral brain tissue) and cerebral ischemia (hypointense); thus, there was

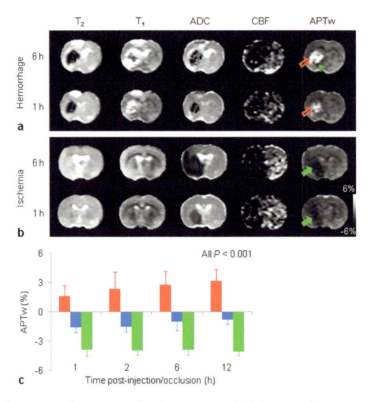

Figure 15.3 Comparison of multi-parametric MRI features of hyperacute ICH and cerebral ischemia. (a) ICH at 1 h and 6 h after the injection of collagenase (red open arrow). (b) Cerebral ischemia at 1 h and 6 h after occlusion of the middle cerebral artery (green solid arrow). The display windows are T_2 (25–75 ms), T_1 (1–2 s), ADC (0.4 to 1.2×10^{-9} m^2/s), blood flow (0–120 ml/100g/min), and APTw (−6% to 6% of the bulk water signal intensity). ADC and CBF maps showed a primarily hypointense signal in both pathologies. Areas of perfusion abnormality are larger than areas of diffusion abnormality. APTw MRI showed hyperintensity in ICH (red open arrow), but hypointensity (green solid arrow) in cerebral ischemia at both time points. Possible perihematomal ischemia (APTw hypointense, green thin arrow) was seen at 6 h post-injection. (c) Comparison of average APTw signal intensities at several time points post-injection/occlusion ($n = 10$ each group). Hyperacute ICH (red bar) and cerebral ischemia (green bar) showed opposite the APTw MRI signals. With the saturation settings used in this study, the APTw signal in the contralateral brain tissue (blue bar) was slightly negative. APTw MRI provided a stark image contrast between ICH and cerebral ischemia at the hyperacute stage. Reprinted with permission from Ref. [79], Copyright 2015, John Wiley and Sons.

a stark contrast in signal intensity between these two pathologies consistently at all of these hyperacute time points.

It has been reported previously [61] that APTw imaging can detect an ischemic lesion that is associated with local tissue acidosis following impaired aerobic metabolism even before an ADC abnormality, which usually shows a low APT-pH MRI signal in the ischemic lesion. In hyperacute hemorrhage, the hematoma following vessel rupture consists of a collection of red blood cells (which are rich in hemoglobin content), white blood cells, platelet clumps, and protein-rich serum [75]. Therefore, the hyperacute hemorrhage APT signal may reflect the presence of abundant mobile proteins and peptides, causing APTw hyperintensity in the lesion. Quantitative analysis (Fig. 15.3c) shows that APT-MRI signal intensities were significantly higher in ICH than in ischemia (all $P < 0.01$) at all time points at the hyperacute stage. Our preclinical results suggest that APT-MRI can accurately detect hyperacute ICH and distinctly differentiate hyperacute ICH from cerebral ischemia, thus opening up the possibility of introducing to the clinic a single MRI scan for the simultaneous visualization and separation of hemorrhagic and ischemic strokes at the hyperacute stage.

15.5 APT Imaging of Brain Tumors

Soon after introducing the APT technology to the imaging of stroke, Zhou et al. [80] also demonstrated the possibility of producing APT contrast in the rat 9L gliosarcoma model. The results (Fig. 15.4a) show that, unlike in ischemic stroke, a high APTw signal signifies viable tumor, compared to the contralateral hemisphere. The APTw image can identify the tumor more accurately than several conventional and advanced imaging approaches, such as T_1w, T_2w, and diffusion imaging, as confirmed by histology. In order to have results that would be more relevant to human studies, they also investigated nude rats implanted with human glioblastoma xenografts (such as U87 and GBM#22) [81]. These xenografts mimic human tumor biogenesis and growth patterns, including tumor necrosis and invasive growth. The initial data on the rat tumor models showed diffuse intensity patterns for many conventional and advanced imaging approaches, and the increases in these conventional and advanced image signals extended over an area larger than that outlined by the APTw hyperintensity and by the histological changes. The additional regions were classified as peritumoral edema. This suggests the possibility of separating tumor from edema using APT imaging.

After showing the existence of APT effects in animal brain tumors, Jones et al. [82] applied the APT technique to patients with brain tumors, in 2006. The data confirmed the preclinical findings, suggesting that there are increased APTw signals in malignant tumors. This and several subsequent studies have clearly shown that APTw imaging for malignant brain tumors has much potential (Fig. 15.4c). For example, APTw imaging has the potential to differentiate between tumor and peritumoral edema, the potential to separate high-grade from low-grade gliomas [37, 83–85], the potential to detect high-grade tumors that do not show gadolinium (Gd) enhancement [86], the potential to assess brain tumors and treatment effects [87], and the potential to differentiate between primary CNS lymphomas and high-grade gliomas [88]. Gd enhancement, although it depicts the disruption of the blood–brain barrier, is not specific for malignant tumor proliferation. Gd-enhanced MRI is limited in that some high-grade gliomas (roughly 10% of GBM and 30% of anaplastic astrocytoma [89, 90]) demonstrate no Gd enhancement. Therefore, it can be difficult to identify the most malignant portions of tumor prior to surgery or local therapies. In addition, Gd enhancement is not always specific for tumor grade, as low-grade gliomas occasionally enhance [91]. However, the prior data suggest that the APTw hyperintensity (compared to the contralateral normal-appearing white matter) is a typical feature of high-grade gliomas, consistent with the increased cellular content of proteins, as revealed by MRI-guided proteomics [59, 60] and in vivo NMR spectroscopy [92].

The clinical applications of APT to tumor detection are promising but have often been limited to single-slice acquisitions. A few technical limitations to this technique are the specific absorption rate (SAR) guidelines and a long scan time. In addition, multi-slice imaging is complicated by APT signal losses due to T_1 relaxation, with respect to the order in which the slices are acquired [93, 94].

APT Imaging of Brain Tumors | 359

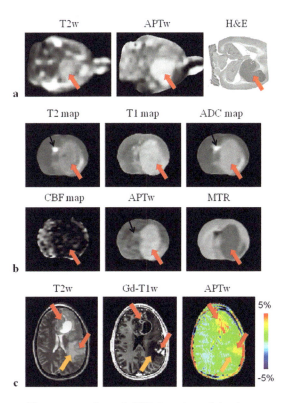

Figure 15.4 Three examples of APT imaging of brain tumors. (a) 9L gliosarcoma in a nude rat (10 days post-implantation). This was the first example of APT imaging of brain tumors. All experimental parameters are the same as those in Fig. 15.2(a). Because many proteins are overexpressed in tumor, the APTw signal would be greater in tumor than in normal brain tissue. Reprinted with permission from Ref. [80], Copyright 2003, John Wiley and Sons. (b) U87MG glioma in a rat (10 days post-implantation). MRI data were acquired on a 4.7 T clinical MRI scanner. Weak continuous wave saturation was used for APT imaging (irradiation power 1.3 μT, irradiation time 4 s, TR 10 s, TE 30 ms, 16 scans, scan time 8 min). The display windows are T_2 (0–150 ms), T_1 (0.5–2.5 sec), ADC (0 to 2×10^{-9} m^2/s), CBF (0–150 ml/100g/min), *MTR* at 2 kHz (0–50% of the bulk water signal intensity), and APTw (−8% to 8% of the bulk water signal intensity). Note: The ventricle (dark arrow) has high T_2 and ADC signals, but a low APTw signal. (c) Multifocal anaplastic astrocytoma (WHO grade 3) in a patient. MRI was acquired on a 3 T clinical MRI scanner. The APT imaging parameters were as follows: saturation time 0.8 s; saturation power 2 μT; TR 3 s. Red solid arrow: tumor core (Gd enhancement, high APTw signal); orange open arrow: peritumoral edema (low signal on APTw). The APTw image can identify the tumor (red arrow) more accurately than the T_2 map, as confirmed by Gd-enhanced MRI.

Several novel three-dimensional (3D) APT-MRI sequences have been designed in the past several years [40, 95]. Notably, Zhu et al. have developed a 3D gradient- and spin-echo (GRASE) APT-MRI technology with a sensitivity-encoding (SENSE) acceleration that allows rapid acquisition on 3T clinical instruments [95]. Based on this sequence, it is feasible to perform APTw imaging of brain tumors with 15 slices of 4.4 mm thickness (Fig. 15.5) [96], which is within a clinically relevant time frame (10 min 42 s) and FDA SAR requirements (1.1 W/kg). When combined with the time-interleaved, multi-channel transmit RF saturation technique [97], 3D APTw imaging would provide the highly sensitive, standard optimized approaches required to translate this new technology into the clinic. This could potentially improve diagnosis and treatment planning for patients with brain tumors by providing more accurate targets for biopsy, tumor resection, radiation therapy, and local chemotherapy.

15.6 Differentiation between Active Glioma and Radiation Necrosis

A major obstacle in both the daily management of patients with malignant gliomas and in the development of new agents for these cancers is the inability to accurately assess tumor response to therapy [98]. Specifically, patients often have changes on their MRI after treatment (new or increased Gd enhancement) that suggest tumor recurrence but are actually found to be radiation necrosis (treatment effects) on biopsy. The MRI appearance of tumor recurrence and radiation necrosis is often identical and indistinguishable, posing a formidable clinical and radiologic dilemma that has remained for decades [99]. The current standard chemo-radiotherapy with temozolomide improves cancer-cell killing and survival rates [100], but this regimen significantly increases the rates of treatment effects, a phenomenon called pseudoprogression [101]. Therefore, most clinicians agree that reliable imaging parameters that can distinguish between true tumor progression and treatment effects are urgently needed, leading to an international, multidisciplinary

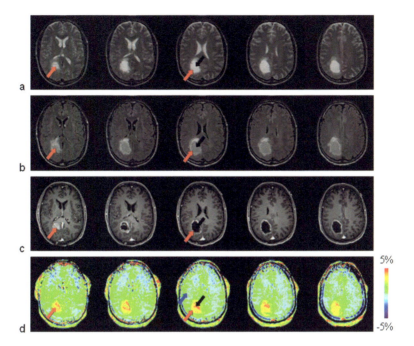

Figure 15.5 Three-dimensional APTw and conventional MR images for a patient with a glioblastoma at 3 T. (a) T_2-weighted image and (b) FLAIR image demonstrate a predominantly cystic mass in the right parietal lobe. (c) Gd-enhanced T_{1w} image shows an enhancing rim with a non-enhancing central area. (d) APTw image shows hyperintensity in the Gd-enhancing area (red arrows) and in the centrally cystic area (back arrows), compared to the contralateral brain area (blue arrow). APT imaging was acquired with the 3D GRASE approach (RF saturation duration, a series of four block saturation pulses, 200 ms each, total t_{sat} 800 ms; inter-pulse delay 10 ms; RF power 2 µT; FOV = 212 × 212 mm^2; 15 slices; thickness 4.4 mm; in-plane resolution 2.2 × 2.2 mm^2; SENSE factor 2, in the right-left direction; TR 3 s). To correct for B_0 inhomogeneity effects, APT imaging was acquired with a six-offset protocol (± 3, ± 3.5, ± 4 ppm from water; two to eight averages; scan time, 10 min 40 s). Five of 15 slices are shown. Reprinted with permission from Ref. [86], Copyright 2013, John Wiley and Sons.

effort to develop new response assessment criteria for malignant gliomas [98].

In a recent preclinical study published in 2011, Zhou et al. applied the APT-MRI approach to various glioma models and models of radiation-induced necrosis in rats [87]. The data showed that radiation-induced brain necrosis (hypointense to isointense) and glioma (hyperintense) could be clearly differentiated with APT-MRI, while these two entities show similar T_2 hyperintensity and Gd-enhancing MRI characteristics (Fig. 15.6). This and a few further studies have shown that the APTw signal would be a novel biomarker of the tumor response to chemotherapy [102, 103], radiotherapy [87, 104, 105], and high-intensity focused ultrasound (HIFU) [106]. Patients with treatment effects are treated conservatively, and the current therapy is usually maintained, whereas tumor recurrence requires a new anti-cancer therapy. Currently, tissue sampling via surgery is the only reliable approach in the clinic. If the initial preclinical results can be validated with appropriately powered studies in patients, APTw imaging would dramatically change the care of roughly 50% of patients with malignant gliomas, who exhibit non-specific MRI changes and currently undergo repeated surgery or spend precious months in an indeterminant diagnostic state.

15.7 Conclusions and Future Directions

An important area of in vivo MR imaging that has yet to be explored is the non-invasive detection of proteins and exchange properties in living tissues. APT imaging has the potential to introduce into the clinic an entirely new molecular MRI methodology that can detect endogenous cellular protein signals non-invasively. The early preclinical and clinical data suggest that APT-based imaging has unique features by which to detect and characterize strokes and brain tumors. In addition, this technology may be relevant for application to other cancers (such as prostate cancer [107], breast cancer [102], lung cancer [43, 108], and head and neck cancer [109]), other disorders (such as multiple sclerosis [110], Parkinson's disease [111], and Alzheimer's disease [112]), and other human diseases (such as pediatric brain developmental delay [113] and traumatic brain injury [114]). This is particularly meaningful in the era of proteomics and molecular imaging. APTw imaging is a safe, non-invasive MRI technology that can be easily implemented for

Conclusions and Future Directions | 363

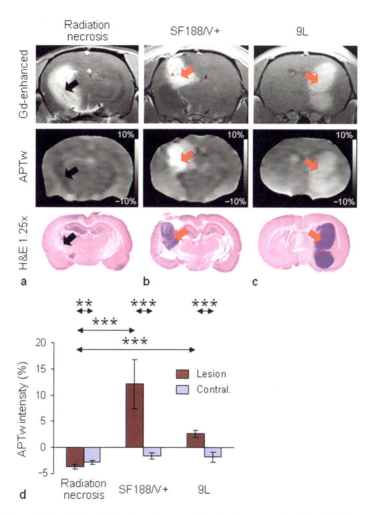

Figure 15.6 MR and histology images of radiation necrosis (a) and gliomas (b,c) in rats. MRI data were acquired on a 4.7 T clinical MRI scanner. The APT imaging parameters were as follows: saturation time 4 s; saturation power 1.3 µT; TR 10 s. Radiation necrosis (black arrow) is revealed by Gd enhancement. APT-MRI is hypointense to isointense in the lesion compared to the contralateral brain tissue. Both SF188/V+ and 9L tumors (red arrow) are hyperintense on the APTw images, corresponding to high cellularity on histology. Reprinted with permission from Ref. [87], Copyright 2010, Nature Publishing Group.

clinical applications. The research is of high potential impact since the validation of the results will result in a new sensitive and specific MRI modality with the potential to aid in the molecular diagnosis and treatment assessment of malignancies and other diseases at the protein level.

Acknowledgments

The authors sincerely thank Drs. Peter van Zijl, Peter Barker, Zaver Bhujwalla, Lindsay Blair, Jaishri Blakeley, Paul Bottomley, Che-Feng Chang, Min Chen, Zhong Chen, Kai Ding, Charles Eberhart, De-Xue Fu, Zongming Fu, Erik Tryggestad, Jochen Keupp, Guang Jia, Matthias Holdhoff, Xiaohua Hong, Shuguang Hu, Jun Hua, Craig Jones, Raymond Koehler, Kun Yan, Bachchu Lal, John Laterra, Richard Leigh, Chunmei Li, Haiyun Li, Qiang Li, Michael Lim, Xiaojie Luo, Bo Ma, Antonella Mangraviti, Yun Peng, Martin Pomper, Alfredo Quinones-Hinojosa, Amandeep Salhotra, Phillip Zhe Sun, Betty Tyler, Jian Wang, Meiyun Wang, Silun Wang, Wenzhu Wang, Xiangyang Wang, Yi-Xiang Wang, Zhibo Wen, Chen Yang, Yang Yu, Jing Yuan, Hong Zhang, Kai Zhang, Yansong Zhao, and Henry Zhu for guidance, cooperation, and help on this project in the past 15 years. The authors also thank Ms. Mary McAllister for editorial assistance. This work was supported in part by grants from the National Institutes of Health (R01EB009731, R01CA166171, R01NS083435, R21EB015555, R01EB015032, and P41EB015909).

References

1. Gillies RJ, Bhujwalla Z, Evelhoch J, Garwood M, Neeman M, Robinson SP, Sotak CH, and van der Sanden B. Applications of magnetic resonance in model systems: Tumor biology and physiology. *Neoplasia*, 2000; 2: 139–151.
2. Ellingson BM, Wen PY, van den Bent MJ, and Cloughesy TF. Pros and cons of current brain tumor imaging. *Neuro-Oncology*, 2014; 16: 2–11.
3. Baird AE and Warach S. Magnetic resonance imaging of acute stroke. *J. Cereb. Blood Flow Metab.*, 1998; 18: 583–609.

4. Mori S and van Zijl PCM. Fiber tracking: Principles and strategies—a technical review. *NMR Biomed.*, 2002; 15: 468–480.
5. Kauppinen RA and Peet AC. Using magnetic resonance imaging and spectroscopy in cancer diagnostics and monitoring: Preclinical and clinical approaches. *Cancer Biol. Ther.*, 2011; 12: 665–679.
6. Weissleder R, Moore A, Mahmood U, Bhorade R, NBenveniste H, Chiocca EA, and Basilion JP. In vivo magnetic resonance imaging of transgene expression. *Nature Med.*, 2000; 6: 351–354.
7. Bulte JW and Kraitchman DL. Iron oxide MR contrast agents for molecular and cellular imaging. *NMR Biomed.*, 2004; 17: 484–499.
8. Wolff SD and Balaban RS. NMR imaging of labile proton exchange. *J. Magn. Reson.*, 1990; 86: 164–169.
9. Guivel-Scharen V, Sinnwell T, Wolff SD, and Balaban RS. Detection of proton chemical exchange between metabolites and water in biological tissues. *J. Magn. Reson.*, 1998; 133: 36–45.
10. Ward KM, Aletras AH, and Balaban RS. A new class of contrast agents for MRI based on proton chemical exchange dependent saturation transfer (CEST). *J. Magn. Reson.*, 2000; 143: 79–87.
11. Zhou J and van Zijl PC. Chemical exchange saturation transfer imaging and spectroscopy. *Progr. NMR Spectr.*, 2006; 48: 109–136.
12. Sherry AD and Woods M. Chemical exchange saturation transfer contrast agents for magnetic resonance imaging. *Annu. Rev. Biomed. Eng.*, 2008; 10: 391–411.
13. van Zijl PCM and Yadav NN. Chemical exchange saturation transfer (CEST): What is in a name and what isn't? *Magn. Reson. Med.*, 2011; 65: 927–948.
14. Goffeney N, Bulte JWM, Duyn J, Bryant LH, and van Zijl PCM. Sensitive NMR detection of cationic-polymer-based gene delivery systems using saturation transfer via proton exchange. *J. Am. Chem. Soc.*, 2001; 123: 8628–8629.
15. Zhang S, Winter P, Wu K, and Sherry AD. A novel europium(III)-based MRI contrast agent. *J. Am. Chem. Soc.*, 2001; 123: 1517–1578.
16. Aime S, Barge A, Delli Castelli D, Fedeli F, Mortillaro A, Nielsen FU, and Terreno E. Paramagnetic lanthanide(III) complexes as pH-sensitive chemical exchange saturation transfer (CEST) contrast agents for MRI applications. *Magn. Reson. Med.*, 2002; 47: 639–648.
17. Schroder L, Lowery TJ, Hilty C, Wemmer DE, and Pines A. Molecular imaging using a targeted magnetic resonance hyperpolarized biosensor. *Science*, 2006; 314: 446–449.

18. Wolff SD and Balaban RS. Magnetization transfer contrast (MTC) and tissue water proton relaxation in vivo. *Magn. Reson. Med.*, 1989; 10: 135–144.
19. Henkelman RM, Stanisz GJ, and Graham SJ. Magnetization transfer in MRI: A review. *NMR Biomed.*, 2001; 14: 57–64.
20. Behar KL and Ogino T. Assignment of resonances in the ^{1}H spectrum of rat brain by two dimensional shift correlated and J-resolved NMR spectroscopy. *Magn. Reson. Med.*, 1991; 17: 285–303.
21. Behar KL and Ogino T. Characterization of macromolecule resonances in the ^{1}H NMR spectrum of rat brain. *Magn. Reson. Med.*, 1993; 30: 38–44.
22. Kauppinen RA, Kokko H, and Williams SR. Detection of mobile proteins by proton nuclear magnetic resonance spectroscopy in the guinea pig brain *ex vivo* and their partial purification. *J. Neurochem.*, 1992; 58: 967–974.
23. Pfeuffer J, Tkac I, Provencher SW, and Gruetter R. Toward an in vivo neurochemical profile: Quantification of 18 metabolites in short-echo-time ^{1}H NMR spectra of the rat brain. *J. Magn. Reson.*, 1999; 141: 104–120.
24. Mori S, Eleff SM, Pilatus U, Mori N, and van Zijl PCM. Proton NMR spectroscopy of solvent-saturable resonance: A new approach to study pH effects *in situ*. *Magn. Reson. Med.*, 1998; 40: 36–42.
25. van Zijl PCM, Zhou J, Mori N, Payen J, and Mori S. Mechanism of magnetization transfer during on-resonance water saturation. A new approach to detect mobile proteins, peptides, and lipids. *Magn. Reson. Med.*, 2003; 49: 440–449.
26. Zhou J, Payen J, Wilson DA, Traystman RJ, and van Zijl PCM. Using the amide proton signals of intracellular proteins and peptides to detect pH effects in MRI. *Nature Med.*, 2003; 9: 1085–1090.
27. Zhou JY and van Zijl PCM. Defining an acidosis-based ischemic penumbra from pH-weighted MRI. *Transl. Stroke Res.*, 2012; 3: 76–83.
28. Zhou J and Hong X. Molecular imaging using endogenous cellular proteins. *Bo Pu Xue Za Zhi*, 2013; 30: 307–321.
29. Vinogradov E, Sherry AD, and Lenkinski RE. CEST: From basic principles to applications, challenges and opportunities. *J. Magn. Reson.*, 2013; 229: 155–172.
30. Kogan F, Hariharan H, and Reddy R. Chemical exchange saturation transfer (CEST) imaging: Description of technique and potential clinical applications. *Curr. Radiol. Reports*, 2013; 1: 102–114.

31. Liu G, Song X, Chan KW, and McMahon MT. Nuts and bolts of chemical exchange saturation transfer MRI. *NMR Biomed.*, 2013; 26: 810–828.
32. Wuthrich K. *NMR of Proteins and Nucleic Acids*, 2 edn. New York: John Wiley & Sons; 1986 p.
33. Englander SW, Downer NW, and Teitelbaum H. Hydrogen exchange. *Annu. Rev. Biochem.*, 1972; 41: 903–924.
34. Zhou JY, Yan K, and Zhu H. A simple model for understanding the origin of the amide proton transfer MRI signal in tissue. *Appl. Magn. Reson.*, 2012; 42: 393–402.
35. Yan K, Fu Z, Yang C, Zhang K, Jiang S, Lee DH, Heo HY, Zhang Y, Cole RN, Van Eyk JE, and Zhou J. Assessing amide proton transfer (APT) MRI contrast origins in 9L gliosarcoma in the rat brain using proteomic analysis. *Mol. Imaging Biol.*, 2015: DOI:10.1007/s11307-11015-10828-11306.
36. Hua J, Jones CK, Blakeley J, Smith SA, van Zijl PCM, and Zhou J. Quantitative description of the asymmetry in magnetization transfer effects around the water resonance in the human brain. *Magn. Reson. Med.*, 2007; 58: 786–793.
37. Zhou J, Blakeley JO, Hua J, Kim M, Laterra J, Pomper MG, and van Zijl PCM. Practical data acquisition method for human brain tumor amide proton transfer (APT) imaging. *Magn. Reson. Med.*, 2008; 60: 842–849.
38. Zaiss M and Bachert P. Chemical exchange saturation transfer (CEST) and MR Z-spectroscopy in vivo: A review of theoretical approaches and methods. *Phys. Med. Biol.*, 2013; 58: R221–R269.
39. Kim J, Wu Y, Guo Y, Zheng H, and Sun PZ. A review of optimization and quantification techniques for chemical exchange saturation transfer MRI toward sensitive in vivo imaging. *Contrast Media Mol. Imaging*, 2015; 10: 163–178.
40. Jones CK, Polders D, Hua J, Zhe H, Hoogduin HJ, Zhou J, Luijten P, and van Zijl PCM. In vivo 3D whole-brain pulsed steady state chemical exchange saturation transfer at 7T. *Magn. Reson. Med.*, 2012; 67: 1579–1589.
41. Zaiss M, Schmitt B, and Bachert P. Quantitative separation of CEST effect from magnetization transfer and spillover effects by Lorentzian-line-fit analysis of z-spectra. *J. Magn. Reson.*, 2011; 211: 149–155.
42. Zaiss M, Xu J, Goerke S, Khan IS, Singer RJ, Gore JC, Gochberg DF, and Bachert P. Inverse Z-spectrum analysis for spillover-, MT-, and T1-corrected steady-state pulsed CEST-MRI: Application to pH-weighted MRI of acute stroke. *NMR Biomed.*, 2014; 27: 240–252.

43. Desmond KL, Moosvi F, and Stanisz GJ. Mapping of amide, amine, and aliphatic peaks in the CEST spectra of murine xenografts at 7 T. *Magn. Reson. Med.*, 2014; 71: 1841–1853.

44. Cai K, Singh A, Poptani H, Li W, Yang S, Lu Y, Hariharan H, Zhou XJ, and Reddy R. CEST signal at 2 ppm (CEST@2ppm) from Z-spectral fitting correlates with creatine distribution in brain tumor. *NMR Biomed.*, 2015; 28: 1–8.

45. Jin T, Wang P, Zong X, and Kim S-G. MR imaging of the amide-proton transfer effect and the pH-insensitive nuclear Overhauser effect at 9.4 T. *Magn. Reson. Med.*, 2013; 69: 760–770.

46. Chappell MA, Donahue MJ, Tee YK, Khrapitchev AA, Sibson NR, Jezzard P, and Payne SJ. Quantitative Bayesian model-based analysis of amide proton transfer MRI. *Magn. Reson. Med.*, 2013; 70: 556–567.

47. Heo H-Y, Zhang Y, Lee D-H, Hong X, and Zhou J. Quantitative assessment of amide proton transfer (APT) and nuclear Overhauser enhancement (NOE) imaging with extrapolated semi-solid magnetization transfer reference (EMR) signals: Application to a rat glioma model at 4.7 T. *Magn. Reson. Med.*, 2015: DOI:10.1002/mrm.25581.

48. Heo HY, Zhang Y, Jiang S, Lee DH, and Zhou J. Quantitative assessment of amide proton transfer (APT) and nuclear Overhauser enhancement (NOE) imaging with extrapolated semisolid magnetization transfer reference (EMR) signals: II. Comparison of three EMR models and application to human brain glioma at 3 Tesla. *Magn. Reson. Med.*, 2015: DOI:10.1002/mrm.25795.

49. Scheidegger R, Vinogradov E, and Alsop DC. Amide proton transfer imaging with improved robustness to magnetic field inhomogeneity and magnetization transfer asymmetry using saturation with frequency alternating RF irradiation. *Magn. Reson. Med.*, 2011; 66: 1275–1285.

50. Lee JS, Regatte RR, and Jerschow A. Isolating chemical exchange saturation transfer contrast from magnetization transfer asymmetry under two-frequency rf irradiation. *J. Magn. Reson.*, 2012; 215: 56–63.

51. Lee JS, Xia D, Ge Y, Jerschow A, and Regatte RR. Concurrent saturation transfer contrast in in vivo brain by a uniform magnetization transfer MRI. *Neuroimage*, 2014; 95: 22–28.

52. Zu ZL, Janve VA, Li K, Does MD, Gore JC, and Gochberg DF. Multi-angle ratiometric approach to measure chemical exchange in amide proton transfer imaging. *Magn. Reson. Med.*, 2012; 68: 711–719.

53. Zu Z, Janve VA, Xu J, Does MD, Gore JC, and Gochberg DF. A new method for detecting exchanging amide protons using chemical exchange rotation transfer. *Magn. Reson. Med.*, 2013; 69: 637–647.
54. Xu J, Yadav NN, Bar-Shir A, Jones CK, Chan KW, Zhang J, Walczak P, McMahon MT, and van Zijl PC. Variable delay multi-pulse train for fast chemical exchange saturation transfer and relayed-nuclear Overhauser enhancement MRI. *Magn. Reson. Med.*, 2014; 71: 1798–1812.
55. Zhou J, Hong X, Zhao X, Gao J-H, and Yuan J. APT-weighted and NOE-weighted image contrasts in glioma with different RF saturation powers based on magnetization transfer ratio asymmetry analyses. *Magn. Reson. Med.*, 2013; 70: 320–327.
56. Zhou J, Wilson DA, Sun PZ, Klaus JA, and van Zijl PCM. Quantitative description of proton exchange processes between water and endogenous and exogenous agents for WEX, CEST, and APT experiments. *Magn. Reson. Med.*, 2004; 51: 945–952.
57. Kintner DB, Anderson ME, Sailor KA, Dienel G, Fitzpatrick JH, Jr., and Gilboe DD. In vivo microdialysis of 2-deoxyglucose 6-phosphate into brain: A novel method for the measurement of interstitial pH using ^{31}P-NMR. *J. Neurochem.*, 1999; 72: 405–412.
58. Griffiths JR. Are cancer cells acidic? *Br. J. Cancer*, 1991; 64: 425–427.
59. Hobbs SK, Shi G, Homer R, Harsh G, Altlas SW, and Bednarski MD. Magnetic resonance imaging-guided proteomics of human glioblastoma multiforme. *J. Magn. Reson. Imag.*, 2003; 18: 530–536.
60. Li J, Zhuang Z, Okamoto H, Vortmeyer AO, Park DM, Furata M, Lee Y-S, Oldfield EH, Zeng W, and Weil RJ. Proteomic profiling distinguishes astrocytomas and identifies differential tumor markers. *Neurology*, 2006; 66: 733–736.
61. Sun PZ, Zhou J, Sun W, Huang J, and van Zijl PCM. Detection of the ischemic penumbra using pH-weighted MRI. *J. Cereb. Blood Flow Metab.*, 2007; 27: 1129–1136.
62. Sun PZ, Benner T, Kumar A, and Sorensen AG. Investigation of optimizing and translating pH-sensitive pulsed-chemical exchange saturation transfer (CEST) imaging to a 3T clinical scanner. *Magn. Reson. Med.*, 2008; 60: 834–841.
63. Sun PZ, Cheung JS, Wang E, and Lo EH. Association between pH-weighted endogenous amide proton chemical exchange saturation transfer MRI and tissue lactic acidosis during acute ischemic stroke. *J. Cereb. Blood Flow Metab.*, 2011; 31: 1743–1750.

64. Jokivarsi KT, Grohn HI, Grohn OH, and Kauppinen RA. Proton transfer ratio, lactate, and intracellular pH in acute cerebral ischemia. *Magn. Reson. Med.*, 2007; 57: 647–653.
65. Jin T, Wang P, Zong XP, and Kim SG. Magnetic resonance imaging of the Amine-Proton EXchange (APEX) dependent contrast. *NeuroImage*, 2012; 59: 1218–1227.
66. Zong XP, Wang P, Kim SG, and Jin T. Sensitivity and source of amine-proton exchange and amide-proton transfer magnetic resonance imaging in cerebral ischemia. *Magn. Reson. Med.*, 2014; 71: 118–132.
67. Wei M, Shen Z, Xiao G, Qiu Q, Chen Y, and Wu R. Study of magnetic resonance imaging at 1.5 Tesla based on pH-sensitive magnetization transfer technology. *Chin. J. Magn. Reson. Imag.*, 2012; 3: 40–43.
68. McVicar N, Li AX, Goncalves DF, Bellyou M, Meakin SO, Prado MA, and Bartha R. Quantitative tissue pH measurement during cerebral ischemia using amine and amide concentration-independent detection (AACID) with MRI. *J. Cereb. Blood Flow Metab.*, 2014; 34: 690–698.
69. Zhao X, Wen Z, Huang F, Lu S, Wang X, Hu S, Zu D, and Zhou J. Saturation power dependence of amide proton transfer image contrasts in human brain tumors and strokes at 3 T. *Magn. Reson. Med.*, 2011; 66: 1033–1041.
70. Tietze A, Blicher J, Mikkelsen IK, Ostergaard L, Strother MK, Smith SA, and Donahue MJ. Assessment of ischemic penumbra in patients with hyperacute stroke using amide proton transfer (APT) chemical exchange saturation transfer (CEST) MRI. *NMR Biomed.*, 2014; 27: 163–174.
71. Tee YK, Harston GW, Blockley N, Okell TW, Levman J, Sheerin F, Cellerini M, Jezzard P, Kennedy J, Payne SJ, and Chappell MA. Comparing different analysis methods for quantifying the MRI amide proton transfer (APT) effect in hyperacute stroke patients. *NMR Biomed.*, 2014; 27: 1019–1029.
72. Harston GW, Tee YK, Blockley N, Okell TW, Thandeswaran S, Shaya G, Sheerin F, Cellerini M, Payne S, Jezzard P, Chappell M, and Kennedy J. Identifying the ischaemic penumbra using pH-weighted magnetic resonance imaging. *Brain*, 2015; 138: 36–42.
73. Kidwell CS and Wintermark M. Imaging of intracranial haemorrhage. *Lancet Neurol.*, 2008; 7: 256–267.
74. Morgenstern LB, Hemphill JC, 3rd, Anderson C, Becker K, Broderick JP, Connolly ES, Jr., Greenberg SM, Huang JN, MacDonald RL, Messe SR, Mitchell PH, Selim M, and Tamargo RJ. Guidelines for the management

of spontaneous intracerebral hemorrhage: A guideline for healthcare professionals from the American Heart Association/American Stroke Association. *Stroke*, 2010; 41: 2108–2129.

75. Smith EE, Rosand J, and Greenberg SM. Hemorrhagic stroke. *Neuroim. Clin. N. Am.*, 2005; 15: 259–272.

76. Schellinger PD, Jansen O, Fiebach JB, Hacke W, and Sartor K. A standardized MRI stroke protocol: Comparison with CT in hyperacute intracerebral hemorrhage. *Stroke*, 1999; 30: 765–768.

77. Kidwell CS, Chalela JA, Saver JL, Starkman S, Hill MD, Demchuk AM, Butman JA, Patronas N, Alger JR, Latour LL, Luby ML, Baird AE, Leary MC, Tremwel M, Ovbiagele B, Fredieu A, Suzuki S, Villablanca JP, Davis S, Dunn B, Todd JW, Ezzeddine MA, Haymore J, Lynch JK, Davis U, and Warach S. Comparison of MRI and CT for detection of acute intracerebral hemorrhage. *JAMA*, 2004; 292: 1823–1830.

78. Siddiqui FM, Bekker SV, and Qureshi AI. Neuroimaging of hemorrhage and vascular defects. *Neurotherapeutics*, 2011; 8: 28–38.

79. Wang M, Hong X, Chang CF, Li Q, Ma B, Zhang H, Xiang S, Heo HY, Zhang Y, Lee DH, Jiang S, Leigh R, Koehler RC, van Zijl PC, Wang J, and Zhou J. Simultaneous detection and separation of hyperacute intracerebral hemorrhage and cerebral ischemia using amide proton transfer MRI. *Magn. Reson. Med.*, 2015; 74: 42–50.

80. Zhou J, Lal B, Wilson DA, Laterra J, and van Zijl PCM. Amide proton transfer (APT) contrast for imaging of brain tumors. *Magn. Reson. Med.*, 2003; 50: 1120–1126.

81. Salhotra A, Lal B, Laterra J, Sun PZ, van Zijl PCM, and Zhou J. Amide proton transfer imaging of 9L gliosarcoma and human glioblastoma xenografts. *NMR Biomed.*, 2008; 21: 489–497.

82. Jones CK, Schlosser MJ, van Zijl PC, Pomper MG, Golay X, and Zhou J. Amide proton transfer imaging of human brain tumors at 3T. *Magn. Reson. Med.*, 2006; 56: 585–592.

83. Wen Z, Hu S, Huang F, Wang X, Guo L, Quan X, Wang S, and Zhou J. MR imaging of high-grade brain tumors using endogenous protein and peptide-based contrast. *NeuroImage*, 2010; 51: 616–622.

84. Togao O, Yoshiura T, Keupp J, Hiwatashi A, Yamashita K, Kikuchi K, Suzuki Y, Suzuki SO, Iwaki T, Hata N, Mizoguchi M, Yoshimoto K, Sagiyama K, Takahashi M, and Honda H. Amide proton transfer imaging of adult diffuse gliomas: Correlation with histopathological grades. *Neuro. Oncol.*, 2014; 16: 441–448.

85. Sakata A, Okada T, Yamamoto A, Kanagaki M, Fushimi Y, Dodo T, Arakawa Y, Schmitt B, Miyamoto S, and Togashi K. Grading glial tumors with amide proton transfer MR imaging: Different analytical approaches. *J. Neuro-Oncol.*, 2015; 122: 339–348.
86. Zhou J, Zhu H, Lim M, Blair L, Quinones-Hinojosa A, Messina AA, Eberhart CG, Pomper MG, Laterra J, Barker PB, van Zijl PCM, and Blakeley JO. Three-dimensional amide proton transfer MR imaging of gliomas: Initial experience and comparison with gadolinium enhancement. *J. Magn. Reson. Imaging*, 2013; 38: 1119–1128.
87. Zhou J, Tryggestad E, Wen Z, Lal B, Zhou T, Grossman R, Wang S, Yan K, Fu D-X, Ford E, Tyler B, Blakeley J, Laterra J, and van Zijl PCM. Differentiation between glioma and radiation necrosis using molecular magnetic resonance imaging of endogenous proteins and peptides. *Nature Med.*, 2011; 17: 130–134.
88. Jiang S, Yu H, Wang X, Lu S, Li Y, Feng L, Zhang Y, Heo HY, Lee DH, Zhou J, and Wen Z. Molecular MRI differentiation between primary central nervous system lymphomas and high-grade gliomas using endogenous protein-based amide proton transfer MR imaging at 3 Tesla. *Eur. Radiol.*, 2015: DOI:10.1007/s00330-00015-03805-00331.
89. Segall HD, Destian S, and Nelson MD. CT and MR imaging in malignant gliomas. In: Apuzzo MLJ (ed), *Malignant Cerebral Glioma*. Park Ridge, IL: American Association of Neurological Surgeons; 1990. p. 63–78.
90. Scott JN, Brasher PM, Sevick RJ, Rewcastle NB, and Forsyth PA. How often are nonenhancing supratentorial gliomas malignant? A population study. *Neurology*, 2002; 59: 947–949.
91. Knopp EA, Cha S, Johnson G, Mazumdar A, Golfinos JG, Zagzag D, Miller DC, Kelly PJ, and Kricheff II. Glial neoplasms: Dynamic contrast-enhanced T2*-weighted MR imaging. *Radiology*, 1999; 211: 791–798.
92. Howe FA, Barton SJ, Cudlip SA, Stubbs M, Saunders DE, Murphy M, Wilkins P, Opstad KS, Doyle VL, McLean MA, Bell BA, and Griffiths JR. Metabolic profiles of human brain tumors using quantitative in vivo ^{1}H magnetic resonance spectroscopy. *Magn. Reson. Med.*, 2003; 49: 223–232.
93. Sun PZ, Murata Y, Lu J, Wang X, Lo EH, and Sorensen AG. Relaxation-compensated fast multislice amide proton transfer (APT) imaging of acute ischemic stroke. *Magn. Reson. Med.*, 2008; 59: 1175–1182.
94. Dixen WT, Hancu I, Ratnakar SJ, Sherry AD, Lenkinski RE, and Alsop DC. A multislice gradient echo pulse sequence for CEST imaging. *Magn. Reson. Med.*, 2009; 63: 253–256.

95. Zhu H, Jones CK, van Zijl PCM, Barker PB, and Zhou J. Fast 3D chemical exchange saturation transfer (CEST) imaging of the human brain. *Magn. Reson. Med.*, 2010; 64: 638–644.
96. Zhou J, Zhu H, Lim M, Blair L, Quinones-Hinojosa A, Wen Z, Wang S, Laterra J, Barker PB, van Zijl PCM, and Blakeley J. Assessment of brain tumors at the protein level with amide proton transfer MRI. In *Proc. 97th Ann. Meeting RSNA*, Chicago, IL, 2011. p. MSVN31–11.
97. Keupp J, Baltes C, Harvey PR, and van den Brink J. Parallel RF transmission based MRI technique for highly sensitive detection of amide proton transfer in the human brain. In *Proc. 19th Annual Meeting ISMRM*, Montreal, Quebec, 2011. p. 710.
98. Wen PY, Macdonald DR, Reardon DA, Cloughesy TF, Sorensen AG, Galanis E, DeGroot J, Wick W, Gilbert MR, Lassman AB, Tsien C, Mikkelsen T, Wong ET, Chamberlain MC, Stupp R, Lamborn KR, Vogelbaum MA, van den Bent MJ, and Chang SM. Updated response assessment criteria for high-grade gliomas: Response assessment in neuro-oncology working group. *J. Clin. Oncol.*, 2010; 28: 1963–1972.
99. Kumar AJ, Leeds NE, Fuller GN, van Tassel P, Maor MH, Sawaya RE, and Levin VA. Malignant gliomas: MR imaging spectrum of radiation therapy- and chemotherapy-induced necrosis of the brain after treatment. *Radiology*, 2000; 217: 377–384.
100. Stupp R, Mason WP, van den Bent MJ, Weller M, Fisher B, Taphoorn MJB, Belanger K, Brandes AA, Marosi C, Bogdahn U, Curschmann J, Janzer RC, Ludwin SK, Gorlia T, Allgeier A, Lacombe D, Cairncross JG, Eisenhauer E, and Mirimanoff RO. Radiotherapy plus concomitant and adjuvant temozolomide for glioblastoma. *N. Engl. J. Med.*, 2005; 352: 987–996.
101. Brandes AA, Tosoni A, Spagnolli F, Frezza G, Leonardi M, Calbucci F, and Franceschi E. Disease progression or pseudoprogression after concomitant radiochemotherapy treatment: Pitfalls in neurooncology. *Neuro-Oncology*, 2008; 10: 361–367.
102. Dula AN, Arlinghaus LR, Dortch RD, Dewey BE, Whisenant JG, Ayers GD, Yankeelov TE, and Smith SA. Amide proton transfer imaging of the breast at 3 T: Establishing reproducibility and possible feasibility assessing chemotherapy response. *Magn. Reson. Med.*, 2013; 70: 216–224.
103. Sagiyama K, Mashimo T, Togao O, Vemireddy V, Hatanpaa KJ, Maher EA, Mickey BE, Pan E, Sherry AD, Bachoo RM, and Takahashi M. In vivo

chemical exchange saturation transfer imaging allows early detection of a therapeutic response in glioblastoma. *Proc. Natl. Acad. Sci. USA*, 2014; 111: 4542–4547.

104. Wang SL, Tryggestad E, Zhou TT, Armour M, Wen ZB, Fu DX, Ford E, van Zijl PCM, and Zhou JY. Assessment of MRI parameters as imaging biomarkers for radiation necrosis in the rat brain. *Int. J. Rad. Onc. Biol. Phys.*, 2012; 83: E431–E436.

105. Hong X, Liu L, Wang M, Ding K, Fan Y, Ma B, Lal B, Tyler B, Mangraviti A, Wang S, Wong J, Laterra J, and Zhou J. Quantitative multiparametric MRI assessment of glioma response to radiotherapy in a rat model. *Neuro. Oncol.*, 2014; 16: 856–867.

106. Hectors S, Jacobs I, Strijkers GJ, and Nicolay K. Amide proton transfer imaging of high intensity focused ultrasound-treated tumor tissue. *Magn. Reson. Med.*, 2014; 72: 1113–1122.

107. Jia G, Abaza R, Williams JD, Zynger DL, Zhou JY, Shah ZK, Patel M, Sammet S, Wei L, Bahnson RR, and Knopp MV. Amide proton transfer MR imaging of prostate cancer: A preliminary study. *J. Magn. Reson. Imaging*, 2011; 33: 647–654.

108. Togao O, Kessinger CW, Huang G, Soesbe TC, Sagiyama K, Dimitrov I, Sherry AD, Gao J, and Takahashi M. Characterization of lung cancer by amide proton transfer (APT) imaging: An in-vivo study in an orthotopic mouse model. *Plos One*, 2013; 8: e77019.

109. Yuan J, Chen S, King AD, Zhou J, Bhatia KS, Zhang Q, Yeung DK, Wei J, Mok GS, and Wang YX. Amide proton transfer-weighted imaging of the head and neck at 3 T: A feasibility study on healthy human subjects and patients with head and neck cancer. *NMR Biomed.*, 2014; 27: 1239–1247.

110. Dula AN, Asche EM, Landman BA, Welch EB, Pawate S, Sriram S, Gore JC, and Smith SA. Development of chemical exchange saturation transfer at 7T. *Magn. Reson. Med.*, 2011; 66: 831–838.

111. Li C, Peng S, Wang R, Chen H, Su W, Zhao X, Zhou J, and Chen M. Chemical exchange saturation transfer MR imaging of Parkinson's disease at 3 Tesla. *Eur. Radiol.*, 2014; 24: 2631–2639.

112. Wang R, Li SY, Chen M, Zhou JY, Peng DT, Zhang C, and Dai YM. Amide proton transfer magnetic resonance imaging of Alzheimer's disease at 3.0 Tesla: A preliminary study. *Chin. Med. J.*, 2015; 128: 615–619.

113. Zhang H, Zhao X, Zhou J, and Peng Y. Amide proton transfer (APT) MR imaging of the brain in Children at 3T. In *Proc. 99th Ann. Meeting RSNA*, Chicago, IL, 2013. p. SSQ17–07.
114. Zhang H, Wang W, Ma B, Peng Y, Wang W, and Zhou J. Molecular MRI detection of traumatic brain injury (TBI) with amide proton transfer (APT) imaging. In *Proc. 100th Ann. Meeting RSNA*, Chicago, IL, 2014. p. SSK13–05.

Chapter 16

Cartilage and Intervertebral Disc Imaging and Glycosaminoglycan Chemical Exchange Saturation Transfer (gagCEST) Experiment

Joshua I. Friedman,[a] Ravinder R. Regatte,[b] Gil Navon,[c] and Alexej Jerschow[a]

[a]*Department of Chemistry, New York University, 100 Washington Square East, New York, NY 10003, USA*
[b]*Radiology Department, New York University Langone Medical Center, 550 First Avenue, New York, NY 10016, USA*
[c]*School of Chemistry, Tel Aviv University, Tel Aviv, Israel*
alexej.jerschow@nyu.edu

16.1 Introduction

Osteoarthritis (OA) is a degenerative joint disease that affects nearly 80% of individuals by the time they reach age 65 [1]. The near universal prevalence of this disorder in older populations coupled with an absence of effective treatments aside from symptom management makes OA one of the costliest diseases to treat, and it is responsible for nearly 4% of US health care related

---

Chemical Exchange Saturation Transfer Imaging: Advances and Applications
Edited by Michael T. McMahon, Assaf A. Gilad, Jeff W. M. Bulte, and Peter C. M. van Zijl
Copyright © 2017 Pan Stanford Publishing Pte. Ltd.
ISBN 978-981-4745-70-3 (Hardcover), 978-1-315-36442-1 (eBook)
www.panstanford.com

expenditures [2]. The disease is phenomenologically described by the degenerative loss of cartilage found at joints, which normally insures that bones move past each other non-abrasively during limb motion. Progression of OA eventually leads to direct bone-on-bone contact and may result in wear-related morphological deformations in bone structure, limited range of motion, and acute pain. The development of new and preventative treatments for OA is hindered by the absence of appropriate radiological tools for assessing the condition and detailed morphology of cartilage tissue, making the development of such radiological techniques an important step in the development of effective treatments.

The current gold-standards for the diagnosis and monitoring of OA are X-ray and tomographic imaging [3,4], but these techniques are incapable of visualizing the soft tissues directly and instead infer the condition of cartilage by its effect on the surrounding denser bones. Such indirect probes cannot provide detailed anatomic insights that are relevant to specific pathologies and are not well suited to detect subtle changes in cartilage morphology or composition that have been shown to be early indicators of diseases [5, 6]. These limitations in observing cartilage make it difficult to detect OA prior to the onset of acute symptoms and to gauge the efficacy of existing or emerging treatment options that might act before irreparable damage has been done to the joint.

The gradual degradation and loss of a particular class of polysaccharide, known as glycosaminoglycan (GAG), from healthy cartilage tissue are known early indicators of OA progression [7–9]. These GAG sugars imbue cartilage with many unique chemical and physical properties that are important to the tissue's function as a lubricant and shock absorber. Because of the GAG's uneven distribution in tissues and high concentration in cartilage, these compounds can serve as important tissue-specific biochemical markers for directly monitoring cartilage extracellular matrix and OA progression using MRI.

Chemical exchange saturation transfer (CEST) is one such MRI technique that is particularly well suited to detect cartilage via the presence of GAG sugars. In this specific application, known as gagCEST, protons on GAG sugars are tagged with non-equilibrium spin magnetization, and the redistribution of this polarization to water is measured. Because GAG sugars have many exchangeable protons and many transfer-labeling steps can occur for each proton prior to detection, gagCEST greatly amplifies GAG signals compared to direct detection using conventional MR spectroscopy. Additionally, the chemical exchange processes are sensitive to parameters such as temperature and pH. GagCEST offers the capability, in principle, to non-invasively probe these internal variables as well. This chapter will first provide a brief cursory overview of cartilage tissue, a discussion of the emerging MRI methodologies that are currently available for imaging cartilage, followed by an overview of recent advances in gagCEST methodologies.

16.2 Composition and Organization of Cartilage

Cartilage is a complex material with many structural features that collectively span many orders of magnitude in scale and that are relevant for detection by magnetic resonance. The tissue has surprisingly few living cells (chondrocytes); their sole function is to build and maintain a vast extracellular matrix around themselves by secreting polysaccharides and proteins that self-assemble into a tough, three-dimensional viscoelastic material, composed of approximately 80% water by weight, providing the physical and mechanical properties of cartilage [10, 11]. At first glance, this extracellular matrix may resemble a randomly assembled jumble of proteins, GAGs, and solvating water.

Both proteins and polysaccharides snake around the chondrocytes forming a tangled mesh-like structure that serves as a basic substrate for building additional levels of structural complexity.

The protein fibrils are primarily composed of collagen bundles and are made up of three protein chains twisted about each other's long axis into a helical structure. Because of this helical construction, collagen fibrils are stretchy and can lengthen in response to applied forces and resist further sheer strain. This effect plays a critical role in solidifying cartilage tissue in response to the compressive loads of load-bearing joints. Unlike the polysaccharide components of the extracellular matrix, however, collagen fibrils are relatively insoluble in water and do little to trap the water molecules that make up the majority of cartilages mass. Enmeshed along with these collagen fibers is hyaluronic acid (HA), which is an unbranched polysaccharide chain that can have up to 25,000 identical sugar subunits. Aside from tying together collagen in a mesh, HA also plays a key role in the next level of the organizational structure. Proteoglycan aggregates—highly glycosylated proteins that are also secreted by chondrocytes and contain specific HA-binding domains— anchor themselves onto the linear contour of the HA substrate.

The proteoglycan aggregates can be thought of as adaptors for attaching a host of different GAG polymers to the HA fibers of the extracellular matrix. These proteoglycans contain multiple instances of the tetrameric amino acid sequence -Ser-Gly-X-Gly- (where X can be any of the 19 biological amino acids). Glycosylating enzymes in the chondrocyte Golgi apparatus recognize this amino acid sequence, and the hydroxyl serine units of proteoglycans are glycosylated with GAG sugars of various lengths. Ultimately, these proteoglycan aggregates resemble brushes with numerous bristles of GAG sugar chains protruding from one end, and a specialized HA-binding domain anchoring them to the HA substrate on the other. The numerous GAG sugars on these aggregate proteins swell with water and become a compressible gel-like material that cushions load-bearing joints and fills voids in the material [11]. The molecular structure of these individual GAG molecules gives rise to many important modalities of MRI.

GAGs encompass a large family of linear, unbranched polysaccharides, made up of a repeating disaccharide unit. These disaccharide units, with the singular exception of keratin, all consist of one sugar with an NH chemical moiety, known as an amino sugar, linked to what is known as a uranic sugar, which contains a carboxylic acid functional group instead. Six kinds of GAG polysaccharides are commonly found in biological tissues: HA, chondroitin sulfate, keratan sulfate (see Fig. 16.1 for molecular structures), dermatan sulfate, heparin, and heparan sulfate.

Apart from these minor differences in polymer length and chemical formula, all GAGs are united by similar set of physical properties, and most importantly, all are highly polar anionic compounds. This charge polarity traps a large number of solvating

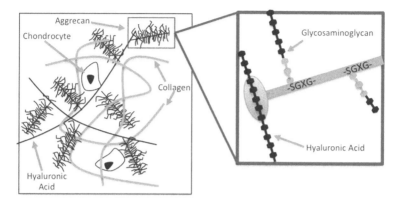

Figure 16.1 Illustration depicting the overall organization of cartilage with detailed insert of an aggrecan showing a proteoglycan bound to a hyaluronic acid polymer (not drawn to scale).

water molecules around the GAGs, thus producing much larger effective molecular assemblies than their chemical formula would otherwise imply. These tightly associated waters are particularly important to the shock absorption and lubricating functions of cartilage.

16.3 Composition and Organization of Intervertebral Disc

The intervertebral disc (IVD) is located between the vertebral bodies and links them together. The main function of the IVD is mechanical, as it provides flexibility to the spine (bending and rotation) and transmits loads from body weight and muscle activity [12]. The IVD is composed of three regions: nucleus pulposus (NP), a gelatinous core in the center of the disc; annulus fibrosus (AF), the outer ring; and the cartilaginous end plates that separate the disc from the vertebral bodies (for a review, see Ref. 13).

The NP contains collagen fibers, which are randomly oriented [14], elastin fibers, proteoglycans (PG), and water (∼80%). The collagen is mainly of type-II and constitutes approximately 20% of the dry weight of the tissue [15]. The PG content of the nucleus is about 50% of the dry weight. The swelling pressure due to the PG is

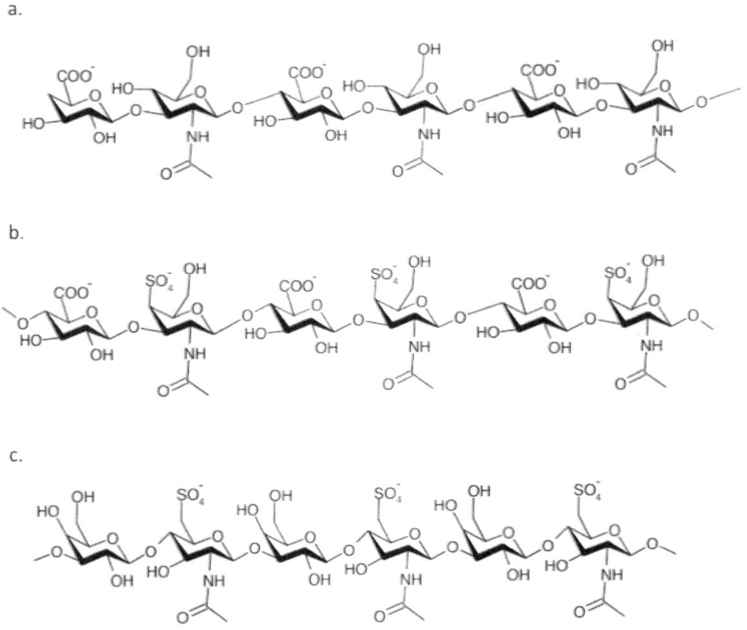

Figure 16.2 The chemical structures of three poly-glycosaminoglycans commonly found in cartilage with the disaccharide constituents of each highlighted: (a) hyaluronic acid, (b) chondroitin-4-sulfate, and (c) keratin-sulfate. Note that each disaccharide is made up of a uric acid sugar ring linked to an amino sugar.

used to support the compressive loads and to maintain the collagen fibers in a state of tension.

The AF is the outer ring of the disc that contains collagen fibers, PG, and water (~70%). The collagen fibers are highly ordered and arranged in a series of concentric lamellae, with the fibers lying parallel within each lamella. The fibers are oriented at approximately 60° to the vertical axis, alternating to the left and right of it in adjacent lamellae [12, 13, 16]. The collagen fibers are mainly of type-I in the outermost regions and type-II in the innermost. They constitute approximately 70% of the dry weight in the AF [15], while the PG constitutes of about 20% of the dry weight.

The main mechanical functions of the disc are governed by the composition and the organization of the two major macromolecular components in the extracellular matrix: (1) the collagen network that provides the tensile strength of the disc and anchors the tissue to the bone, and (2) the negatively charged PGs that are responsible for the strong water-binding ability that generates a swelling pressure [17].

Since the overall disc structure is very similar to the structure of articular cartilage, many detection and diagnostic strategies are similar for both tissues. The exception is probably the relative isolation of disc tissue, and not accessible to contrast agents, which is why the dGEMRIC method cannot be used for the disc.

Disc degeneration is known to be accompanied by decreases in GAG concentration and a loss of the mechanical properties and structure of the disc. The symptoms are often low back pain, and the general disorder is termed degenerative disc disease (DDD).

For simplicity, we will use cartilage for many illustrations and explanations, with the understanding that similar statements can be made for the IVD as well.

16.4 MRI Techniques for Measuring GAG (Other than CEST)

Studies based on T_1 and T_2 contrast alone have been shown to be insufficient to assess cartilage health. Gradual losses of GAG sugars may decrease the water content of cartilage tissue, but this does not necessarily affect the observable relaxation rates of the remaining water. While T_2-weighted images of cartilage generally show good contrast with surrounding tissues, the T_2 constants are poorly correlated with GAG concentration due to the strong dependencies of T_2 on magnetic field inhomogeneity, magic angle effect, and chemical exchange with non-GAG solutes as well [18, 19]. Indeed, clinical findings indicate that these MRI-based assessments of cartilage defects systematically underestimate the amount of damaged tissue compared to measurements derived from invasive surgeries, and that anatomical MRI is not well suited for quantitating defects [20, 21]. A number of alternative MRI detection methods exist, however, which are largely dependent on the strong negative charge of GAGs and which greatly improve the diagnostic potential of

MRI experiments. A recent review compared quantitative radiologic imaging techniques, including sodium MRI, dGEMRIC, and gagCEST [22].

16.4.1 Gadolinium-Enhanced Imaging

Gd-DTPA(2-) can be used to selectively label non-cartilage tissue and increase its contrast relative to surrounding tissues in T_1-weighted images [23, 24]. This agent is used for cartilage imaging because the strong negative charge of the proteoglycan aggregates in cartilage electrostatically inhibits the similarly charged Gd-DTPA(2-) from distributing into these tissues following injection [24, 25] to various degrees, and an inverse relationship results between Gd-DTPA(2-) concentration and GAG concentration. As a consequence, the T_1 values in cartilage are longer than in surrounding tissues [26]. Together, these properties effectively make it possible to correlate local variations in electric charge with the contrast observed by T_1-weighted MRI [27, 28]. A large number of in vivo studies have demonstrated the ability of the Gd-DTPA(2-) contrast agent to detect abnormal changes in GAG concentration prior to the appearance of OA [25, 27–31]. Though powerful, this approach has some clinical disadvantages. Injection of Gd-DTPA(2-) is discouraged for routine use in the general population, and not all tissues are equally well permeated by Gd-DTPA(2-), making it difficult to quantitatively attribute subtle differences in the partitioning of Gd to local electric fields and GAG concentrations.

16.4.2 Sodium Imaging

Sodium imaging exploits the electric charge interactions between highly negatively charged GAGs and sodium. Sodium is an abundant spin 3/2 nucleus in the body, with a concentration of up to 154 mM in neutral fluids and up to 300 mM in cartilage and in the disc. Also in contrast to Gd agents, Na ions are positively charged, which causes them to preferentially accumulate close to the proteoglycan aggregates of cartilage and correlate with the negative charge imparted by GAGs [32].

The (negative) fixed charge density of cartilage can be derived from sodium concentration via the Donnan equilibrium as

$$\text{FCD} = \frac{[\text{Na}^+]_b^2}{[\text{Na}^+]_t} - [\text{Na}^+]_t \quad (16.1)$$

where $[\text{Na}^+]_t$ is the concentration of sodium in the tissue of interest and $[\text{Na}^+]_b$ is the concentration of "bath" sodium, or sodium in the surrounding fluids (typically assumed to be 154 mM) [32]. This concentration can then be translated into GAG concentration via the number of charges per repeat unit.

Sodium MRI has a significant sensitivity penalty with respect to proton imaging, and sodium T_2 relaxation times are very short due to quadrupolar relaxation. Nevertheless, because sodium MRI is non-invasive and it provides an independent measure of GAG concentration, it is an important technique that could also be used for cross-validation of gagCEST and other techniques.

Sodium imaging has been shown to be sensitive to proteoglycan concentration and has been proposed as a quantitative probe of cartilage tissue health [32–36].

16.4.3 $T_{1\rho}$ Contrast

Spin-lattice relaxation in the rotating frame is characterized by the time constant ($T_{1\rho}$) that defines the magnetic relaxation of spins under the influence of a radio frequency field. This mechanism has been used extensively to investigate low-frequency interactions between macromolecules and bulk water [37, 38]. $T_{1\rho}$ MRI has been shown to be sensitive to changes in proteoglycans of cartilage [39]. Consequently, $T_{1\rho}$ has been shown to reflect early degenerative changes to cartilage [40, 41]. Limitations of $T_{1\rho}$ MRI include the fact that relatively high specific absorption rate (SAR) levels are needed, and that exchange effects that are prominent in cartilage and in the IVD are measured non-specifically.

16.5 GagCEST

GAG sugars contain two species of labile protons that typically satisfy the fundamental requirements for CEST under physiological conditions: the amide proton of the amino sugar and the various hydroxyl protons of the sugar rings. GagCEST refers to the use of CEST when targeting the exchangeable protons from GAG for image contrast. Since there are identical copies of each sugar in the GAG chains, the CEST effect scales with the number of repeat units. Apart from the exchange effects from hydroxyl and amide protons, it has been observed that nuclear Overhauser effect (NOE) transfers are visible in the Z-spectra and CEST curves. NOE-mediated polarization transfer is based on through–space dipole–dipole interactions. Because nuclei are physically limited in terms of how closely they can approach each other under physiological conditions, the observed rates of NOE transfer fall into a relatively narrow range of values and are typically close to $1/T_1$ of water. The transfer via NOE is hence much slower than for typical chemical exchange dominated resonances in tissues.

Ex vivo studies on isolated samples of cartilage tissue and purified GAG sugars have been used to assign these groups. This work used conventional multidimensional nuclear magnetic resonance (NMR) assignment techniques on ex vivo samples to assign proton resonances at +3.2 ppm downfield of water to the amide protons of the GAG amino sugars, and those spanning from +0.9 to +1.9 ppm downfield of water to the hydroxyl protons in the sugar rings [42, 43]. Hydroxyl protons, in particular, are highly labile and undergo acid- or base-catalyzed chemical exchange processes with water protons on the order of 100 to 1000 s^{-1} in vivo, providing a significant opportunity for CEST-mediated signal amplification, and are preferred targets for gagCEST [44]. Ideally, the rate of this magnetization transfer (MT) must be on the same order as, or greater than the $1/T_1$ water, but less than the fast exchange limit of the NMR timescale defined as $k_{ex} \leq \Delta\omega/2\pi$, where $\Delta\omega/2\pi$ is the resonance frequency difference between the chemical species of interests and the solvent water peak in units of Hz. The faster the polarization transfer between solute and water occurs, the greater the potential signal amplification that CEST experiments can achieve. Figure 16.3 shows a ^1H spectrum and a Z-spectrum of a bovine cartilage sample, with the –NH, –OH, and NOE sites indicated.

Additional CEST-type spectral features of cartilage tissue are commonly observed between −2.6 ppm and −1.0 ppm upfield of

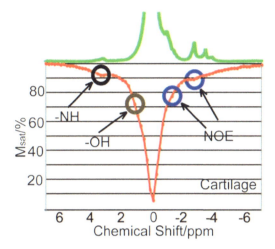

Figure 16.3 Spectrum of a cartilage specimen (top, green), and CEST Z-spectrum (bottom, red). Amide and hydroxyl CEST spectral features are labeled along with less well-characterized resonances between the -1 and -3 ppm that are dominated by NOE-mediated transfer mechanisms [42].

water, which have been attributed to the CH and N-acetyl chemical moieties of the GAG sugars. These upfield signals are located in spectroscopically crowded regions and are difficult to separate from those of other common biological solutes and cellular components [42, 43, 45]. These sites typically do not display significant rates of chemical exchange. The polarization transfer of these signals detected by the CEST experiment likely originates from slow NOE-mediated transfer mechanisms, thus making them less desirable as gagCEST targets.

Figure 16.4 shows ^{1}H spectra and Z-spectra of the two regions of the IVD, the AF, and the NP, demonstrating that spectral assignments similar to those of cartilage can be made.

Validations of the gagCEST experiment using known concentrations of the purified GAG polymer, chondroitin sulfate (see Fig. 16.2b), have demonstrated excellent quantitative correspondence ($R^2 > 0.99$) between chondroitin sulfate sample concentration and the gagCEST hydroxyl signals [46]. Ex vivo plugs of cartilage and the NP have also been used to validate gagCEST experiments. To create

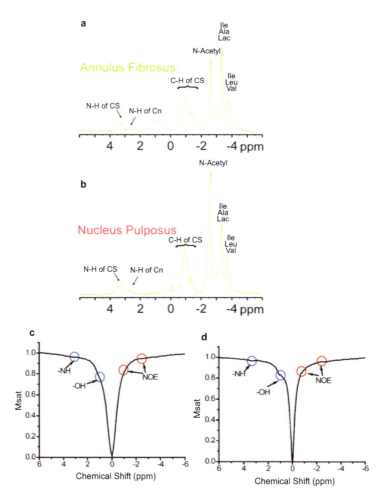

Figure 16.4 Correlation between gagCEST effect and fixed charge density (FCD) as well as N-acetyl concentration. Reprinted with permission from Ref. [18], Copyright 2011, John Wiley and Sons.

these tissue plugs with different GAG concentrations, the tissue samples were treated with trypsin protease for a variable amount of time to simulate the depletion of proteoglycan aggregates occurring during the development of OA (or DDD). The resultant hydroxyl gagCEST signals of these samples were then confirmed to be strongly correlated with both Na imaging data [43] (see Fig. 16.5), the degree of proteoglycan degradation [18], as well as with the concentration of the GAG's N-acetyl group (as detected spectroscopically, see Fig. 16.2). It is noted that the ex vivo experiments are often harder to implement under well-defined conditions, especially for such a fluid tissue as the NP, where it is hard to maintain exact osmotic pressure and water content, which is even more difficult after enzymatic treatment. These conditions are thought to be at the root of the major uncertainties in the ex vivo measurements, as shown in Fig. 16.5.

The pH dependence of the gagCEST effect was also evaluated for IVD specimens ex vivo at 7 T [47]. The pH was measured by calculating a CEST ratio of Iopromide, an FDA-approved CT/X-ray agent, that was injected into the tissue. The dependence of the CEST ratio of Iopromide on pH was independently calibrated. Inside the tissue, pH was adjusted by injecting sodium lactate. The gagCEST effect was found to peak at approximately pH 7.2 and decrease by approximately 30–50% toward pH 5.7 or pH 8.2. This study can serve as a basis for quantifying gagCEST variations in response to in vivo pH changes [47].

Since the first in vivo demonstration of gagCEST on cartilage [42], a number of practical challenges of this approach were identified, which are common to many CEST implementations. The susceptibility of the experiment to B_0 inhomogeneities can be quite strong, since small changes in the magnetic fields can lead to significant false positives when calculating the *MTR* asymmetry. This effect is especially pronounced when using resonances for CEST that are close to water, such as the hydroxyl groups. More recent studies all use a B_0 correction based on WASSR or its derivatives [48]. In many CEST techniques that use hydroxyl groups, the frequency separation between the exchangeable site and the water resonance is relatively small, thus complicating the selective saturation. A smaller frequency separation also increases the likelihood of an intermediate or fast exchange regime, thus further reducing the ability to selectively saturate the exchangeable site.

Typical measurement parameters in vivo include a saturating field of 0.5–1.5 µT for the saturating field at an irradiation time of up to 1 s, and often a slice-selection spoiled segmented gradient-echo sequence is used for image readout.

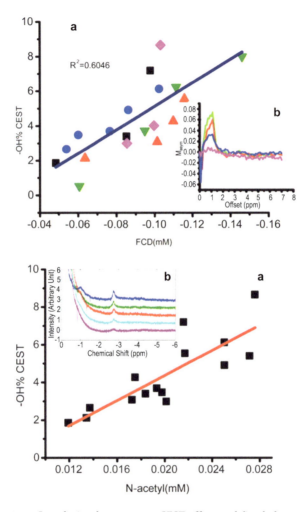

Figure 16.5 Correlation between gagCEST effect and fixed charge density (FCD) as well as N-acetyl concentration. Reprinted with permission from Ref. [18], Copyright 2011, John Wiley and Sons.

Singh et al. have examined gagCEST at both 3 T and 7 T in vivo [9]. In these studies, gagCEST effects that had been observed at a 3 T field strength disappeared after corrections for B_0 inhomogeneity were made [48], and at 7 T, GAG signals dropped from nearly 35% of the total solvent signal to approximately 6% following the same corrections. This finding suggests that (1) GAG hydroxyl groups may typically be in a fast exchange with water on the MRI timescale at 3 T, and that gagCEST is easier to perform at a 7 T field strength or higher where intermediate to slow exchange conditions are more likely to hold, and (2) careful B_0 referencing is critical for in vivo quantification of gagCEST [9].

Others have seen, however, noticeable cartilage gagCEST effects even at 3 T from cartilage [49] after B_0 correction. A WASSR-type B_0 correction method was also compared with a gradient-echo method, and the latter was found to be feasible for a 3 T implementation [49].

The dGEMRIC, gagCEST, and T_2 mapping of cartilage were also compared recently [50]. The conclusion of this study was that CEST was able to detect normal and damaged cartilage and was non-inferior in distinguishing between them. It was further suggested

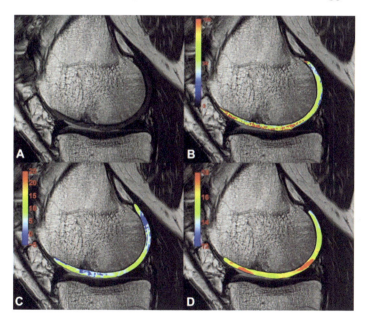

Figure 16.6 Comparison of MRI-based cartilage detection on a human knee in vivo with colored overlays representing different contrast techniques in the articular cartilage. (a) Anatomical proton density image; (b) T_2-weighted image of cartilage, with color representing relaxation time in ms; (c) gagCEST-weighted image; (d) sodium imaging. Reprinted from Ref. [51], Copyright 2012, with permission from Elsevier.

that some variation of the CEST contrast is related to complex remodeling processes and may also relate to changes in T_2. It was also indicated that a T_2 correction algorithm could be suitable for enhancing accuracy.

GagCEST of cartilage was also compared with sodium MRI at 7 T (Fig. 16.6) [51, 52]. It is noted, however, that the sodium pulse sequences had relatively long TE values and thus did not sample the short-lived cartilage-sodium signals very well.

For the IVD, several studies evaluated gagCEST at 3 T [19, 46, 53]. In these studies, it was easy to distinguish NP from AF in those studies. A turbo spin echo was used to suppress bowel movements in Ref. 46. Since dGEMRIC is not applicable in the disc, and sodium MRI is difficult to perform for the disc as well, it is important to identify suitable alternative markers for GAG in cartilage, such that gagCEST could be validated in this tissue.

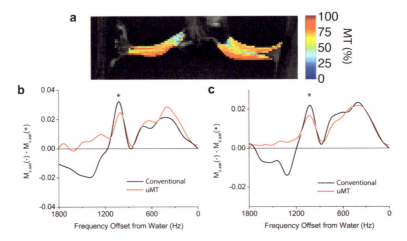

Figure 16.7 Removal of asymmetric MT effects from CEST curves in cartilage tissue via the uniform MT (uMT) CEST method [55]. (a) A uMT contrast map from the right knee of a volunteer, superimposed onto the anatomical reference image. MT asymmetry curves averaged over the segments on the lateral (b) and medial (c) femorotibial cartilage, which consist of the pixels with the uMT contrast more than 50%. Black and red lines are MT asymmetry curves, respectively, from the conventional and uMT CEST acquisitions. The MT asymmetry curves from the uMT CEST method were inverted and shifted for ease of comparison with the conventional CEST method.

Recently Liu et al. [54] demonstrated the feasibility of low back pain detection employing the ratio of $T_{1\rho}$ dispersion and CEST for pH-level dependent imaging of the IVD while eliminating the effect of labile proton concentration. The technique was validated by numerical simulations on phantoms, ex vivo porcine spines, and healthy and low back pain patients on a 3.0 Tesla MR scanner and correlated with reference standard discography.

In recent work, it was found that the contribution of asymmetric MT could be an important factor in gagCEST quantification [55]. The so-called uniform MT CEST method could be used to measure gagCEST with minimized contributions from asymmetric MT (Fig. 16.7). The method is described in detail in Chapter 7.

16.6 Conclusion

We have outlined here the structure of cartilage and the IVD tissues, and how these could be assessed via CEST. The main usable exchangeable component in cartilage comes from the hydroxyl groups of the GAG saccharide chains. The method, hence called gagCEST, has the potential to measure or qualitatively assess GAG concentrations in vivo, which have also been linked to the onset of OA and degenerative disc disease. B_0 inhomogeneity and insufficient frequency dispersion (at fields of 3 T and below) can present some challenges, because the exchangeable sites could not be saturated efficiently without affecting the main water resonance. Nonetheless, important proof-of-principle demonstrations have already been performed and the method was compared to dGEMRIC and sodium MRI. It can be expected that gagCEST will become useful at fields of 3 T or greater. The non-invasiveness of the method and the potential of specificity in cartilage and disc tissues are of advantage.

References

1. Bradley JD, Brandt KD, Katz BP, Kalasinski LA, and Ryan SI. Comparison of an antiinflammatory dose of ibuprofen, an analgesic dose of ibuprofen, and acetaminophen in the treatment of patients with

osteoarthritis of the knee. *New England Journal of Medicine*, 1991; 325(2): 87–91.

2. Torio CM and Andrews RM. National inpatient hospital costs: The most expensive conditions by payer, 2011: Statistical Brief #160. *Healthcare Cost and Utilization Project (HCUP) Statistical Briefs*. Rockville (MD): Agency for Health Care Policy and Research (US); 2006-2013 Aug.

3. Kanis JA. Diagnosis of osteoporosis and assessment of fracture risk. *Lancet*, 2002; 359(9321): 1929–1936.

4. Kanis JA, Melton LJ, Christiansen C, Johnston CC, and Khaltaev N. The diagnosis of osteoporosis. *Journal of Bone and Mineral Research*, 1994; 9(8): 1137–1141.

5. Chan WP, Lang P, Stevens MP, et al. Osteoarthritis of the knee: Comparison of radiography, CT, and MR imaging to assess extent and severity. *American Journal of Roentgenology*, 1991; 157(4): 799–806.

6. Lang P, Noorbakhsh F, and Yoshioka H. MR imaging of articular cartilage: Current state and recent developments. *Radiologic Clinics of North America*, 2005; 43(4): 629–639.

7. Malemud CJ. Changes in proteoglycans in osteoarthritis: Biochemistry, ultrastructure and biosynthetic processing. *Journal of Rheumatology*, 1991; 27: 60–62.

8. Juras V, Bittsansky M, Majdisova Z, et al. In vitro determination of biomechanical properties of human articular cartilage in osteoarthritis using multi-parametric MRI. *Journal of Magnetic Resonance*, 2009; 197(1): 40–47.

9. Singh A, Haris M, Cai K, et al. Chemical exchange saturation transfer magnetic resonance imaging of human knee cartilage at 3 T and 7 T. *Magnetic Resonance in Medicine*, 2012; 68(2): 588–594.

10. Poole CA. Review. Articular cartilage chondrons: Form, function and failure. *Journal of Anatomy*, 1997; 191(1): 1–13.

11. Pearle AD, Warren RF, and Rodeo SA. Basic science of articular cartilage and osteoarthritis. *Clinics in Sports Medicine*, 2005; 24(1): 1–12.

12. Urban JP and Roberts S. Degeneration of the intervertebral disc. *Arthritis Research and Therapy*, 2003; 5(3): 120–130.

13. Eyre DR. Biochemistry of the intervertebral disc. *International Review of Connective Tissue Research*, 1979; 8: 227–291.

14. Inoue H. Three-dimensional architecture of lumbar intervertebral discs. *Spine*, 1981; 6(2): 139–146.

15. Eyre DR and Muir H. Quantitative analysis of types I and II collagens in human intervertebral discs at various ages. *Biochimica et Biophysica Acta*, 1977; 492(1): 29–42.

16. Hardy PA. Intervertebral disks on MR images: Variation in signal intensity with the disk-to-magnetic field orientation. *Radiology*, 1996; 200(1): 143–147.
17. Urban JP and McMullin JF. Swelling pressure of the intervertebral disc: Influence of proteoglycan and collagen contents. *Biorheology*, 1985; 22(2): 145–157.
18. Saar G, Zhang B, Ling W, Regatte RR, Navon G, and Jerschow A. Assessment of glycosaminoglycan concentration changes in the intervertebral disc via chemical exchange saturation transfer. *NMR in Biomedicine*, 2012; 25(2): 255–261.
19. Haneder S, Apprich SR, Schmitt B, et al. Assessment of glycosaminoglycan content in intervertebral discs using chemical exchange saturation transfer at 3.0 Tesla: Preliminary results in patients with low-back pain. *European Radiology*, 2013; 23(3): 861–868.
20. Figueroa D, Calvo R, Vaisman A, Carrasco MA, Moraga C, and Delgado I. Knee chondral lesions: Incidence and correlation between arthroscopic and magnetic resonance findings. *Arthroscopy: The Journal of Arthroscopic and Related Surgery*, 2007; 23(3): 312–315.
21. Campbell AB, Knopp MV, Kolovich GP, et al. Preoperative MRI underestimates articular cartilage defect size compared with findings at arthroscopic knee surgery. *The American Journal of Sports Medicine*, 2013; 41(3): 590–595.
22. Oei EH, van Tiel J, Robinson WH, and Gold GE. Quantitative radiologic imaging techniques for articular cartilage composition: Toward early diagnosis and development of disease-modifying therapeutics for osteoarthritis. *Arthritis Care and Research*, 2014; 66(8): 1129–1141.
23. Bashir A, Gray ML, Boutin RD, and Burstein D. Glycosaminoglycan in articular cartilage: In vivo assessment with delayed Gd(DTPA)(2-)-enhanced MR imaging. *Radiology*, 1997; 205(2): 551–558.
24. Williams A, Gillis A, McKenzie C, et al. Glycosaminoglycan distribution in cartilage as determined by delayed gadolinium-enhanced MRI of cartilage (dGEMRIC): Potential clinical applications. *American Journal of Roentgenology*, 2004; 182(1): 167–172.
25. Tiderius CJ, Olsson LE, Leander P, Ekberg O, and Dahlberg L. Delayed gadolinium-enhanced MRI of cartilage (dGEMRIC) in early knee osteoarthritis. *Magnetic Resonance in Medicine*, 2003; 49(3): 488–492.
26. Koenig SH. From the relaxivity of Gd (DTPA) 2–to everything else. *Magnetic Resonance in Medicine*, 1991; 22(2): 183–190.
27. Bashir A, Gray ML, and Burstein D. Gd-DTPA2–as a measure of cartilage degradation. *Magnetic Resonance in Medicine*, 1996; 36(5): 665–673.

28. Trattnig S, Mlyńarik Vr, Breitenseher M, et al. MRI visualization of proteoglycan depletion in articular cartilage via intravenous administration of Gd-DTPA. *Magnetic Resonance Imaging*, 1999; 17(4): 577–583.
29. Bashir A, Gray ML, Hartke J, and Burstein D. Nondestructive imaging of human cartilage glycosaminoglycan concentration by MRI. *Magnetic Resonance in Medicine*, 1999; 41(5): 857–865.
30. Kim Y-J, Jaramillo D, Millis MB, Gray ML, and Burstein D. Assessment of early osteoarthritis in hip dysplasia with delayed gadolinium-enhanced magnetic resonance imaging of cartilage. *The Journal of Bone and Joint Surgery*, 2003; 85(10): 1987–1992.
31. Eckstein F, Wyman BT, Buck RJ, et al. Longitudinal quantitative MR imaging of cartilage morphology in the presence of gadopentetate dimeglumine (Gd-DTPA). *Magnetic Resonance in Medicine*, 2009; 61(4): 975–980.
32. Shapiro EM, Borthakur A, Gougoutas A, and Reddy R. ^{23}Na MRI accurately measures fixed charge density in articular cartilage. *Magnetic Resonance in Medicine*, 2002; 47(2): 284–291.
33. Reddy R, Insko EK, Noyszewski EA, Dandora R, Kneeland JB, and Leigh JS. Sodium MRI of human articular cartilage in vivo. *Magnetic Resonance in Medicine*, 1998; 39(5): 697–701.
34. Borthakur A, Shapiro EM, Beers J, Kudchodkar S, Kneeland JB, and Reddy R. Sensitivity of MRI to proteoglycan depletion in cartilage: Comparison of sodium and proton MRI. *Osteoarthritis and Cartilage*, 2000; 8(4): 288–293.
35. Wheaton AJ, Borthakur A, Shapiro EM, et al. Proteoglycan loss in human knee cartilage: Quantitation with sodium MR imaging—feasibility study 1. *Radiology*, 2004; 231(3): 900–905.
36. Borthakur A, Mellon E, Niyogi S, Witschey W, Kneeland JB, and Reddy R. Sodium and $T_{1\rho}$ MRI for molecular and diagnostic imaging of articular cartilage. *NMR in Biomedicine*, 2006; 19(7): 781–821.
37. Duvvuri U, Charagundla SR, Kudchodkar SB, et al. Human knee: In vivo T_{1rho}-weighted MR imaging at 1.5 T—preliminary experience. *Radiology*, 2001; 220(3): 822–826.
38. Regatte RR, Akella SV, Borthakur A, Kneeland JB, and Reddy R. In vivo proton MR three-dimensional T_{1rho} mapping of human articular cartilage: Initial experience. *Radiology*, 2003; 229: 269–274.
39. Akella SV, Regatte RR, Gougoutas AJ, Borthakur A, Shapiro EM, and Kneeland JB. Proteoglycan-induced changes in T_{1rho}-relaxation of

articular cartilage at 4T. *Magnetic Resonance in Medicine*, 2001; 46: 419–423.

40. Regatte RR, Akella SV, Wheaton AJ, et al. 3D-T_{1rho}-relaxation mapping of articular cartilage: In vivo assessment of early degenerative changes in symptomatic osteoarthritic subjects. *Academic Radiology*, 2004; 11: 741–749.

41. Li X, Benjamin Ma C, Link TM, et al. In vivo $T_{(1rho)}$ and $T_{(2)}$ mapping of articular cartilage in osteoarthritis of the knee using 3 T MRI. *Osteoarthr Cartilage*, 2007; 15: 789–797.

42. Ling W, Regatte RR, Navon G, and Jerschow A. Assessment of glycosaminoglycan concentration in vivo by chemical exchange-dependent saturation transfer (gagCEST). *Proceedings of the National Academy of Sciences USA*, 2008; 105(7): 2266–2270.

43. Ling W, Regatte RR, Schweitzer ME, and Jerschow A. Characterization of bovine patellar cartilage by NMR. *NMR in Biomedicine*, 2008; 21(3): 289–295.

44. Hills BP, Cano C, and Belton PS. Proton NMR relaxation studies of aqueous polysaccharide systems. *Macromolecules*, 1991; 24(10): 2944–2950.

45. Schiller J, Naji L, Huster D, Kaufmann J, and Arnold K. ^{1}H and ^{13}C HR-MAS NMR investigations on native and enzymatically digested bovine nasal cartilage. *Magnetic Resonance Materials in Physics, Biology and Medicine*, 2001; 13(1): 19–27.

46. Liu Q, Jin N, Fan Z, et al. Reliable chemical exchange saturation transfer imaging of human lumbar intervertebral discs using reduced-field-of-view turbo spin echo at 3.0 T. *NMR in Biomedicine*, 2013; 26(12): 1672–1679.

47. Melkus G, Grabau M, Karampinos DC, and Majumdar S. Ex vivo porcine model to measure pH dependence of chemical exchange saturation transfer effect of glycosaminoglycan in the intervertebral disc. *Magnetic Resonance in Medicine*, 2014; 71(5): 1743–1749.

48. Kim M, Gillen J, Landman BA, Zhou J, and van Zijl P. Water saturation shift referencing (WASSR) for chemical exchange saturation transfer (CEST) experiments. *Magnetic Resonance in Medicine*, 2009; 61(6): 1441–1450.

49. Wei W, Jia G, Flanigan D, Zhou J, and Knopp MV. Chemical exchange saturation transfer MR imaging of articular cartilage glycosaminoglycans at 3 T: Accuracy of B_0 Field Inhomogeneity corrections with gradient echo method. *Magnetic Resonance Imaging*, 2014; 32(1): 41–47.

50. Rehnitz C, Kupfer J, Streich NA, et al. Comparison of biochemical cartilage imaging techniques at 3 T MRI. *Osteoarthritis and Cartilage*, 2014; 22(10): 1732–1742.
51. Krusche-Mandl I, Schmitt B, Zak L, et al. Long-term results 8 years after autologous osteochondral transplantation: 7 T gagCEST and sodium magnetic resonance imaging with morphological and clinical correlation. *Osteoarthritis and Cartilage*, 2012; 20(5): 357–363.
52. Schmitt B, Zbyn S, Stelzeneder D, et al. Cartilage quality assessment by using glycosaminoglycan chemical exchange saturation transfer and (23)Na MR imaging at 7 T. *Radiology*, 2011; 260(1): 257–264.
53. Kim M, Chan Q, Anthony MP, Cheung KM, Samartzis D, and Khong PL. Assessment of glycosaminoglycan distribution in human lumbar intervertebral discs using chemical exchange saturation transfer at 3 T: Feasibility and initial experience. *NMR in Biomedicine*, 2011; 24(9): 1137–1144.
54. Liu Q, Tawackoli W, Pelled G, et al. Detection of low back pain using pH level-dependent imaging of the intervertebral disc using the ratio of $R1\rho$ dispersion and $-$OH chemical exchange saturation transfer (RROC). *Magnetic Resonance in Medicine*, 2014; 73(3): 1196–1205.
55. Lee J-S, Regatte RR, and Jerschow A. Isolating chemical exchange saturation transfer contrast from magnetization transfer asymmetry under two-frequency rf irradiation. *Journal of Magnetic Resonance*, 2012; 215: 56–63.

Chapter 17

GlucoCEST: Imaging Glucose in Tumors

Francisco Torrealdea, Marilena Rega, and Xavier Golay

*Institute of Neurology, University College London, Queen Square,
London WC1N 3BG, UK*
x.golay@ucl.ac.uk

17.1 Introduction

Glucose plays a central role in biology. Almost every organism, from bacteria to humans, use it as an energy and biomass source to sustain their metabolic demands. In the human body, a single glucose molecule can provide up to 36 ATP molecules via complete aerobic respiration, but it can also be transformed into several carbon scaffolds for biosynthetic reactions [3, 10, 16]. Cancer cells avidly consume glucose at rates of up to 20 times faster than their healthy counterparts. The elevated glucose uptake enables the cells to meet the energetic requirements for fast cell proliferation and to produce many intermediate biosynthetic precursors involved in biomass duplication [8, 9]. Upregulated glycolysis is arguably the single most common feature in nearly all primary and metastatic cancers, a phenomenon known as the Warburg effect. Even under well-oxygenated conditions, cancer cells generally metabolize glucose to produce lactate by aerobic glycolysis followed by reaction with lactate dehydrogenase. The aberrant consumption of glucose by tumors has been widely exploited in the diagnosis of cancer with the use of fluorodeoxyglucose positron emission tomography (^{18}FDG-PET) in nuclear medicine. Similarly, using glucose chemical exchange saturation transfer (glucoCEST) magnetic resonance imaging (MRI), tumors are studied by looking at the concentration of natural glucose in the tissue, which can be detected through the chemical exchange of hydroxyl groups. This chapter explains the principles and rationale behind the glucoCEST technique and presents a summary of the most recent developments in the field.

17.2 Cancer Metabolism and the Warburg Effect

In contrast to normal cells, which grow and divide under tightly controlled biological conditions, cancer cells have the ability to evade the body's regulatory processes and multiply in an uncontrolled manner. This ability to alter physiological pathways to their advantage is wired deep in the genetic code of the cancer cells. Overall, these genetic mutations give rise to a wide range of phenotypes, which makes both the study and therapy of malignant neoplastic diseases difficult. Toward the middle of the twentieth century, Otto Warburg [33] observed that tumor cells use conversion of glucose to lactate for energy production instead of relying on the much more efficient mitochondrial oxidation. This was an unexpected finding, as tumor cells need, in principle, more energy than normal differentiated cells. Warburg's opinion was that this respiratory defect, conversion of glucose to lactate together with the cancellation of strong reduction of the Krebs cycle in the mitochondria, was the most characteristic change of tumor cells. The reader is directed to the paper by Ganapathy et al. [4] for a description of the main differences in transport and metabolism between normal and tumor cells. Although hypoxia in the tumoral tissue due to abnormal tumor microvasculature is sometimes given as a possible contributing factor to mitochondrial malfunction, mitochondrial function is partially suppressed in most tumor cells even in the abundant presence of oxygen [4]. According to Gosalvez [6], the mitochondrial malfunction might be due to an alteration in the tumoral mitochondria's filamentation–defilamentation cycle, which reduces its affinity to ADP below the affinity of the pyruvate kinase isoenzymes. Thus, in a competition for ADP, which occurs between mitochondria and glycolysis, the latter is favored. Glycolytic metabolism of glucose to lactate under normoxic conditions, a process sometimes referred to as "aerobic glycolysis," is accompanied by the release of lactic acid to the extracellular compartment. Therefore, aerobic glycolysis in tumors drives extracellular pH (pHe) to become acidic [23]. Recent studies observed that this acidic environment promotes invasion and enhances metastasis, which offers cancer a selective evolutionary advantage [27, 28]. Even though glycolysis yields only two ATP per molecule of glucose, tumors obtain energy by enhancing the glycolysis rate to at least 25-fold higher than in normal cells, which is more than compensating for the glycolytic energy inefficiency. Obviously, this high rate of glycolysis cannot be reached with normal enzymatic levels. To get such an extraordinary high rate of glucose conversion in aerobic glycolysis, cancer cells must enhance their glucose transport to match the elevated glycolysis rate. Tumor cells must activate not only glucose transporters but also genes encoding necessary glycolytic enzymes to increase their expression at the required levels to enable the metabolism of an amount of glucose 25 times higher than normal, hence the term *metabolic reprogramming* [9]. In addition to these changes in expression levels, it is worth noting that enzymatic activity enhancement can also be achieved by several post-translational modifications as well. Fast proliferation requires abundant energy as well as a constant supply of building blocks to enable anabolism. As Warburg himself noticed, the respiration process in tumors is small compared to glucose consumption, but not small relative to the respiration in normal tissues [34]. During periods of rapid proliferation, many cell types ranging from microbes to lymphocytes resort to aerobic glycolysis, which suggests that this process may play a crucial role in supporting normal cell growth. Heiden et al. [10, 16] argue that the main function of aerobic glycolysis is to sustain high levels of glycolytic intermediates as the building blocks for anabolic reactions in cells. An alternative to plain energetic considerations, this argument provides an explanation as to why increased glucose metabolism is chosen in proliferating cells throughout nature.

17.3 Imaging Methods Targeting Metabolism

For more than three decades, diagnosis in oncology has exploited the elevated glucose uptake of tumors by using positron emission tomography (PET) in combination with 2-deoxy-D-glucose labeled with an ^{18}F radiotracer. Based on the relative uptake of tissues, ^{18}F-FDG PET has enabled the distinction of areas of active tumor from non-tumor or necrotic regions. Furthermore, ^{18}F-FDG PET has been correlated with tumor grade in a wide range of cancers, with an intense PET signal associated with fast proliferating malignant cells [7, 15, 21, 24]. In the field of nuclear magnetic resonance (NMR), spectroscopic techniques have successfully identified and characterized mostly primary brain tumors (gliomas) based on their increased anaerobic pathway (glycolysis) of glucose, by measuring increases in lactate and choline concentrations together with a reduction of N-acetyl aspartate (NAA) in cancerous areas. Again, these NMR profiles correlate with glioma grades, displaying high choline and lactate levels in the most common and aggressive Grade IV gliomas, also known as glioblastoma multiforme (GBM) [2, 22]. In this context glucoCEST shows up as a novel, potentially promising technique for the study of tumors by exploiting their aberrant metabolism. It is worth pointing out that increased lactate is produced as a direct consequence of the elevated glycolysis, together with activation of the overexpression of lactate dehydrogenase. This excess of lactate gets secreted from the cell, which in turn creates an acidic environment in the extracellular space (ECS). In this context, cancer cells present an unusual situation in which extracellular pH becomes more acidic than the intracellular one. This anomaly, sometime referred to as "reverse pH gradient," could also play an important role in CEST-based experiments, as the exchange rate is greatly influenced by the pH of the milieu. In this sense, image contrast could be enhanced based on local pH variation around tumors.

17.4 GlucoCEST: The Concept

CEST is extremely sensitive to a number of chemical species and metabolites present in the body (see Chapter 8). All these species exchange magnetization simultaneously resulting in a combined effect that produces a complex shaped Z-spectrum. Moreover, in vivo transverse relaxation times are shorter than in aqueous solutions, which leads to broader absorption bandwidths of the solutes and to overlapping resonant frequencies. Accurate deconvolution of the Z-spectrum into its individual contributors is, therefore, very difficult and a major area of research in the CEST field. Owing to these limitations, exact quantification of glucose content in tissues is unrealistic at present.

The approach in glucoCEST, however, is not to measure the intrinsic glucose content of tissues but the concentration increase following an external administration of glucose. For that purpose, two CEST images are acquired, one before and another after the injection of glucose. The final glucoCEST signal, named glucoCEST enhancement (GCE) by Walker-Samuel et al. [32], arises from the difference of the two images (see Fig. 17.1). In this way, a substantial increase in glucose following administration leads to a strong glucoCEST signal, while no signal is observed in the case of absence of glucose uptake. The approach taken in glucoCEST avoids many of

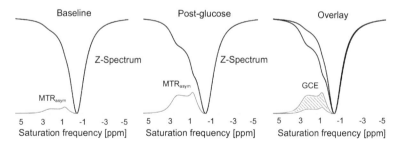

Figure 17.1 Simulated Z-spectra of three hydroxyl group resonances alongside the magnetization transfer ratio asymmetry (MTR_{asym}) before (left) and after (center) glucose administration. The signal from the hydroxyl groups resonating at 1.2, 2.1, and 2.9 ppm from water increases. The GCE is defined as the change in the area under the MTR_{asym} curve from baseline (right).

the complications involved in the interpretation of CEST data. The first pre-glucose image serves as a reference from which changes can be measured; therefore, no deconvolution methods or sophisticated multi-pool fittings are required in order to identify the source of the signal. The method picks up regions of higher glucose concentration with relatively little manipulation of the data. In this way, glucoCEST may allow the use of natural sugar as a potential biomarker of metabolism. This makes it a very attractive and readily applicable technique for a number of metabolism-related disorders.

17.4.1 Advantages

Like all CEST-based methods, glucoCEST offers an enhanced sensitivity to labile protons, glucose hydroxyl groups in this case, which allows in vivo and non-invasive imaging of glucose distribution in tissues. In vivo Z-spectra display complex profiles deriving from various independent processes, i.e., spillover, macromolecular MT, chemical exchange from amide, amines and hydroxyl groups, aliphatic nuclear Overhauser effect (NOE), and distortions from B_1 inhomogeneities. A major challenge in endogenous CEST is the lack of specificity to separate and weigh the individual contribution of each process that shapes the Z-spectrum. While under certain conditions, clever methods of data analysis can yield consistent values for in vivo concentrations of labile protons, reliable quantification is still very problematic, particularly for those proton species resonating in a frequency range close to water (~1 ppm), as it is the case here. Glucose administrated exogenously alters the basal in vivo CEST profile, which allows easier determination of its specific contribution to the CEST signal by a subtraction of the baseline image acquired prior glucose administration.

17.4.2 Drawbacks

Currently, the main limitation of glucoCEST is arguably its low signal-to-noise ratio (SNR). Similar to the blood-oxygen-level dependent (BOLD) contrast used in functional MRI (fMRI), the CEST contrast is generated from the difference of two quasi-identical CEST images. Considering that these CEST images have a low signal to begin with (especially at high saturation powers where effects of direct saturation are dominant), the final glucoCEST image is prone to elevated noise.

Hence, glucoCEST images tend to compromise spatial resolution in favor of higher SNR. Another hurdle associated with glucoCEST imaging arises from the potential changes in the physiology during acquisition. As a biologically active compound, administration of glucose can change the physiological conditions of the inspected tissues. Altered blood flow, pH, oxygen, and CO_2 levels can lead to significant variations in the observed CEST profile. If not controlled, these changes can lead to misreading of the influence of glucose in the Z-spectrum and consequently to the misinterpretation of the glucoCEST data.

Finally, the limitation on the applicable B_1 power for saturation also diminishes the efficiency of glucoCEST, in particular for clinical applications. Intense and prolonged saturation RF pulses can easily lead to specific absorption rate (SAR) depositions above the regulated safe levels. This may constrain the detection sensitivity for fast exchangeable protons in humans. Safety considerations aside, even within the allowed SAR limits, modern clinical MRI systems have very strict allowed B_1 power levels and the use of solid state RF amplifiers inherently limits their duty cycle.

17.5 GlucoCEST: State of the Art

Over the last 5 years, the number of studies involving CEST MRI has largely expanded due to its attractiveness for in vivo imaging of important physiological traits. The method in glucoCEST differs from endogenous CEST modalities in that it relies on the detection of an exogenously administered CEST agent (or on its physiological effects). As such the analysis of the glucoCEST data becomes more straightforward, as endogenous CEST images serve as control upon which new contrast is generated. In the following, a brief review of the works published around the concept of glucoCEST and the detection of glucose though chemical exchange is presented.

17.5.1 *The Origins: GlycoCEST*

In 2007, van Zijl et al. [31] published a seminal work demonstrating the feasibility of in vivo imaging of glycogen content in the mouse liver based on CEST MRI. The method, named glycoCEST, opened an opportunity to study glycogen metabolism and liver disorders with MRI. By applying a 1 s long RF pulse (3 µT) 1 ppm downfield from water (higher frequency), protons from hydroxyl groups of glycogen get saturated, producing a dip in the Z-spectra as they exchange with the hydroxyl protons in water. While the principles in glycoCEST are similar to other CEST methods, the publication set a precedent in the CEST community as for the first time it was possible to image fast exchanging –OH groups using a non-invasive technique. The glycogen levels in the mouse liver were tracked in time by comparing CEST images irradiated at opposite sides of the water spectrum ($MTR_{asym}(\delta) = MTR(\delta) - MTR(-\delta)$, with $\delta = 1$ ppm), acquired with a [Saturations + Spin-Echo] sequence design on a 4.7 T field scanner. The progressive breakdown of glycogen was sustained by constant infusion of glucagon and validated in parallel experiments by ^{13}C magnetic resonance spectroscopy (MRS), the gold standard technique for non-invasive determination of glycogen concentration in vivo. The CEST measurements showed a gradual reduction in the MTR_{asym} for a period of 2 h from the start of the infusion, after which the MTR_{asym} stabilized, which marked the glycogen depletion point. The time for depletion of glycogen matched with the data found using ^{13}C MRS. The results were somewhat unexpected as the proton exchange rate between glycogen and water protons was supposed to be too fast compared to the small chemical shift (1 ppm) between the two species. While generally protons in –OH groups exchange too rapidly for optimum CEST contrast ($\kappa > \delta$), hydroxyls in glycogen macromolecules exchange at a relatively slow rate ($\kappa < 1000$ Hz), due to their restricted mobility and accessibility within the molecular structure.

This paper and a 2005 review [36] also showed the ability of CEST MRI to detect D-glucose in solution, but this work was not expanded upon. The demonstration of the possibility to detect exchangeable hydroxyl groups led researchers in different labs to start studying ways of exploiting chemical exchange processes to

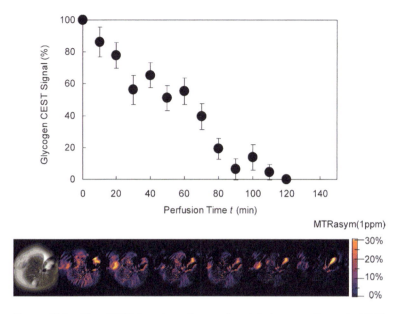

Figure 17.2 GlycoCEST imaging of a perfused fed-mouse liver at 4.7 T and 37°C. The first image (grayscale) marks the beginning of perfusion ($t = 0$) with glucose-free media containing 500 pg/ml glucagon. The colorized glycoCEST images as a function of time during perfusion show the relative CEST intensity [MTR_{asym} (1 ppm)] of liver tissue as a function of perfusion time. With time, as glycogen disappears, the CEST images become more uniformly dark blue, corresponding to depletion of glycogen. The corresponding glycogen reduction for a homogeneous region of interest is quantified in the graph ($n = 4$). Reprinted with permission from Ref. [31], Copyright 2007, National Academy of Sciences.

image biomolecules at the fast end of the intermediate exchange regime. A notable example of this has been the study of glucose and other sugar analogues as CEST agents, fueled by a big clinical interest due to their central role in a number of conditions where metabolism is disrupted.

17.5.2 Cancer Studies

Alteration of glucose metabolism is a ubiquitous feature in cancer. As such, glucoCEST has the qualities to become an ideal method to investigate the abnormally elevated glycolytic flux (Warburg effect) of proliferating tumors. Similar to ^{18}FDG-PET, glucoCEST aims to differentiate healthy tissue from tumors based on their glucose concentration, which has the potential to provide direct information on both its underlying metabolism and aggressiveness. Unlike ^{18}FDG-PET though, glucoCEST is free of the burdens associated with the use of radio-tracers, i.e., high production costs, limited accessibility, and the safety implications of radioactive exposure. Currently, three independent studies have been published on the use of natural glucose as CEST agent for the study of cancer in vivo: the work by Chan et al. [1] from The Johns Hopkins University, the work by Walker-Samuel et al. [32] from University College London, and the study by Rivlin et al. [25] from Tel Aviv University. The Johns Hopkins University group looked at the glucoCEST signal in two different human breast tumor xenograft models (MDA-MB-231 and the less aggressive MCF-7 cell lines) and compared the data to ^{18}FDG-PET and Gd-DTPA contrast-enhanced MRI scans, performed in the same mice cohort. Results of both tumor types displayed elevated signal for glucoCEST and PET with just partial Gd-enhancement. Furthermore, among the three techniques explored, only glucoCEST showed significant separation between cancer cell lines, suggesting the possibility of the characterization of tumors based on their glucoCEST profile.

Interestingly, the glucoCEST signal was found to be consistently lower in the more aggressive phenotype tumors. These results, together with the idea that once inside the cell, glucose is very rapidly converted into lactate, and the fact that acidic extracellular pH in tumors is likely to provide a more favorable exchange rate for the detection of hydroxyl groups in glucose, led researchers at Hopkins to conclude that glucoCEST does not directly inform on metabolism but mainly on interstitial and intravascular glucose levels.

Nonetheless, the debate of the precise origin of the glucoCEST signal was far from settled.

In the study presented by the University College London group [32], the glucoCEST technique was compared to ^{18}FDG autoradiography in two mouse models of colorectal cancer (cell lines SW1222 and LS174T) with distinct metabolic characteristics. Animals were

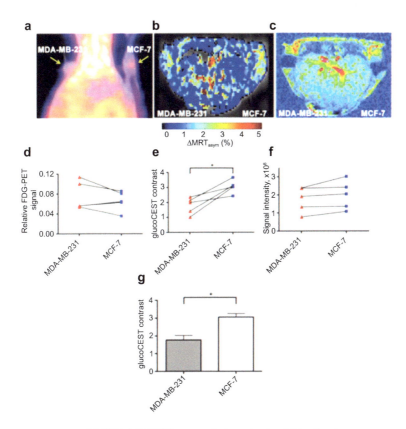

Figure 17.3 (a) ^{18}FDG-PET/CT coronal view obtained 1 h after intravenous injection of ^{18}FDG showing high signal in both tumors. (b) GlucoCEST ΔMTR_{asym} map (infusion, pre-infusion). (c) T_1-weighted difference image (injection, pre-injection) showing Gd-enhanced regions, mainly at the edges of the tumors. (d–f) Comparison of signal intensities ($n = 5$) for the three modalities using ROIs comprising both tumors. Even though some trends appear visible for PET and contrast-enhanced MRI, significant differences ($p < 0.05$; paired student's t-test) between tumors could be detected only by glucoCEST. (g) Bar graph showing average glucoCEST contrast for MDA-MB-231 and MCF-7 tumors. Reprinted with permission from Ref. [1], Copyright 2012, John Wiley and Sons.

410 | GlucoCEST

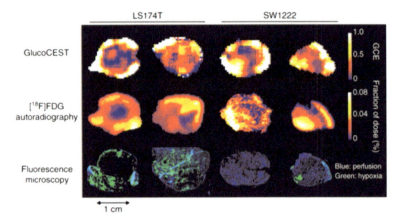

Figure 17.4 The fluorescence microscopy images show perfused (blue) and hypoxic (green) regions corresponding to Hoechst 33342 and pimonidazole staining, respectively. Reprinted with permission from Ref. [32], Copyright 2013, Nature Publishing Group.

scanned with glucoCEST pre- and post-administration of glucose (1 g/kg intraperitoneal [IP] bolus to provide a more sustained glucose delivery than intravascular [IV] bolus), and 24 h later, ^{18}FDG autoradiography experiments were performed on the same animal cohort. The study also included parallel Gd-enhancement MRI, ^{13}C spectroscopic analysis (using uniformly labeled ^{13}C glucose) aimed to track the metabolic path of glucose in the tumors, as well as perfusion and hypoxia fluorescence microscopy images (using Hoechst 33342 and pimonidazole staining, respectively).

In agreement with the study by Chan et al., results showed that glucoCEST is sensitive to tumor glucose uptake and can distinguish tumor types with different metabolism and pathophysiologies.

Results from the ^{13}C MRS study showed ratios of glucose 6-phosphate (G6P) to glucose concentration close to 40% in both tumor models (42% and 38% in SW1222 and LS174T, respectively). In addition, glucose was observed to be metabolized into lactate and a number of amino acids, such as glutamine, glutamate, and alanine. In vitro assessment of the CEST properties of the glycolytic intermediates and amino acids found in the tumors suggested that intracellular stages of glucose metabolism could be detected with glucoCEST, with G6P giving an approximately equal CEST contrast

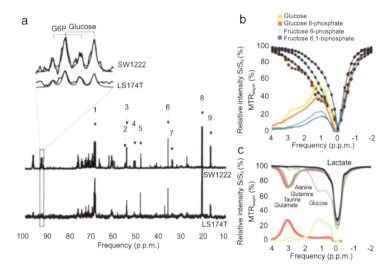

Figure 17.5 (a) Example ^{1}H-decoupled ^{13}C NMR spectra from SW1222 and LS174T tumors after administration of [U-^{13}C]glucose. The peak assignments are as follows: (1) lactate C2; (2) glutamate C2; (3) glutamine C2; (4) alanine C2; (5) taurine C1; (6) taurine C2; (7) glutamate C4; (8) lactate C3; and (9) alanine C3. An expansion of the C1 multiplet is shown, which corresponds to doublets from glucose and glucose-6-phosphate (chemically shifted by 0.13 ppm from the glucose doublet). Fitted Lorentzian peaks are overlaid. (b) Z-spectra and MTR_{asym} spectra from glucose, glucose 6-phosphate, fructose 6-phosphate, and fructose 6,1-biphosphate. In vitro, glucose and glucose-6-phosphate show similar CEST effects, whereas fructose-6-phosphate and fructose 6,1-biphosphate display a smaller effect. Units on the vertical axis are signal intensity, S, normalized to a reference measurement, S_0, at 200 ppm from the peak of the water resonance. (c) Z-spectra and MTR_{asym} spectra from glucose, lactate, glutamine, glutamate, alanine, and taurine. Glucose shows a strong CEST effect from hydroxyl proton exchange, whereas the amino acids show a CEST effect through amide proton exchange. Lactate shows a minimal effect. Reprinted with permission from Ref. [32], Copyright 2013, Nature Publishing Group.

as glucose [32]. This was a surprising result as it seems to contradict previous studies using ^{13}C-labeled glucose in several organs and might be related to the particular cancer cells used in that study.

Provided that the metabolic activity of cancer cells increases in response to a sudden rise in glucose availability, it is expected that the glycolytic intermediates and amino acids from fast catabolism of glucose will contribute to the observed glucoCEST signal.

The lack of correlation between either FDG or glucoCEST signal and perfusion (measured with Hoechst 33342 staining and Gd-DTPA) indicated that the measured glucoCEST contrast was not limited to the vascular delivery of glucose. In fact, based on the observed correlation between FDG and glucoCEST signal and the elevated presence of glucose intermediates found inside the cells, we argued that a significant contribution to the glucoCEST contrast might be attributed to the intracellular compartment. In the third work published on glucoCEST in cancer, the team from Tel Aviv University used the glucose analogues 2-deoxyglucose (2DG) and 2-fluoro-deoxyglucose (FDG, non-radioactive stable isotope ^{19}F) to study the glucoCEST response of a poorly differentiated mammary adenocarcinoma mouse model (cell line DA3-D1-DMBA-3). Both glucose analogues are taken up into the cells by the GLUT transporters and are phosphorylated by hexokinase, after which the catabolic process cannot continue. The molecules accumulate inside the cells until they are slowly cleared through dephosphorylation of 2-deoxyglucose-6-phosphate (2DG6P), or though isomerization to 2-deoxyglucose-1-phosphate (2DG1) to form glycogen [19, 20]. Unlike natural glucose, which quickly metabolizes, the glucoCEST signal from the analogue molecules built up in the tumors over the 2 h that followed the IP administration of FDG/2DG. This results in an increased and long-lasting CEST contrast. However, toxic effects of FDG and 2DG limit their use in clinical studies, where tolerated dose is around 60 mg/kg/day IV (or 250 mg/kg orally), which is an order of a magnitude below the dosage tested in the study [25]. In a recent publication [26], the team from Tel Aviv have reported the use of another less toxic glucose analogue 3-O-methyl-D-glucose (3OMG) as an alternative glucoCEST agent. Once inside the cell, 3OMG does not undergo phosphorylation by hexokinase and can readily return to the blood stream (through the bidirectional GLUT transporters), which consequently is fully excreted by the kidneys. By testing it on the same cancer model as in the previous study, they observed an intense glucoCEST signal, which slowly faded, retaining 70% of the maximum contrast 1 h after IP injection. While further tolerance tests are needed to assess safety dosage of the compound

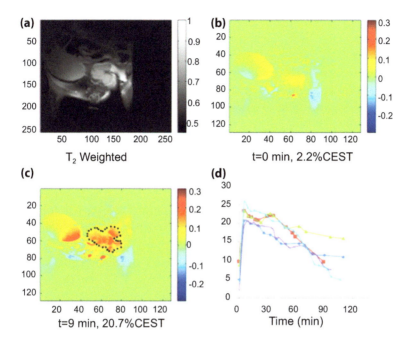

Figure 17.6 GlucoCEST MRI kinetic measurements in the tumor at different times following injection of 3OMG, 1.5 g/kg ($B_1 = 2.5$ μT, $B_0 = 7$ T). (a) A T_2-weighted image before the administration of the agent. (b) A CEST image before the administration of the agent. (c) A CEST image 9 min after the injection. The marked ROI was used for the CEST signal calculation. (d) The time series of the %CEST for the five mice tested. Reprinted with permission from Ref. [26], Copyright 2014, John Wiley and Sons.

in humans, initial data suggest that 3OMG can be potentially useful for the detection of tumors and monitoring of treatment in the clinic.

17.5.3 Brain Studies

Moving now to the field of neuroimaging, an area where, traditionally, MRI excels, two published works on glucoCEST exploring the metabolism of the healthy brain can be found. In the first study, Nasrallah et al. [18] present a thorough study of the evolution of the glucoCEST signal in rat brain under different conditions. The study shows a significantly elevated glucoCEST signal following administration of both 2DG and natural D-glucose. Similar to the results in other studies, signal from 2DG was shown to be nearly twice as intense and lasting longer than that from natural D-glucose. Even with a sustained high blood D-glucose concentration via IV infusion, the signal did not reach levels observed with 2DG, which points to an accumulation of 2DG6P inside the cell (rather than to the extracellular glucose concentration) as an important contributor to the glucoCEST contrast using this tracer. This observation agrees with similar changes observed in 2DG6P levels measured by in vivo ^{31}P NMR, suggesting that glucoCEST reflects the rate of glucose assimilation.

Importantly, a dramatic attenuation of signal was observed when high levels of anesthesia (2% isoflurane) were used, further supporting a link between glucoCEST contrast and the metabolic state of the brain.

Moreover, contribution from the intravascular compartment was ruled out by two independent experiments. First, no change in glucoCEST signal was measured with increased cerebral blood flow (CBF) under hypercapnic conditions (achieved with 1.8% CO_2 in the gas mixture, which doubled basal CBF). Second, administration of L-glucose, which does not leave the vascular compartment in the brain, produced no glucoCEST signal.[a] These results led the research team to conclude that glucoCEST provides an interesting alternative MRI method to image the uptake and metabolism of glucose/2DG in the brain. The second MRI study exploring glucose uptake in the brain was published by Tao Jin et al. [12] in 2014. This study is not, strictly speaking, based on CEST but on a variant dubbed CESL (chemical exchange spin-lock). Both on-resonance and off-resonance spin-lock sequences have indeed been proposed as alternative and more sensitive approach for the detection of fast exchanging protons, like the hydroxyl protons in glucose. These techniques rely on the measurement of $T_{1\rho}$ (the spin-lattice relaxation time in the rotating frame), which ultimately depends on the concentration of the exchangeable pool, just as in CEST. Using an on-resonance

---

[a] L-glucose is the enantiomer (L-isomer) of the naturally occurring D-glucose. In vitro, it displays the same CEST signature as D-glucose. In vivo, however, the Levo chirality prevents it from binding to the GLUT transporters (nor does it serve as substrate for glycolytic enzymes); therefore, its permeability across the blood–brain barrier is very restricted [29].

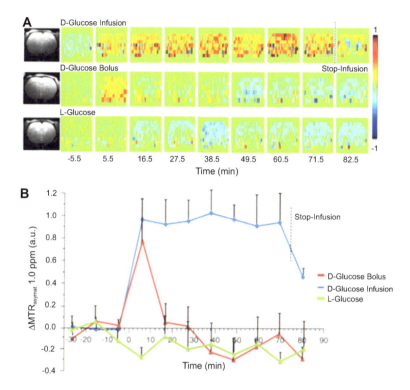

Figure 17.7 (A) Time series of glucoCEST images of rat brain under constant infusion of D-glucose (top), after bolus injection of 1 g/kg D-glucose (middle), and after bolus injection of 1 g/kg L-glucose (bottom). The image intensity represents relative glucoCEST signal change from the baseline. (B) Time courses of glucoCEST signal under the above three injection conditions in (A). The signal represents the difference of MTR_{asym} integral around 1.0 ppm from the baseline signal. Reprinted with permission from Ref. [18], Copyright 2013, SAGE Publications.

spin-lock technique (spin-lock at the water Larmor's frequency), Jin et al. measured a linear glucoCESL response to the glucose dose in the rat brain. Similar to what was reported in the study by Nasrallah et al. [18], the signal measured following the administration of 2DG was more intense (~2.2 times) and prolonged in time than with natural glucose, which is consistent with the different metabolic characteristics previously discussed. Compared to the glucoCEST technique, advantages of on-resonance glucoCESL

416 | *GlucoCEST*

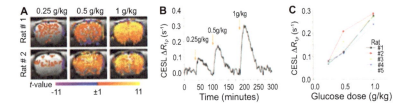

Figure 17.8 Rat-brain glucoCESL studies at 9.4 T showing near-linear contrast for intravenously administered D-glucose doses of 0.25, 0.5, and 1.0 g/kg, and robust detection for doses down to 0.25 g/kg. (A) The t-maps for each dose for two of the animals show highest t-values in the cortex where surface coil sensitivity is higher. Color scale: t-value. (B) Average of time courses for all animals ($n = 5$, mean ± s.e.m.) showing the increase in brain $\Delta R_{1\rho}$ with glucose. Arrows indicate time of injection. (C) The nearly linear dependence of peak brain $\Delta R_{1\rho}$ on glucose appears for each individual animal. Reprinted with permission from Ref. [12], Copyright 2014, SAGE Publications.

include (1) enhanced detection sensitivity for fast exchange regimes ($\kappa/\delta > 1$), which has been numerically demonstrated using two-pool model Bloch McConnell equations [12]; (2) high temporal resolution, as the duration of the spin-locking irradiation is shorter (in the order of $T_{1\rho}$ as opposed to T_1 in CEST); and (3) inherent robustness for B_0 inhomogeneities with no asymmetry analysis required.

Importantly, the sensitivity enhancement of glucoCESL is expected to be higher at low B_0 field as the ratio between the exchange rate and the chemical shift (κ/δ) is increased due to the smaller chemical shift. This makes on-resonance spin-lock particularly interesting for clinical MRI systems. $T_{1\rho}$ is modulated by all the simultaneously occurring chemical exchange processes; therefore, the signal measured with on-resonance spin-lock lacks specificity to any particular type of labile proton. While in vivo studies have shown $T_{1\rho}$ to be fairly insensitive to vascular oxygenation [17], changes in vascular volume and CBF are known to cause variations in $T_{1\rho}$ [11, 13]. Therefore, in glucoCESL, measurements of the contribution of glucose and its potential derivatives have to be carefully weighed against possible changes in tissue vasculature. Finally, although no journal publications have been documented exploring the possibilities of glucoCEST for the study of brain tumors, studies from the research groups at The Johns Hopkins University and University College London have already been presented in various scientific conferences. These ongoing studies aim to provide insight into relevant metabolic aspects of brain tumors as well as to explore the possibility of using natural glucose for detection and characterization of tumor aggressiveness and even for monitoring therapy based on their glucoCEST response. The following example presented by Torrealdea et al. [30] shows follow-up images of a mouse brain bearing a human glioblastoma xenograft, scanned within 2 weeks from each other. Correlation of glucoCEST contrast at day 220 with T_2-weighted spin-echo image and immunohistochemistry at day 235 suggests that glucoCEST could identify tumors at an earlier stage than conventional MRI. The observed glucoCEST contrast could be a reflection of the high glucose uptake in fast proliferating cancer cells, which manifests before significant changes in T_2 occur due to disruption of tissue structure.

17.5.4 *Alternative Technique for Glucose Detection*

As seen in this chapter, CEST and CESL have shown to be promising MRI techniques to investigate glucose uptake and metabolism in vivo, yet a more conventional method has recently been proposed as a means to detect a rise in glucose concentration. Acknowledging the effect that chemical exchange processes have in the relaxation times of water, Yadav et al. [35] from The Johns Hopkins University reported in vivo detection of changes in glucose concentration by measuring the spin–spin relaxation constant T_2, similar to other earlier observations reported by Gore et al. [5].

Using standard T_2-weighted images, the researchers demonstrated contrast generation in mouse kidneys arising at the time of a glucose infusion (a bolus of 0.15 mL of 0.5 M glucose solution). The contrast in the kidneys lasted just over a minute, after which it returned to initial levels, within 2 min after the infusion.

Two of the greatest advantages of this technique are its simplicity (T_2 relaxometry sequences are readily available on all clinical

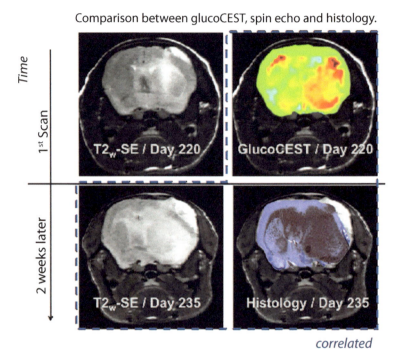

Figure 17.9 At day 220 after injection of cancer cells, the T_{2w} images do not show the full spread of the tumor, while glucoCEST images display features that will be detectable 15 days later. Pixel-by-pixel analysis of glucoCEST signal at day 220 versus T_{2w} image at day 235 shows a weak correlation ($R^2 = 0.13$) but no correlation ($R^2 = 0.027$) at day 220. The overlay of MRI and immunochemistry slice at day 235 shows tumor cells highlighted in brown (vimentin staining) and non-invaded areas of the brain in blue (hematoxylin counterstain) [30].

MRI systems) and the high temporal resolution, which allows for statistical averaging of the signal to boost the effective sensitivity. However, alike CEST, $T_{1\rho}$- and T_2-based methods have a reduced sensitivity in tissues with high intrinsic R20 (exchange-independent spin–spin relaxation constant). In practice, this might become a sensitivity-limiting factor, as very accurate measurements of the respective relaxation times are required.

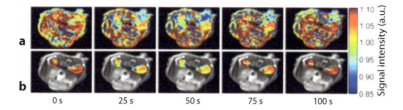

Figure 17.10 Transverse relaxation data from a dynamic in vivo glucose infusion experiment on a mouse. T_2-weighted images of mouse kidneys from the time of a glucose bolus infusion are displayed (time indicated below each image). (a) Effects in all regions displayed. (b) Only the kidney signal is displayed in color on grayscale image to highlight effects. After infusion, the signal in the kidneys drops by approximately 10% and then recovers over the course of 50–100 s. Reprinted with permission from Ref. [35], Copyright 2014, John Wiley and Sons.

17.6 GlucoCEST: Good Practices

17.6.1 *Main Magnetic Field Drifts*

Corrections for the potential shifts of the water resonant frequency (B_0 drifts) over the course of the experiments are essential for reliable asymmetry analysis and meaningful subtractions of the initial baseline image. Therefore, special attention has to be given to making sure every CEST image acquired during the experiment can be adequately corrected. This can be done using different methods.

In a preclinical setup, high field strengths produce sharp Z-spectra. Moreover, less stringent time constrains in animal studies allow for a denser packing of the off-resonant frequencies in the Z-spectrum. In this situation, correction of B_0 inhomogeneities can be successfully achieved by simply interpolating the experimental Z-spectra (typically with a non-parametric algorithm, such as smoothing spline or cubic interpolating spline) and repositioning the water frequency to the position of the minimum intensity (where water is fully saturated). By doing so on a pixel-by-pixel basis, a map of the B_0 shift is obtained, which tells how much each pixel needs to be shifted (in the frequency dimension) before proceeding with any asymmetry analysis. Alternatively, one can obtain B_0 maps parallel to the CEST acquisitions (using the WASSR [14] method or by a standard double TE gradient echo technique) from which to correct for B_0 irregularities in the image slab. This last approach is predominant in clinical systems where, due to lower field and shorter scan times, the Z-spectra tend to be less defined around the water peak. If the chosen method for B_0 correction is to acquire an independent B_0 map, multiple instances of it should be run over the course of the experiment in order to track the evolution of B_0 in time.

17.6.2 Timing of Frequency Offsets

It is important to avoid time delays between offset frequencies downfield and upfield from water. During the relatively long acquisition time of a full Z-spectrum dataset, perceptible changes in the tissue glucose concentrations and drifts in the magnetic field homogeneity may occur. A good practice to minimize the impact of these variations on the asymmetry analysis is to sample the offset frequencies in an alternating pattern, swapping from downfield to upfield frequencies centered at the water resonance. In this way, symmetrical offset data points are acquired with a minimum time gap between them.

17.6.3 Offset and Integration Range

Choosing the right saturation parameters that maximize the CEST signal is not a trivial matter. Generally, fast exchanging hydrogen protons (as in glucose) are expected to produce higher CEST contrast when saturated with elevated RF power, whereas slower exchanging molecules (like amide groups) require less power. However, the SNR also gets affected at high B_1 levels. Intense spillover effect widens the Lorentzian absorption profile reducing the observable signal and resulting in lower SNR close to water. Moreover, strong spillover effects shift the *MTR* asymmetry profile away from water. Hence, the optimum range of offset frequencies that contain the CEST information also depends on the applied saturation power. Figure 17.11 illustrates the glucoCEST enchantment (GCE) signal modulated both in amplitude and frequency by the B_1 irradiation power. Precise optimization of the saturation power will depend on the characteristics of the tissue inspected, such as water content, relaxation times T_2 and T_1, and pH.

17.7 Conclusion: Remaining Open Questions

The data presented in this chapter provide tangible proof of the potential use of glucoCEST for the study of cancer in vivo. Indeed, as shown in the study by Walker-Samuel et al. [32], glucoCEST and FDG autoradiography imaging methods might offer closely related information. Aside from the physical principles of both imaging

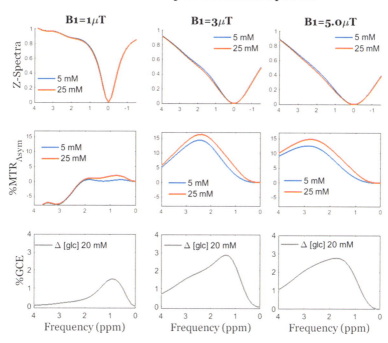

Figure 17.11 Simulation of glucoCEST experiments for 5 and 25 mM glucose concentrations (blue and orange, respectively) at 9.4 T field strength. GCE profile (bottom row) varies both in intensity and frequency with different B_1 irradiation powers (1, 3, and 5 µT, respectively). Simulations run with a multi-pool Bloch-McConnell model with $T_1 = 1850$ ms and $T_2 = 40$ ms, mimicking relaxation in brain tissue.

modalities, there is an important difference between the glucoCEST and PET (or autoradiography) methods, which lies in the marker-molecule used. While ^{18}F-labeled FDG (used in PET) is an analogue of glucose that cannot be metabolized, glucoCEST uses naturally occurring sugar, which is free to go through all the steps of the metabolic chain. This means that the glucose absorbed by the cells does not necessarily build up, as it is the case for FDG. This leads to the question of the compartmental origin of the glucoCEST signal. To this date, there is no consensus among researchers on whether the glucoCEST signal originates from glucose only, or if other sugars along the glycolytic pathway contribute to it as well. If glucoCEST was only sensitive to glucose, it is reasonable to think that it would predominately reflect on blood perfusion in tissues. On the other hand, an intracellular origin of glucoCEST could open an exciting possibility to study the kinetics of cancer metabolism with MRI. Future studies will no doubt shed some light on this ongoing debate.

Acknowledgments

Francisco Torrealdea was supported by a UCL Grand Challenge studentship and Marilena Rega by an Impact studentship in collaboration with Agilent.

References

1. Chan, K. W. Y., McMahon, M. T., Kato, Y., Liu, G., Bulte, J. W. M., Bhujwalla, Z. M., Artemov, D., and van Zijl, P. C. M. Natural D-glucose as a biodegradable MRI contrast agent for detecting cancer. *Magn Reson Med*, 2012; 68(6): 1764–1773.
2. Di Costanzo, A., Scarabino, T., Trojsi, F., Popolizio, T., Catapano, D., Giannatempo, G. M., Bonavita, S., Portaluri, M., Tosetti, M., d'Angelo, V. A., Salvolini, U., and Tedeschi, G. Proton MR spectroscopy of cerebral gliomas at 3 T: Spatial heterogeneity, and tumour grade and extent. *Eur Radiol*, 2008; 18(8): 1727–1735.
3. Fields, R. D. and Burnstock, G. Purinergic signalling in neuron-glia interactions. *Nat Rev Neurosci*, 2006; 7(6): 423–436.

4. Ganapathy, V., Thangaraju, M., and Prasad, P. D. Nutrient transporters in cancer: Relevance to Warburg hypothesis and beyond. *Pharmacol Ther*, 2009; 121(1): 29–40.

5. Gore, J. C., Brown, M. S., Mizumoto, C. T., and Armitage, I. M. (1986). Influence of glycogen on water proton relaxation times. *Magn Reson Med*, 1986; 3(3): 463–466.

6. Gosalvez, M. Mitochondrial filamentation: A therapeutic target for neurodegeneration and aging. *Am J Alzheimers Dis Other Demen*, 2013; 28(5): 423–426.

7. Gulyas, B. and Halldin, C. New PET radiopharmaceuticals beyond FDG for brain tumor imaging. *Q J Nucl Med Mol Imaging*, 2012; 56(2): 173–190.

8. Hanahan, D. and Weinberg, R. A. The hallmarks of cancer. *Cell*, 2000; 100(1): 57–70.

9. Hanahan, D. and Weinberg, R. A. Hallmarks of cancer: The next generation. *Cell*, 2011; 144(5): 646–674.

10. Heiden, M. G. V., Cantley, L. C., and Thompson, C. B. Understanding the Warburg effect: The metabolic requirements of cell proliferation. *Science*, 2009; 324(5930): 1029–1033.

11. Hulvershorn, J., Borthakur, A., Bloy, L., Gualtieri, E. E., Reddy, R., Leigh, J. S., and Elliott, M. A. $T_{1\rho}$ contrast in functional magnetic resonance imaging. *Magn Reson Med*, 2005; 54(5): 1155–1162.

12. Jin, T., Mehrens, H., Hendrich, K. S., and Kim, S.-G. Mapping brain glucose uptake with chemical exchange-sensitive spin-lock magnetic resonance imaging. *J Cereb Blood Flow Metab*, 2014; 34(8): 1402–1410.

13. Jin, T., Wang, P., Zong, X., and Kim, S.-G. MR imaging of the amide-proton transfer effect and the pH-insensitive nuclear Overhauser effect at 9.4 T. *Magn Reson Med*, 2013; 69(3): 760–770.

14. Kim, M., Gillen, J., Landman, B. A., Zhou, J., and van Zijl, P. C. M. Water saturation shift referencing (WASSR) for chemical exchange saturation transfer (CEST) experiments. *Magn Reson Med*, 2009; 61(6): 1441–1450.

15. Lin, T.-J., Huang, C.-C., Wang, I.-J., Lin, J.-W., Hung, K.-S., Fan, L., Tsao, H.-H., Yang, N.-S., and Lin, K.-J. Validation of an animal FDG PET imaging system for study of glioblastoma xenografted mouse and rat models. *Ann Nucl Med Sci*, 2010; 23: 77–83.

16. Lunt, S. Y. and Vander Heiden, M. G. Aerobic glycolysis: Meeting the metabolic requirements of cell proliferation. *Annu Rev Cell Dev Biol*, 2011; 27(1): 441–464.

17. Magnotta, V. A., Heo, H.-Y., Dlouhy, B. J., Dahdaleh, N. S., Follmer, R. L., Thedens, D. R., Welsh, M. J., and Wemmie, J. A. Detecting activity-evoked pH changes in human brain. *Proc Natl Acad Sci USA*, 2012; 109(21): 8270–8273.
18. Nasrallah, F. A., Pagès, G., Kuchel, P. W., Golay, X., and Chuang, K.-H. Imaging brain deoxyglucose uptake and metabolism by glucoCEST MRI. *J Cereb Blood Flow Metab*, 2013; 33(8): 1270–1278.
19. Newman, G. C., Hospod, F. E., Maghsoudlou, B., and Patlak, C. S. Simplified brain slice glucose utilization. *J Cereb Blood Flow Metab*, 1996; 16(5): 864–880.
20. Newman, G. C., Hospod, F. E., and Patlak, C. S. Kinetic model of 2-deoxyglucose metabolism using brain slices. *J Cereb Blood Flow Metab*, 1990; 10(4): 510–526.
21. Nihashi, T., Dahabreh, I. J., and Terasawa, T. PET in the clinical management of glioma: Evidence map. *Am J Roentgenol*, 2013; 200(6): W654–W660.
22. Pedersen, P. L. Tumor mitochondria and the bioenergetics of cancer cells. *Prog Exp Tumor Res*, 1978; 22: 190–274.
23. Prescott, D. M., Charles, H. C., Poulson, J. M., Page, R. L., Thrall, D. E., Vujaskovic, Z., and Dewhirst, M. W. The relationship between intracellular and extracellular pH in spontaneous canine tumors. *Clin Cancer Res*, 2000; 6(6): 2501–2505.
24. Riedl, C. C., Slobod, E., Jochelson, M., Morrow, M., Goldman, D. A., Gonen, M., Weber, W. A., and Ulaner, G. A. Retrospective analysis of ^{18}F-FDG PET/CT for staging asymptomatic breast cancer patients younger than 40 years. *J Nucl Med*, 2014; 55(10): 1578–1583.
25. Rivlin, M., Horev, J., Tsarfaty, I., and Navon, G. Molecular imaging of tumors and metastases using chemical exchange saturation transfer (CEST) MRI. *Sci Rep*, 2013; 3: 3045.
26. Rivlin, M., Tsarfaty, I., and Navon, G. Functional molecular imaging of tumors by chemical exchange saturation transfer MRI of 3-O-methyl-D-glucose. *Magn Reson Med*, 2014; 72(5): 1375–1380.
27. Robey, I. F., Baggett, B. K., Kirkpatrick, N. D., Roe, D. J., Dosescu, J., Sloane, B. F., Hashim, A. I., Morse, D. L., Raghunand, N., Gatenby, R. A., and Gillies, R. J. Bicarbonate increases tumor pH and inhibits spontaneous metastases. *Cancer Res*, 2009; 69(6): 2260–2268.
28. Robey, I. F. and Nesbit, L. A. Investigating mechanisms of alkalinization for reducing primary breast tumor invasion. *Biomed Res Int*, 2013; 2013: 1–10.

29. Silvani, A., Asti, V., Berteotti, C., Bojic, T., Cianci, T., Ferrari, V., Franzini, C., Lenzi, P., and Zoccoli, G. Sleep-related brain activation does not increase the permeability of the blood–brain barrier to glucose. *J Cereb Blood Flow Metab*, 2005; 25(8): 990–997.
30. Torrealdea, F., Rega, M., Thomas, D. L., Brandner, S., Walker-Samuel, S., and Golay, X. GlucoCEST for the detection of human xenografts glioblastoma at early stage. *Proc Intl Soc Mag Reson Med*, 2013; 21: 0424.
31. van Zijl, P. C. M., Jones, C. K., Ren, J., Malloy, C. R., and Sherry, A. D. MRI detection of glycogen in vivo by using chemical exchange saturation transfer imaging (glycoCEST). *Proc Natl Acad Sci USA*, 2007; 104(11): 4359–4364.
32. Walker-Samuel, S., Ramasawmy, R., Torrealdea, F., Rega, M., Rajkumar, V., Johnson, S. P., Richardson, S., Gonçalves, M., Parkes, H. G., Arstad, E., Thomas, D. L., Pedley, R. B., Lythgoe, M. F., and Golay, X. In vivo imaging of glucose uptake and metabolism in tumors. *Nat Med*, 2013; 19(8): 1067–1072.
33. Warburg, O. On the origin of cancer cells. *Science*, 1956; 123(3191): 309–314.
34. Warburg, O. H. New methods of cell physiology: Applied to cancer, photosynthesis, and mechanism of x-ray action. Science, 1962; 137(3523): 30–31.
35. Yadav, N. N., Xu, J., Bar-Shir, A., Qin, Q., Chan, K. W., Grgac, K., Li, W., McMahon, M. T., and van Zijl, P. C. M. Natural D-glucose as a biodegradable MRI relaxation agent: Glucose as an MRI relaxation agent. *Magn Reson Med*, 2014; 72(3): 823–828.
36. Zhou, J. and van Zijl, P. C. M. Chemical exchange saturation transfer imaging and spectroscopy. *Prog Nucl Magn Reson Spectrosc*, 2006; 48(2–3): 109–136.

Chapter 18

Creatine Chemical Exchange Saturation Transfer Imaging

Catherine DeBrosse,[a] Feliks Kogan,[b] Mohammad Haris,[c] Kejia Cai,[d] Anup Singh,[e] Ravi P. R. Nanga,[a] Mark Elliott,[a] Hari Hariharan,[a] and Ravinder Reddy[a]

[a]*Center for Magnetic Resonance and Optical Imaging, University of Pennsylvania, Philadelphia, PA 19104, USA*
[b]*Stanford University, Stanford, CA 94305, USA*
[c]*Sidra Medical and Research Center, Doha PO Box 26999, Qatar*
[d]*University of Illinois at Chicago, Chicago, IL 60607, USA*
[e]*Indian Institute of Technology, Delhi 110016, India*
krr@mail.med.upenn.edu

18.1 Introduction

Creatine (Cr) is a metabolite that plays a vital role in bioenergetics. Cellular energy, during increases in instantaneous demand, is supplied by generation of adenosine triphosphate (ATP) through the creatine kinase (CK) reaction (Eq. 18.1) [1]. During the CK reaction, ATP is produced from the conversion of phosphocreatine (PCr) and adenosine diphosphate (ADP) to Cr [2]. When cells undergo processes that require an increase in energy, like muscle contraction, PCr is depleted to maintain the available supply of ATP. This mechanism allows cells access to an immediate energy supply, without requiring storage of large amounts of ATP [3]. Once the instantaneous energy demand is reduced, generation of ATP through Krebs cycle or glycolysis leads to a subsequent increase in PCr levels. In muscular diseases and metabolic disorders, CK kinetics are often affected [4]. Oxidative phosphorylation (OXPHOS), the process by which mitochondria produce ATP, is often impaired in such diseases, indicated by delayed recovery of PCr.

$$Cr + ATP \xrightleftharpoons{CK} ADP + PCr + H^+ \qquad (18.1)$$

Metabolism of exercising muscle and myocardium has been studied extensively through non-invasive magnetic resonance (MR) techniques. The most commonly used technique has been phosphorus magnetic resonance spectroscopy (^{31}P MRS) [5], which can measure signals from several metabolites of the CK reaction. Although ^{31}P MRS has been critical to understanding muscle energetics, it suffers from poor spatial resolution and low sensitivity.

As interest continues to rise in applying these techniques to study muscular diseases and secondary conditions that also affect muscle energy metabolism, a more robust method for studying the CK reaction in vivo would be invaluable. Such a technique would aid in the study of metabolic abnormalities in various muscular disorders.

In this chapter, the recent developments in chemical exchange saturation transfer (CEST) [6] for imaging of creatine will be discussed, along with applications in skeletal muscle and myocardium and in the brain in vivo.

18.2 Study of Energy Metabolism: ^{31}P MRS

18.2.1 ^{31}P Magnetic Resonance Spectroscopy

For many decades, ^{31}P MRS has been the standard technique for the study of phosphorus compounds [7]. It is an extremely useful tool due to certain properties of phosphorus: its natural abundance (100%), spin (1/2), and large range of chemical shifts [5, 8].

Inorganic compounds had been studied using ^{31}P MRS since the early 1950s [9, 10], and during the 1970s, it was applied to the study of biological phosphorus compounds [11]. Since then, it has been an invaluable technique for the study of muscle metabolism.

In muscle tissue, ^{31}P MRS can non-invasively measure the phosphate components of the CK reaction [12]. Concentrations of these metabolites (PCr, ATP, inorganic phosphates (Pi), phosphomonoesters (PME), and phosphodiesters (PDE)) can be obtained from ^{31}P spectra. Additionally, changes in intracellular pH can be estimated by changes in spectral distance of the inorganic phosphate (P$_i$) peak [13].

During exercise in healthy muscle tissue, the CK reaction maintains a constant concentration of ATP by the conversion of PCr to Cr. ^{31}P MRS spectra of exercising skeletal muscle show the depletion of PCr and any changes in intracellular pH [14]. Conditions of muscle fatigue [15, 16]. or muscle injury can be studied through ^{31}P MRS. Muscle-related disorders, such as muscular dystrophy, mitochondrial myopathies [17, 18], and cardiac disease [19] have also been studied using ^{31}P MRS.

18.2.2 ^{31}P MRS versus CEST Imaging

With critical developments already made in the area of muscle research through ^{31}P MRS, what could be contributed by an imaging modality such as CEST? Although ^{31}P MRS has greatly expanded the knowledge of muscle metabolism, it suffers from poor spatial resolution, which is characteristic of most spectroscopy techniques. Phosphorus spectroscopy suffers from low sensitivity compared to ^{1}H MRS, due to the low gyromagnetic ratio of ^{31}P [8]. On the other hand, ^{1}H MRS methods can measure total creatine (PCr + Cr), but struggle to differentiate PCr from Cr and thus have had limited application toward muscle energetics.

In an effort to address some of the shortcomings of ^{31}P MRS, the CEST technique has been applied to measure free Cr utilizing the exchangeable amine (–NH$_2$) protons of creatine with those of bulk water [20]. Imaging of free creatine is relevant for the muscle disorders discussed earlier, as well as in the study of brain metabolites.

18.3 Development of Creatine CEST

18.3.1 Definition of Exchangeable CK Amine Protons and Their Exchange Rates

During initial development of creatine CEST (CrCEST), limited information was available about the exchange rates of the CK metabolites at physiological conditions. The resonances from the CK components with exchangeable amine protons (Cr, PCr, and ATP) were identified through high-resolution ^{1}H NMR spectra (Fig. 18.1).

In its ionic form, creatine has two groups of amine protons (imide: $=NH_2^+$ and amine: $-NH_2$) that show a single resonance at 1.8 ppm away from water. Phosphocreatine has one amide group (2.5 ppm) and one amine group (1.8 ppm), while the amine group of ATP resonates at 2.0 ppm.

At physiological temperature (37°C) and pH (7.0), line broadening of creatine's exchangeable amine protons occurs to a much greater extent than those of PCr or ATP protons. Line broadening indicates that the amine protons are in faster exchange with bulk water. When the exchange rates were determined using the method described by McMahon et al. [21], it was discovered that Cr did indeed have a much faster exchange rate (950 ± 100 s^{-1}) than the other components ($<140 \pm 50$ s^{-1}) of the CK reaction.

Another key piece of information required for CrCEST imaging was the pK_a values for the CK metabolites. Using ^{1}H MRS, the pK_a values were measured for each metabolite at 37°C. Under physiological conditions, creatine has a lower pK_a than both PCr and ATP, suggesting a faster dissociation, which leads to faster exchange with bulk water spins. For the purposes of imaging at 3 T, the exchange rate of Cr is in the slow-intermediate exchange regime ($k \leq \Delta\omega$, chemical shift between water and Cr amine protons (in radians)). On the other hand, the exchange rates of amine protons on PCr and ATP are in a slow exchange regime at 3 T ($k \ll \Delta\omega$). Based on the exchange rate information, it was determined that CEST imaging could be performed on CK metabolites.

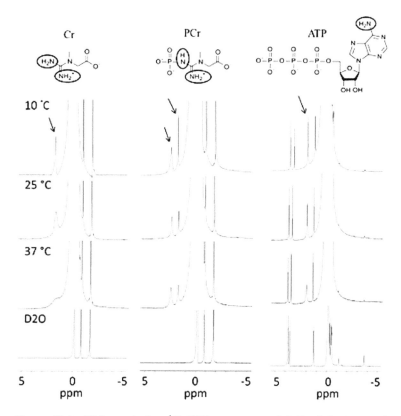

Figure 18.1 High-resolution ^{1}H NMR spectra at 9.4 T of Cr, PCr, and ATP, from Haris et al. (2012) [23]. Chemical structures are shown for each metabolite above their spectra. The water peak was referenced to 0 ppm. Spectra were acquired for each sample at 10°C, 25°C, and 37°C. At 37°C, line broadening was observed due to increased exchange with bulk water. In order to confirm that the resonances were from exchangeable protons, a hydrogen–deuterium exchange experiment was performed (bottom spectra).

18.3.2 CrCEST Phantom Imaging

CEST imaging of CK metabolites was first performed at 3 T on "phantoms": NMR tubes each containing components of the CK reaction (Cr, PCr, ATP, and ADP) at their physiological concentrations (Fig. 18.2). A preliminary "Z-spectrum" [22] was gathered on a 30 mM Cr phantom, showing the broad resonance centered at ~1.8

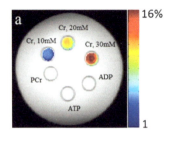 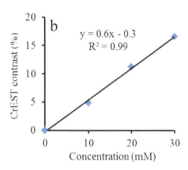

Figure 18.2 CrCEST maps of the CK metabolites (a) at physiological conditions, by Haris et al. [23]. There is a linear dependence of CrCEST on Cr concentration (b).

ppm, attributed to the (–NH$_2$) protons, also observed in the earlier ^{1}H NMR experiments.

The observed contrast in CEST experiments is strongly dependent on the combination of exchange rates, saturation pulse parameters, metabolite concentrations, and relaxation times. For CrCEST imaging of phantoms at 3 T, the optimal saturation pulse amplitude (B_{1rms}) was 3.64 µT. A specially optimized saturation pulse train was developed. It consisted of 20 Hanning windowed rectangular pulses of 49 ms duration, each with a 1 ms delay between them, for a total saturation duration of 1 s. The saturation pulse excitation bandwidth was 5 Hz for 1 s pulse, with a 1% bandwidth of 20 Hz (∼0.15 ppm at 3 T). With these parameters, no appreciable CEST is observed from the other components of the CK reaction, which is likely due to their slower exchange rates.

CEST maps for CrCEST are computed using Eq. 18.2, where $M_{sat}(-\Delta\omega)$ and $M_{sat}(+\Delta\omega)$ are the B_0-corrected MR signals acquired while saturating at -1.8 ppm and $+1.8$ ppm from water resonance. For normalization, $M_{sat}(-\Delta\omega)$ was used instead of M_0, to account for the loss of magnetization due to the direct saturation effect. The CEST contrast map is further corrected for B_1 inhomogeneity [23].

$$\text{CEST}_{asym}(\pm 1.8\,\text{ppm}) = \frac{M_{sat}(-1.8\,\text{ppm}) - M_{sat}(+1.8\,\text{ppm})}{M_{sat}(-1.8\,\text{ppm})}$$

(18.2)

As discussed previously, one of the expected benefits of using CEST instead of traditional spectroscopy is an increase in sensitivity. This was observed using single voxel point resolved spectroscopy (PRESS) on the same 30 mM Cr phantom. The sensitivity of CrCEST is approximately three orders of magnitude greater than ^{31}P MRS, due to the low receptivity of ^{31}P compared to ^{1}H. Increased sensitivity of this magnitude was predicted to be able to detect relatively small changes in Cr levels at high resolution (voxel size of 0.02 cc) [23].

18.3.3 In Vivo CrCEST Studies of Skeletal Muscle Exercise at Ultra-High Field

After establishing that CrCEST images could be obtained from in vitro phantom experiments, the next step was implementation in vivo. An important step for in vivo optimization was maximizing both the CrCEST$_{asym}$ and the signal-to-noise (SNR) ratio. It was discovered that increasing the B_{1rms} could increase the CEST$_{asym}$; however, this led to a decrease in SNR. To maximize both of these parameters, a saturation pulse train with a B_{1rms} of 123 Hz (2.9 µT) and 500 ms duration was chosen for CrCEST in vivo. Another consideration for in vivo imaging was the fat content of the muscle tissue. In order to reduce the fat signal, a chemical shift selective, fat saturation pulse was applied before image readout.

CrCEST of exercising muscle was performed on human calf muscles at 7 T using a 28-channel ^{1}H knee coil [24]. For these studies, it was important to account for B_0 and B_1 inhomogeneities. To correct for any inhomogeneities, water saturation shift reference images [25] and B_1 maps were collected along with CEST maps. Using higher order shimming, B_0 inhomogeneities could be kept within 0.3 ppm ($\Delta B_0 < 0.3$ ppm) and thus CrCEST images were collected in a chemical shift range of ± 1.5 to ± 2.1 ppm. As Cr amine protons have a chemical shift of 1.8 ppm downfield from water, this allows for adequate B_0 inhomogeneity correction in calculating CrCEST$_{asym}$.

Figure 18.3 shows the major muscles of the lower leg and a baseline CrCEST asymmetry map of a healthy human volunteer overlaid on an anatomical image. In resting muscle tissue, CrCEST maps showed a uniform CrCEST$_{asym}$.

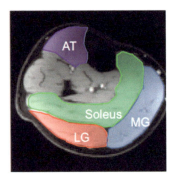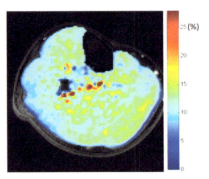

Figure 18.3 Anatomical image showing an axial slice of the human calf muscle (left; AT: anterior tibialis, MG: medial gastrocnemius, LG: lateral gastrocnemius). Baseline, resting CrCEST$_{asym}$ map is overlaid in the figure on the right [24].

After optimizing imaging parameters and establishing baseline CrCEST$_{asym}$ in resting muscle, exercise studies were performed. Human subjects were asked to perform plantar flexion, a type of exercise known to utilize the muscles in the posterior compartment of the lower leg. Exercise was performed in the magnet using an MR compatible, pneumatically controlled foot pedal.

For initial studies of CrCEST in vivo, only mild exercise was performed in order to maintain a consistent pH. During strenuous exercise, the pH of the muscle can drop, due to breakdown of glycogen and production of lactate [26]. A pH decrease would consequently decrease the exchange rate between Cr amine protons and water protons, leading to a lower CEST contrast [27]. Two minutes of mild exercise were sufficient enough to induce creatine changes in exercising muscle without changing the pH of the muscle tissue. CrCEST maps, as well as ^{31}P MRS, were obtained pre- and post-exercise for healthy volunteers (Fig. 18.4).

Following plantar flexion exercise, creatine recovery kinetics as measured from CrCEST were strongly correlated to PCr recovery kinetics measured from ^{31}P MRS. At 7 T, the change in CrCEST is dependent on the change in creatine concentration, with 0.84% CrCEST$_{asym}$ observed per mM Cr. An exciting finding was the ability of CrCEST to distinguish between specific muscle group utilization while exercising. Plantar flexion exercise primarily uses

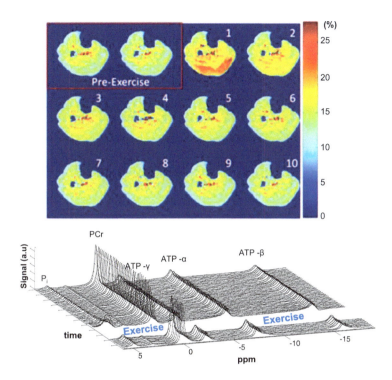

Figure 18.4 Representative CrCEST maps for pre- and post-exercise of the human calf muscle at 7 T. Increase in Cr immediately following exercise is observed, with a subsequent recovery to baseline within about 2 min. ^{31}P data show the corresponding decrease in PCr. Reprinted with permission from Ref. [24], Copyright 2013, John Wiley and Sons.

the soleus muscle and the medial (MG) and lateral (LG) heads of the gastrocnemius muscles. Following exercise, creatine changes were observed in the posterior compartment of the lower leg, primarily the gastrocnemius and soleus muscles. The CrCEST method could distinguish between varying levels of muscle utilization between subjects (Fig. 18.5). Other muscle groups did not show a significant increase in Cr post-exercise. Less than 1% increase in CrCEST$_{asym}$ was observed in the tibialis anterior (AT) muscle, which is predominantly involved in dorsiflexion and was thus used as a control in the study. This further demonstrates the precise spatial resolution of CEST (Figs. 18.4 and 18.5). This method of observing

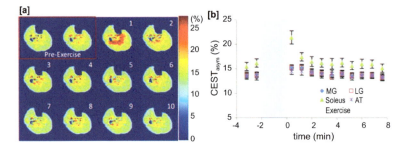

Figure 18.5 (a) CrCEST images of plantar flexion exercise in a subject who has utilized the soleus muscle. (b) Average CrCEST values from four different muscles in the calf (MG: medial gastrocnemius, LG: lateral gastrocnemius, AT: anterior tibialis) as a function of time, pre- and post-exercise. Reprinted with permission from Ref. [24], Copyright 2013, John Wiley and Sons.

the CK reaction in vivo has excellent spatial resolution, which is an important improvement over ^{31}P MRS.

Enhanced sensitivity was observed in in vivo CrCEST compared to spectroscopy. As noted previously, ^{31}P has low sensitivity due to the lower gyromagnetic ratio of ^{31}P (17.235 MHz/T) compared to that of ^{1}H (42.576 MHz/T), which leads to a receptivity of ^{31}P that is only 0.066 that of ^{1}H. The water of muscle tissue has a proton concentration of 82.5 M. Based on the 0.84% CEST observed per mM Cr, 1 mM Cr signal leads to ~700 mM change in water. Therefore, CrCEST has about three orders of magnitude higher sensitivity than ^{31}P MRS.

Phantom experiments showed that ATP, ADP, and PCr do not contribute to the CrCEST effect at 1.8 ppm. During the exercise study, there was no observed change in pH, T_2, or MTR, and the CEST effects from other CK reaction metabolites were negligible. Therefore, using the CrCEST technique in vivo provides an accurate measure of the changes in muscle Cr concentration following exercise [24]. In dynamic studies, this allows for the measurement of OXPHOS capacity, which is reflected in the recovery of Cr post-exercise.

18.3.4 Implementation of CrCEST at Clinical-Strength Field

Numerous disease states can benefit from study by CrCEST. In order for any imaging technique to be applicable to patient populations, it must be validated at clinical-strength fields (3 T).

There were several challenges in translating CrCEST from 7 T to 3 T. At lower fields, the chemical shift ($\Delta\omega$) between Cr amine protons and free water protons is decreased (~540 Hz at 7 T versus ~225 Hz at 3 T). This means that at 3 T, the selective saturation pulse is applied much closer to the water resonance than at 7 T. This results in increased direct saturation of bulk water at lower fields. Direct water saturation results in a lower optimal B_1, consequently decreasing the labile proton saturation efficiency, and decreasing the CrCEST$_{asym}$. Increased direct water saturation also leads to a loss of SNR in the CEST images.

Despite these drawbacks, CrCEST is possible at 3 T because of the optimal exchange rate of Cr amine protons, which is still in the slow-to-intermediate regime at 3 T. The dependence of the change in CrCEST on the change in creatine concentration following exercise at 3 T (0.45% CrCEST$_{asym}$/mM Cr) is approximately half of that observed at 7 T (0.84% CrCEST$_{asym}$/mM Cr). The T_2 relaxation times in muscle are longer at 3 T, ($T_2(3\ T) = 29.3$ ms, $T_2(7\ T) = 23.0$ ms). This somewhat lessens the SNR lost due to increased direct water saturation at 3 T.

There are also advantages to CrCEST imaging at 3 T, including the shorter T_1 relaxation times at 3 T than at 7 T. This allows for increased temporal resolution, since the time necessary to allow for complete longitudinal relaxation between saturation pulses (Shot TR) is directly correlated to the T_1. Temporal resolution is an important consideration when applying CrCEST to studies of creatine kinetics, which have a time course on the order of 30 s to a few minutes following exercise. When implementing the CrCEST sequence at 3 T, the TR was minimized to optimize the temporal resolution while maintaining adequate SNR.

Similar to the experiments performed at 7 T, changes in CrCEST in the calf muscles of healthy human volunteers were first mapped by Kogan et al. at 3 T following plantar flexion exercises (Fig. 18.6) [28]. Changes in PCr levels under the same exercise conditions were

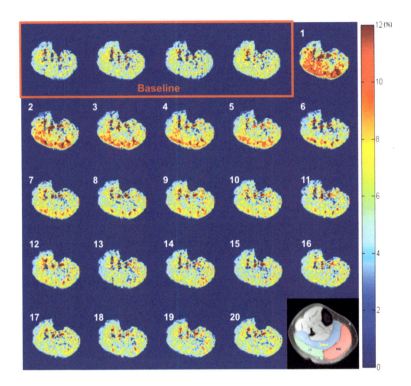

Figure 18.6 CrCEST at 3 T. Increase in Cr post-exercise, as well as recovery to baseline, is observed in the gastrocnemius muscle (images 1 through 20).

mapped using ^{31}P MRS, which showed complementary recovery kinetics to CrCEST under the same exercise conditions. This work demonstrated that CrCEST was feasible on clinical scanners for the study of muscle energetics [28]. CrCEST has many potential applications in the imaging of neuromuscular disorders, such as measuring impaired OXPHOS due to mitochondrial dysfunction, and in monitoring therapeutic attempts to treat creatine transporter defects [29].

18.3.5 Application of CrCEST in Imaging of Myocardial Metabolism

Although the studies of healthy muscle validated the CrCEST imaging method in vivo, the ultimate goal is to apply the technique

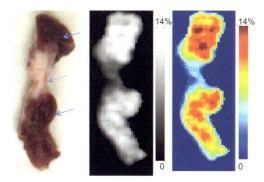

Figure 18.7 Ex vivo CrCEST of non-infarcted and infarcted myocardial tissue from swine. An infarcted region (shown with a dotted arrow) is clearly visible in the ex vivo tissue. CEST maps overlaid onto an anatomical image show lower creatine in the infarcted region.

to disease states. Myocardial metabolism can be disrupted due to an event such as cardiac ischemia or infarction, making this a suitable candidate for CrCEST.

CrCEST imaging and spectroscopy were first performed on non-infarcted, excised lamb heart tissue [30]. Ex vivo tissue was used in order to optimize the imaging parameters for myocardial tissue, and total Cr concentration was measured using ^{1}H MRS. The Cr concentration was validated using the perchloric acid (PCA) extraction method, in which a positive correlation was observed between the imaging method and the biochemical method. For this tissue, the mean sensitivity ($n = 26$) of CrCEST is 0.8% CrCEST contrast per 1 mM of Cr at 37°C.

For a comparison of infarcted tissue to non-infarcted tissue, CrCEST maps were acquired from ex vivo swine myocardial tissue. Infarcted tissue has lower CrCEST contrast (4.7 ± 1.2%) than non-infarcted tissue (10.4 ± 2.0%) (Fig. 18.7).

In vivo CrCEST for myocardial imaging was performed on swine and sheep with infarctions, using healthy animals as a control (Fig. 18.8) [30]. CrCEST maps from two healthy swine displayed uniform CrCEST contrast in myocardium. Infarcted swine and sheep had lower CrCEST contrast in the infarcted regions compared to the normal myocardial regions. CrCEST imaging gave consistent

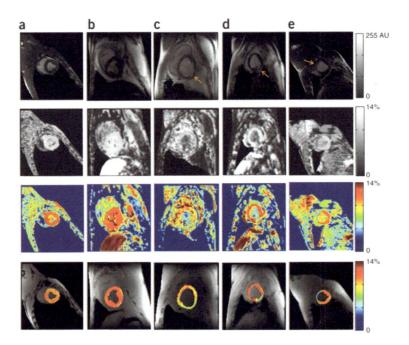

Figure 18.8 In vivo CrCEST data from Haris et al. [30] Columns (a) and (b) show normal swine hearts. Three infarcted animals are shown in two sheep (c, d) and one swine (e). Images from top to bottom show: anatomical CEST-weighted images of the left ventricle, grayscale CrCEST maps, color-coded CrCEST maps, and overlaid color CrCEST maps from all five animals.

regional (infarcted versus non-infarcted) contrast values across all animals studied.

In myocardial imaging, there may be changes in CrCEST in the ischemic and infarcted regions of myocardium due to changes in T_2, MTR, and water content. In these tissues, there is the potential for an increase in water content. This leads to a reduction in Cr concentration and hence CrCEST. However, there is also a decrease in MTR and an increase in water T_2, which would cause an increased CrCEST. Thus, the effects of changes in water content, MTR, and T_2 compete with each other and have no appreciable contribution to the CrCEST$_{asym}$.

CrCEST contrast may also be affected by pH. The CrCEST contrast in infarcted regions of myocardium may be underestimated, because infarcted tissue is expected to have a pH between 6.5 and 7.0. Consequently, the lower CrCEST contrast observed in the infarcted myocardium may be the result of an integrated effect of both lower exchange rate resulting from the decreased pH and lower Cr concentration compared to those in healthy tissue. Thus, although pH may influence the magnitude of CrCEST, it would only enhance the capability of CrCEST in discriminating between healthy and infarcted myocardium [30].

18.3.6 *CrCEST Application in Brain Imaging*

In addition to the study of muscle metabolism, CrCEST has the potential to be used for the study of creatine levels in the brain [27]. There is increasing evidence for creatine depletion in the brain of those with inborn errors of metabolism, which leads to

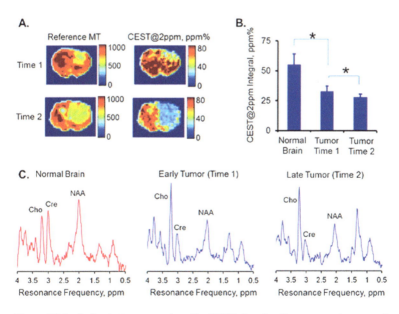

Figure 18.9 In brain tumor region, the CEST signal at 2 ppm was decreased compared with normal brain tissue and further reduced with tumor progression (a, b). MRS results show a similar trend for the creatine change (c). (Cho: choline, Cre: creatine, NAA: N-acetylaspartic acid). Reprinted with permission from Ref. [27], Copyright 2014, John Wiley and Sons.

clinical features of movement disorders, epilepsy, and autism [31]. ^{1}H MRS shows low creatine concentration in the brain of those with known creatine deficiency syndromes [32]. The improved spatial resolution of CrCEST compared to conventional spectroscopy techniques makes it an ideal tool for studying these disorders. Furthermore, recent work has shown CEST imaging of brain creatine in a rat brain tumor model [27] that characterized the multiple exchangeable components through a simplified Z-spectral fitting method. The CEST peak observed at +2 ppm was reduced in brain tumor tissue compared to normal tissue (Fig. 18.9) and continued to decrease with tumor progression. The 2 ppm CEST peak also correlated with ^{1}H MRS derived concentrations of creatine. This non-invasive way to look at Cr levels in the brain will aid in future developments for imaging of cancer and other brain diseases in which creatine metabolism is disrupted.

18.4 Summary

Imaging of creatine using CEST has widespread application. It has already been used to study basic exercise metabolism in humans, as well as cardiac infarction in animal models. The high sensitivity and excellent spatial and temporal resolution make it an excellent tool for assessing creatine metabolism in vivo. It is easily translatable to other disease states in which muscle metabolism is affected, such as mitochondrial dysfunction, myocardial ischemia, and various neuromuscular diseases and cancer.

Acknowledgments

Projects described in this chapter were supported by National Institute of Biomedical Imaging and Bioengineering (Grant Number P41-EB015893) and National Institute of Neurological Disorders and Stroke (Grant Number R01NS087516) of the National Institutes of Health.

References

1. Sweeney HL. The importance of the creatine-kinase reaction: The concept of metabolic capacitance. *Medicine and Science in Sports and Exercise*, 1994; 26(1): 30–36.
2. Cain DF and Davies RE. Breakdown of adenosine triphosphate during a single contraction of working muscle. *Biochemical and Biophysical Research Communications*, 1962; 8(5): 361–366.
3. Wallimann T, Wyss M, Brdiczka D, Nicolay K, and Eppenberger HM. Intracellular compartmentation, structure and function of creatine-kinase isoenzymes in tissues with high and fluctuating energy demands: The phosphocreatine circuit for cellular-energy homeostasis. *Biochemical Journal*, 1992; 281: 21–40.
4. Bottomley PA, Lee YH, and Weiss RG. Total creatine in muscle: Imaging and quantification with proton MR spectroscopy. *Radiology*, 1997; 204(2): 403–410.
5. Gorenstein DG. *Phosphorous-31 NMR: Principles and Applications*, Academic Press; 1984.
6. Sherry AD and Woods M. Chemical exchange saturation transfer contrast agents for magnetic resonance imaging. *Annual Review of Biomedical Engineering*, 2008; 10: 391–411.
7. Lee JH, Komoroski RA, Chu WJ, and Dudley JA. Methods and applications of phosphorus NMR spectroscopy in vivo. In: Webb GA, ed. *Annual Reports on NMR Spectroscopy*, Vol 75. San Diego: Elsevier Academic Press Inc; 2012: 115–160.
8. Hoult DI, Busby SJW, Gadian DG, Radda GK, Richards RE, and Seeley PJ. Observation of tissue metabolites using P-31 nuclear magnetic resonance. *Nature*, 1974; 252(5481): 285–287.
9. Dickinson WC. The time average magnetic field at the nucleus in nuclear magnetic resonance experiments. *Physical Review*, 1951; 81(5): 717–731.
10. Cohn M and Hughes TR. Nuclear magnetic resonance spectra of adenosine di- and triphosphate. II. Effect of complexing with divalent metal ions. *Journal of Biological Chemistry*, 1962; 237(1): 176.
11. Burt CT, Glonek T, and Barany M. Analysis of living tissue by phosphorus-31 magnetic resonance. *Science*, 1977; 195(4274): 145–149.
12. Kemp GJ, Meyerspeer M, and Moser E. Absolute quantification of phosphorus metabolite concentrations in human muscle in vivo by ^{31}P MRS: A quantitative review. *NMR in Biomedicine*, 2007; 20(6): 555–565.

13. Moon RB and Richards JH. Determination of intracellular pH by P-31 magnetic-resonance. *Journal of Biological Chemistry*, 1973; 248(20): 7276–7278.
14. Burt CT, Glonek T, and Barany M. Analysis of phosphate metabolites, intracellular pH, and state of adenosine-triphosphate in intact muscle by phosphorous nuclear magnetic-resonance. *Journal of Biological Chemistry*, 1976; 251(9): 2584–2591.
15. Bendahan D, Giannesini B, and Cozzone PJ. Functional investigations of exercising muscle: A noninvasive magnetic resonance spectroscopy-magnetic resonance imaging approach. *Cellular and Molecular Life Sciences*, 2004; 61(9): 1001–1015.
16. Dawson MJ, Gadian DG, and Wilkie DR. Muscular fatigue investigated by phosphorous nuclear magnetic-resonance. *Nature*, 1978; 274(5674): 861–866.
17. Argov Z and Bank WJ. Phosphorous magnetic-resonance spectroscopy (P-31 MRS) in neuromuscular disorders. *Annals of Neurology*, 1991; 30(1): 90–97.
18. Chance B, Eleff S, Leigh JS, Sokolow D, and Sapega A. Mitochondrial regulation of phosphocreatine inorganic-phosphate ratios in exercising human-muscle: A gated P-31 NMR-study. *Proceedings of the National Academy of Sciences of the United States of America-Biological Sciences*, 1981; 78(11): 6714–6718.
19. Yabe T, Mitsunami K, Inubushi T, and Kinoshita M. Quantiative measurements of cardiac phosphorous metabolites in coronary-artery disease by P-31 magnetic-resonance spectroscopy. *Circulation*, 1995; 92(1): 15–23.
20. van Zijl PC and Yadav NN. Chemical exchange saturation transfer (CEST): What is in a name and what isn't? *Magnetic Resonance in Medicine*, 2011; 65(4): 927–948.
21. McMahon MT, Gilad AA, Zhou JY, Sun PZ, Bulte JWM, and van Zijl PC. Quantifying exchange rates in chemical exchange saturation transfer agents using the saturation time and saturation power dependencies of the magnetization transfer effect on the magnetic resonance imaging signal (QUEST and QUESP): pH calibration for poly-L-lysine and a starburst dendrimer. *Magnetic Resonance in Medicine*, 2006; 55(4): 836–847.
22. Ward KM, Aletras AH, and Balaban RS. A new class of contrast agents for MRI based on proton chemical exchange dependent saturation transfer (CEST). *Journal of Magnetic Resonance*, 2000; 143(1): 79–87.

23. Haris M, Nanga RPR, Singh A, et al. Exchange rates of creatine kinase metabolites: Feasibility of imaging creatine by chemical exchange saturation transfer MRI. *NMR in Biomedicine*, 2012; 25(11): 1305–1309.
24. Kogan F, Haris M, Singh A, et al. Method for high-resolution imaging of creatine in vivo using chemical exchange saturation transfer. *Magnetic Resonance in Medicine*, 2014; 71(1): 164–172.
25. Kim M, Gillen J, Landman BA, Zhou J, and van Zijl PC. Water saturation shift referencing (WASSR) for chemical exchange saturation transfer (CEST) experiments. *Magnetic Resonance in Medicine*, 2009; 61(6): 1441–1450.
26. Iotti S, Lodi R, Frassineti C, Zaniol P, and Barbiroli B. In vivo assessment of mitochondrial functionality in human gastrocnemius-muscle by P-31 MRS: The role of pH in the evaluation of phosphocreatine and inorganic-phosphate recoveries from exercise. *NMR in Biomedicine*, 1993; 6(4): 248–253.
27. Cai K, Singh A, Poptani H, et al. CEST signal at 2 ppm (CEST@2ppm) from Z-spectral fitting correlates with creatine distribution in brain tumor. *NMR in Biomedicine*, 2015; 28(1): 1–8.
28. Kogan F, Haris M, Debrosse C, et al. In vivo chemical exchange saturation transfer imaging of creatine (CrCEST) in skeletal muscle at 3T. *Journal of Magnetic Resonance Imaging*, 2014; 40(3): 596–602.
29. Tarnopolsky MA and Parise G. Direct measurement of high-energy phosphate compounds in patients with neuromuscular disease. *Muscle and Nerve*, 1999; 22(9): 1228–1233.
30. Haris M, Singh A, Cai K, et al. A technique for in vivo mapping of myocardial creatine kinase metabolism. *Nature Medicine,* 2014; 20(2): 209–214.
31. Nasrallah F, Feki M, and Kaabachi N. Creatine and creatine deficiency syndromes: Biochemical and clinical aspects. *Pediatric Neurology*, 2010; 42(3): 163–171.
32. Nouioua S, Cheillan D, Zaouidi S, et al. Creatine deficiency syndrome: A treatable myopathy due to arginine-glycine aminotransferase (AGAT) deficiency. *Neuromuscular Disorders*, 2013; 23: 670–674.

Chapter 19

Iodinated Contrast Media as pH-Responsive CEST Agents

Dario Longo and Silvio Aime

Molecular and Preclinical Imaging Centers, Department of Molecular Biotechnology and Health Sciences, University of Torino, Via Nizza 52, 10126, Torino, Italy
Institute of Biostructure and Bioimaging—CNR (National Research Council of Italy), Torino, Italy
dario.longo@unito.it, silvio.aime@unito.it

Iodinated contrast agents used for radiographic procedures are unique pharmaceuticals. They are available as highly concentrated solutions that are administered at high doses; this is possible because of their high safety profile and special physicochemical properties (high hydrophilicity and inertness). The currently used iodinated contrast agents can be divided into two classes based on their chemical structure. Monomeric agents are composed of a single triiodinated benzene ring, whereas dimeric agents consist of two covalently linked triiodinated benzene rings [1]. The passage from ionic to neutral systems has markedly improved their safety profile, making adverse reaction to their administration rare events [2]. Besides their capability to induce contrast in computed tomography (CT) images, owing to the presence of high X-ray adsorbance iodine atoms, the concomitant presence of amide groups has driven an interest in their exploitation as magnetic resonance imaging (MRI)-

---

Chemical Exchange Saturation Transfer Imaging: Advances and Applications
Edited by Michael T. McMahon, Assaf A. Gilad, Jeff W. M. Bulte, and Peter C. M. van Zijl
Copyright © 2017 Pan Stanford Publishing Pte. Ltd.
ISBN 978-981-4745-70-3 (Hardcover), 978-1-315-36442-1 (eBook)
www.panstanford.com

chemical exchange saturation transfer (CEST) agents. In seeking routes to accelerate the entry of CEST agents into clinical practice, it has been rather straightforward to consider chemicals already approved for human use and potentially able to generate CEST contrast. On this basis, a systematic study of currently used X-ray contrast agents containing mobile protons (e.g., Iopamidol–Isovue® Bracco Imaging SpA, Italy; Iopromide–Ultravist® Bayer-Schering AG, Germany; and Iobitridol–Xenetix® Guerbert, France) has been undertaken. The in vivo results obtained so far appear very encouraging for clinical translation of these CEST agents.

19.1 Iopamidol as a diaCEST Agent in Preclinical Studies

The currently used iodinated contrast agents were discovered about 50 years ago, and since then they are important tools in clinical routine for radiographic procedures. Recently, it has been suggested that the presence of amide and alcoholic groups in these molecules may be exploited for their direct visualization in MRI as CEST agents. Iopamidol (Isovue®–Bracco Imaging, Italy) is characterized by the presence of several types of functionalities (alcoholic and amide groups) that may be exploited for the generation of CEST contrast (Fig. 19.1). In particular, the amide groups that are more shifted from the bulk water signal (at 4.2 and 5.5 ppm, respectively) have been considered for generating a CEST response [3]. Moreover, as the exchange rate of the two amide proton pools is markedly pH-dependent (Fig. 19.2A), a ratiometric method for pH assessment has been set up based on comparison of the saturation transfer effects induced by the selective irradiation of the two resonances [4]. Thus, the setup of the ratiometric approach allows to remove the requirement of the knowledge of the local concentration of the contrast agent for molecules such as Iopamidol and Iopromide, which have two types of exchanging protons endowed with different pH dependences (Fig. 19.2B). By exploiting this ratiometric approach, it is possible to measure the pH value of the solution surrounding the pH-responsive contrast agent, thus obtaining accurate pH measurements inside the physiological range (Fig. 19.2C).

A thorough study by Sun et al., by solving the multi-site exchange properties of Iopamidol, confirmed that the Iopamidol ratiometric

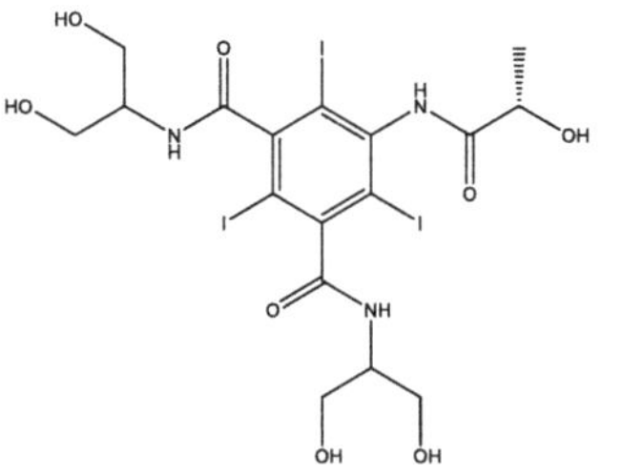

Figure 19.1 Chemical structure of iodinated contrast media exploited as MRI-CEST agents.

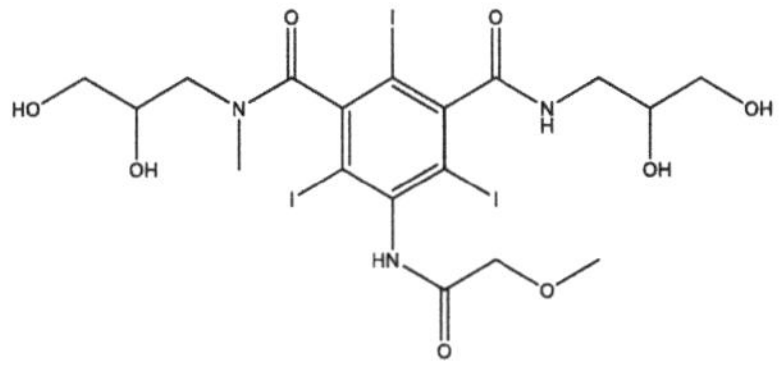

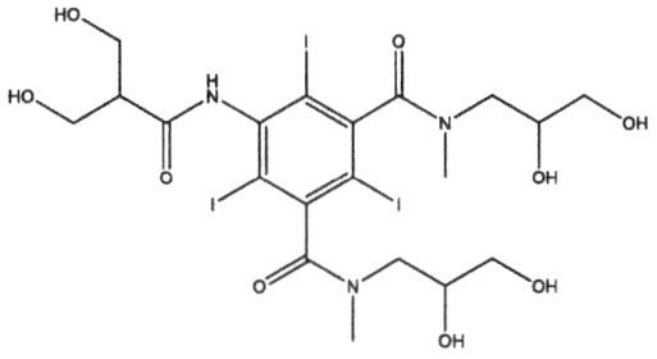

approach enables the measurement of pH from 6 to 7.5 with an accuracy of 0.1 pH unit at 4.7 T even if T_1 and T_2 values vary substantially with pH and concentration [5]. As shown in Fig. 19.3B, the CEST effect increases with pH, peaking at pH 7 for the resonance at 5.5 ppm and at pH 7.5 for the resonance at 4.3 ppm, respectively. The different pH dependencies of the CEST contrast at 4.3 and 5.5 ppm enable the ratiometric quantification of pH. Figure 19.3C shows that the ratiometric curve (rCESTR) increases from 0.6 ± 0.1 to 2.3 ± 0.1 for pH at 6 and 7.5, respectively. A very good correlation between potentiometrically measured pH values and pH values determined by the Iopamidol-CEST method was found, independent of the concentration of Iopamidol in solution (Fig. 19.3D). Upon comparing solutions at different pH values and Iopamidol concentrations, it has been found that the standard deviation was higher at lower pH values and Iopamidol concentration, as a consequence of the lower signal-to-noise ratio (SNR) of the CEST MRI measurement. The observed behavior accounts for the smaller exchange rate at lower pH; hence the CEST

450 | *Iodinated Contrast Media as pH-Responsive CEST Agents*

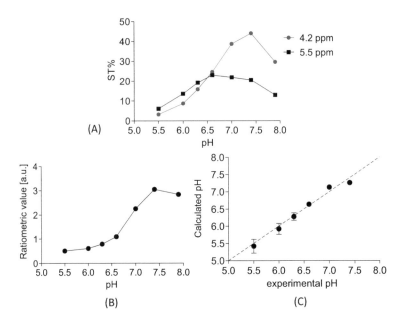

Figure 19.2 Dependence of the saturation transfer (ST%) effect for the two amide proton sets (square, 4.2 ppm; circle, 5.5 ppm) on the solution pH for a 30 mM Iopamidol solution dissolved in PBS upon saturation pulse of 3 µT × 5 s, 7 T, 310 K (A), corresponding ratiometric curve obtained by ratioing the ST% effects at 4.2 and 5.5 ppm (B) and calculated pH values versus experimental pH values showing the good accuracy in pH determination by using Iopamidol MRI-CEST pH mapping (C).

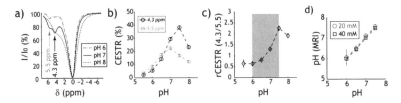

Figure 19.3 (a) Z-spectra for representative pH of 6, 7, and 8 ($B_1 = 2.5$ µT, TS = 5 s) at room temperature. (b) CEST ratio (CESTR) calculated from the asymmetry analysis as a function of pH. CESTR at 4.3 ppm peaks at pH 7.5, while CESTR of 5.5 ppm peaks at pH 7. (c) Ratiometric CEST analysis is sensitive to pH ranging from 6 to 7.5. (d) pH determined from Iopamidol pH MRI versus titrated pH for 20 mM (circles) and 40 mM (squares) Iopamidol phantoms. Reprinted from Ref. [5], by permission of IOP Publishing. All rights reserved.

effect and SNR decrease with pH and Iopamidol concentration. It follows that signal averaging, use of improved hardware and novel pulse sequence designs [6] will lead to enhanced pH accuracy for the ratiometric Iopamidol MRI approach. In addition, understanding pH-dependent exchange properties could help guide experimental optimization and quantification of Iopamidol pH imaging and should aid clinical translation of Iopamidol pH imaging.

The first in vivo pH mapping experiments were carried out on the kidney regions upon i.v. administration of Iopamidol (0.75 g iodine per kg body weight) into healthy mice, exploiting the renal excretion of the molecule and its accumulation within the kidneys. The ratiometric procedure applied to the ST% maps acquired after irradiation at 4.2 and 5.5 ppm (corresponding to the absorption frequencies of the two non-chemically equivalent amide proton pools) yielded in vivo parametric pH maps of the kidney regions (Fig. 19.3) [4]. At 5 min after injection of Iopamidol, the pH values calculated in control mice show a mean cortical pH of 7.0 ± 0.11, a mean medullary pH of 6.85 ± 0.15, and a mean calyx pH of 6.6 ± 0.20. Looking at the pH maps calculated at different time points after i.v. injection of Iopamidol, a progressive pH decrease inside the kidneys, likely due to the activation of H^+ pumps that release protons in the extracellular space as a consequence of the filtration process, has been observed. The obtained renal pH values are in agreement with those obtained with another pH-responsive MRI procedure based on the sequential injection of two Gd-based agents [7].

Several pathological conditions are associated with pH alterations, such as inflammation processes, ischemia, cancer, and kidney diseases. Therefore, pH may be considered a biomarker of kidney injuries and might find considerable clinical relevance. In particular, pathologically altered renal physiology resulting from acute kidney injury (AKI) or tubular acidosis is associated with a perturbation of renal pH [8]. The currently used clinical biomarkers of kidney damage, such as blood urea nitrogen and serum creatinine, apparently report on kidney damage only after a significant loss (50%) of renal function has occurred [9]. We demonstrated the use of Iopamidol to monitor the disease evolution "in vivo" by imaging pH variations in a glycerol-induced model of AKI, as well as in an

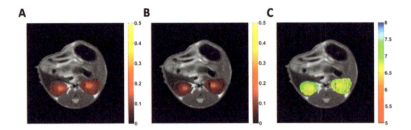

Figure 19.4 ST maps obtained after i.v. injection of Iopamidol 0.75 mg iodine per gram body weight bolus by irradiating the amide proton pools at 4.2 ppm (A) and at 5.5 ppm (B), respectively. Corresponding pH map obtained using the ratiometric curve of Fig. 19.3C. The parametric maps are overimposed to the anatomical T_{2w} image.

ischemia-reperfusion injury model [10]. In contrast to control mice, which showed pH values around 6.7 after Iopamidol injection, we observed in the glycerol-induced AKI model a constant increase in the pH values peaking after 3 days (mean pH values 7.1 and 7.3 after 1 day and 3 days, respectively), and then from week 1 to 3, a steady recovery to control values (pH = 6.7 after 3 weeks) was observed (see Fig. 19.5). This evolution closely followed the time evolution blood urea nitrogen (BUN) levels as well as the damage scoring obtained by histological methods. These results highlight the role that responsive contrast agents may acquire as an in vivo functional imaging tool to assess the evolution of renal damage.

An intrinsic advantage of the proposed methodology relies on the visualization and evaluation of different renal compartments in a longitudinal manner. The commonly used clinical biochemical approaches can assess renal impairment but not distinguish if single kidney functionality is involved or if a particular renal region is compromised. Among the different physiological parameters that are possible to assess in vivo with MRI, such as renal perfusion and glomerular filtration rate, pH is an important addition, as it is directly related to the primary kidney function of extracting protons from blood. We applied the Iopamidol pH mapping procedure for monitoring the alteration and recovery of the renal function following a unilateral ischemia/reperfusion (I/R) injury in mouse

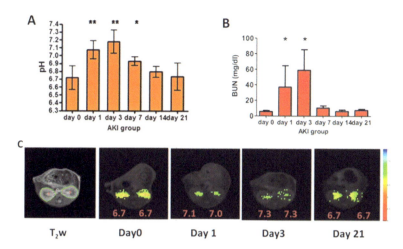

Figure 19.5 Mean renal pH evolution (A) and blood urea nitrogen (BUN) levels (B) after administration of 8 ml/kg body weight of 50% glycerol solution (AKI group). Results are expressed as mean ±SD. (*$p < 0.05$, **$p < 0.01$ in comparison to day 0 values). Representative pH maps obtained 15 min after Iopamidol injection at indicated time points in AKI group after glycerol administration (C). Reprinted with permission from Ref. [10], Copyright 2014, John Wiley and Sons.

kidney in vivo [11]. A flank incision was made in Balb/C mice to expose the left renal pedicle to induce 30 min ($n = 6$) of ischemia, which was followed by reperfusion, and CEST pH images were acquired at 7 T before, and 24 h, 48 h, and 1 week after the I/R injury upon i.v. administration of a clinical dose of Iopamidol (0.75 g iodine per kg body weight). We observed a significant increase in pH values at 1 day and 2 days for the clamped kidney after the I/R injury, in comparison to pH values for the control kidney (Fig. 19.6).

The accuracy of pH determinations provided by Iopamidol pH mapping has also been tested in a mouse model of lung fibrosis. In this study, a dedicated respiration-gated CEST protocol has been designed for measuring extracellular pH in lung lesions [12]. The average extracellular pH of lung lesions was found to be inversely related to lesion size, likely due to a reduced lactic acid production as the lesions decreased in size.

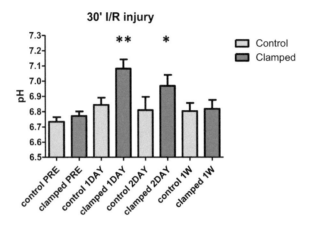

Figure 19.6 Mean renal pH evolution after 30 min of unilateral ischemia/reperfusion injury. Results are expressed as mean ±SD. (*$p < 0.05$, **$p < 0.01$ in comparison to control kidney values).

19.2 Iopamidol as diaCEST Agent on a Clinical MRI Scanner (3 T)

Most CEST studies employing iodinated molecules have been performed at high magnetic field strengths (7 T), due to the advantages that high fields offer to the CEST modality, including the larger dynamic range of chemical shifts in absolute units of Hz, which can improve selective labeling of the mobile proton pools, as well as longer tissue T_1 values that allow for higher SNR during the readout step. In light of a possible translation to human studies, the Iopamidol-CEST pH mapping has been demonstrated to work also at lower field strengths on clinical scanner (3 T), by exploiting a breathing motion-compensated sequence with a long RF saturation that is divided into short saturation elements of 100 ms (pulsed). The feasibility of motion-compensated CEST pH mapping in the kidneys of a rat was demonstrated using a clinical 3 T scanner by measuring the pH time course in the pelvic region [13].

The application of this method on many clinical scanners requires the adoption of pulse train presaturation schemes, in contrast to continuous wave (CW) saturation pulses, due to technical and hardware constraints and to specific absorption rate (SAR) limitations. Despite these limitations, feasibility studies of clinical translation of Iopamidol-based pH mapping in kidney and urinary bladder of human volunteers have been recently demonstrated on a clinical 3 T scanner [13, 14]. Muller-Lütz et al. performed

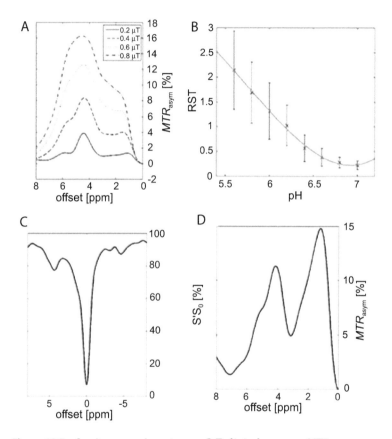

Figure 19.7 In vitro experiments on a 3 T clinical scanner. MTR_{asym} curves of the Iopamidol solution measured with CW amplitude equivalent pulsed saturation schemes at different B_1 strengths (A). Ratiometric curve (RST) obtained from ratioing MTR_{asym} contrast at 4.2 and 5.5 ppm as a function of pH (B). In vivo experiments on a 3 T clinical scanner were performed following Iopamidol injection in a human volunteer upon CT examination. Z-spectrum (C) and the corresponding MTR_{asym} curve (D) are shown for the urinary bladder region. The pH value determined was 6.65 ± 0.34. Reprinted from Ref. [14], Copyright 2014, with permission of Springer.

CEST presaturation using a pulse train scheme that still achieved high-saturation transfer for both the two amide functionalities and a ratiometric saturation transfer value showing a similar pH dependence (for the pH range 5.6–7.0) to that obtained at the field strength of 7 T (Fig. 19.7). This work addressed the measurements of pH in the human bladder (6.65 ± 0.34) and was in good accord with the pH determined in the urine by a pH meter (6.72 ± 0.01). These preliminary results, even though limited to only two volunteers, support the view that Iopamidol pH imaging can become feasible on clinical 3 T MRI scanners.

19.3 Iopromide as a diaCEST Agent in Preclinical Studies

The Iopromide molecule, having a similar chemical structure with two non-equivalent amide proton pools, has similar capabilities to Iopamidol as a pH-responsive contrast agent for the MRI-CEST procedure [15]. The Z-spectrum of this agent shows CEST effects from each amide (with MR frequencies at 4.2 and 5.6 ppm) that can be selectively detected (Fig. 19.8A). Calibration of the ratiometric measurements with pH within the range from 6.3 to 7.2 units yields an excellent linear correlation with a precision of 0.07 pH units (Fig. 19.8B).

Since most solid tumors during growth show a glycolytic phenotype, where high rates of glucose uptake and glycolysis generate an increased lactate production that contributes to an acidic extracellular pH, several MRI-based pH-responsive probes have been proposed for measuring tumor acidosis [16]. The extracellular pH in solid tumors typically ranges between ∼6.5 and ∼7.2, which requires pH measurements with excellent accuracy and precision together with a high spatial resolution to assess tumor acidosis in this narrow pH range [17]. In particular, for Iopromide, a thorough investigation has been performed by the Pagel group for optimizing the dose and route of administration and by measuring the concentration of this agent within the tumor by micro CT modality. This advantage of a direct quantification of the CEST agent by means of CT resulted in the optimization of an i.v. infusion protocol with a dose close to 4 g iodine per kg body weight, which

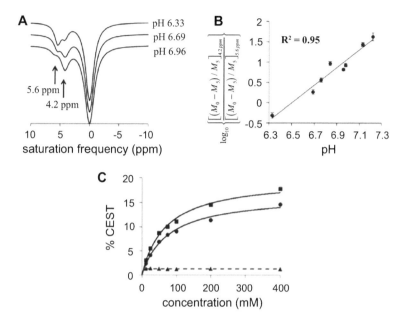

Figure 19.8 The CEST effects of Iopromide are sensitive to pH. The CEST spectra of fitted Lorentzian line shapes are vertically offset to aid the viewing (A). A log10 ratio of the two CEST effects is linearly correlated with pH from 6.3 to 7.2 (B). CEST spectra were acquired with 200 mM Iopromide at 37°C with saturation applied at 2 μT for 5 s. The CEST effects at 5.6 ppm (squares) and 4.2 ppm (circles) measured at pH 6.44 were dependent on concentration, but the ratio (triangles) was independent of concentration (C). Reprinted with permission from Ref. [15], Copyright 2014, John Wiley and Sons.

could deliver a sufficient amount of Iopromide for CEST detection. They demonstrated reliable tumor extracellular pH measurements in a subcutaneous MDA-MB-231 mammary carcinoma tumor model with a range of extracellular acidities from 6.5 to 7.2, therefore showing great heterogeneity that reflects large differences in the glycolysis in these tumors. In addition, it was shown that bicarbonate treatment yielded a net alkalinization of the more acidic tumors, but it did not change the extracellular pH for tumors that were classified as pH-neutral (Fig. 19.9).

In another study, three human B-cell lymphoma tumor xenograft models were investigated and extracellular tumor pH values were

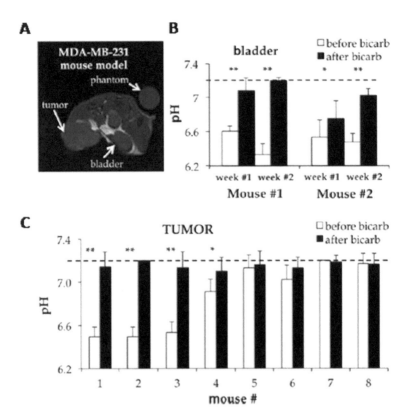

Figure 19.9 The effect of bicarbonate treatment on tissue pHe. The MR image shows the location of the tumor and bladder (A). The extracellular pH in the bladder increased an average of 0.53 pH units 24 h after providing 200 mM bicarbonate in drinking water (B). The initial tumor pHe ranged from 6.5 to 7.2. The tumor pHe increased to ≥ 7.1 pH units after bicarbonate treatment (C). Error bars represent the standard deviation of 4–12 measurements within 30 min. **$P < 0.01$, *$P < 0.02$. Reprinted with permission from Ref. [15], Copyright 2014, John Wiley and Sons.

measured, with tumor pH values between 6.7 and 6.9 [18]. Within these tumor models, a correlation between tumor acidic fractions and carbonic anhydrase IX (CAIX) expression was demonstrated. Interestingly, the same group demonstrated that imaging tumor extracellular pH may be a useful tool for monitoring early response to inhibitors of tumor metabolism. In fact, upon treatment of

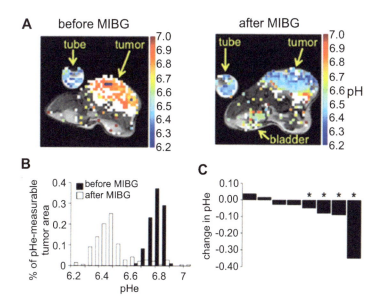

Figure 19.10 AcidoCEST MRI evaluations of mitochondrial poisoning of the Raji tumor model. (A) pHe map of a mouse bearing a Raji xenograft tumor before and after treatment with MIBG. Colored pixels have acidic pHe values <7.0 that correspond to the color bar. White pixels represent tumor regions with only a single CEST effect at 4.2 ppm, which were considered to have neutral pHe values >7.0. (B) Cumulative pixel distribution before and after MIBG treatment from the mouse shown in (A). (C) Waterfall plot of change in average tumor pHe in Raji tumors ($n = 8$) following treatment with MIBG. Mice with statistically significant changes are marked with an asterisk ($p < 0.01$). Reprinted with permission from Ref. [18], Copyright 2014, Taylor and Francis Ltd.

the Raji tumor model with a mithocondrial poison (MIBG) that induces a shift from intracellular oxidative metabolism to glycolysis metabolism, they observed a marked decrease in average tumor extracellular pH in six of eight mice, 4 h later (Fig. 19.10).

19.4 Iobitridol as a diaCEST Agent in Preclinical Studies

As shown above for Iopamidol and Iopromide, MRI-CEST responsive contrast agents are probes able to report on environmental parameters (e.g., pH) in a way that their responsiveness is made concentration independent for in vivo measurements. This task has been fulfilled by using molecules containing multiple sets of magnetically non-equivalent protons through the setup of a ratiometric approach [19]. Unfortunately, only a limited number of CEST systems fulfill this condition; therefore, the capability to act as in vivo responsive agents is restricted to a small number of molecules [4, 15, 20–23]. Moreover, a basic requirement for this method relies on the difference in chemical shift between the two resonances, which has to be large enough for their selective labeling. Recently, it has been shown that a ratiometric approach can be applied to molecules possessing even a single set of mobile protons by exploiting the dependence of the proton exchange rate upon the power of the applied RF saturation field strength (B_1).

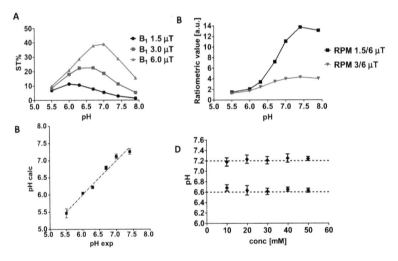

Figure 19.11 Iobitridol (30 mM solution) MRI-CEST contrast (ST%) depends on pH at three representative RF saturation powers 1.5, 3, and 6 µT and with an irradiation time of 5 s (A). Ratiometric curves provide pH-sensitive measurements by ratioing ST% between 1.5/6 µT and 3/6 µT (B). Calculated pH values versus titrated pH values showing high pH accuracy from the ratiometric curve obtained by ratioing 1.5/6 µT (C). Mean pH values calculated for several concentration upon ratioing 1.5/6 µT (D) showing that pH mapping is independent of Iobitridol concentration. (All experiments have been performed at 7 T, 310 K). Reprinted with permission from Ref. [24], Copyright 2014, American Chemical Society.

Thus, instead of calculating the ratio of the ST effects from two resonances in the same molecule at the same RF power, the effect of changing the RF power on the observed ST is exploited. We tested this novel ratiometric power approach on another FDA-approved iodinated contrast agent, Iobitridol (Fig. 19.11), possessing only a single amide proton pool, to test its pH responsiveness both in vitro and in vivo [24]. By saturating the single amide proton pool with two different saturation powers (1.5/6 µT), it is possible to set up a new ratiometric method (Fig. 19.11A) for a concentration-independent pH assessment. This novel ratiometric approach showed high sensitivity and accuracy in pH determinations (Fig. 19.11B,C) as well as concentration independence (Fig. 19.11D) comparable with the Iopamidol and Iopromide measurements described earlier.

The method has been evaluated in vivo upon i.v. injection of Iobitridol (dose of 4 g iodine per kg body weight) in a murine model of breast adenocarcinoma and by acquiring successive CEST images at two different B_1 powers (1.5 and 6 µT). The setup of the new ratiometric method allowed measurement of an extracellular tumor pH of 6.5 ± 0.4 in the investigated murine xenografted tumor model (Fig. 19.12). Therefore, the former prerequisite of the standard ratiometric approach, i.e., the presence of multiple magnetically non-equivalent resonances on the same molecules, is overcome by this new proposed ratiometric approach, which yields accurate pH

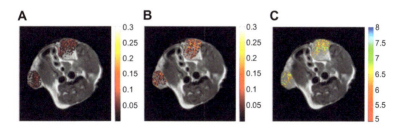

Figure 19.12 CEST-MR images of xenografted tumor bearing mouse following Iobitridol i.v. injection (dose 4 g iodine per kg body weight) overimposed onto T_{2w} anatomical image. CEST contrast difference map between pre- and post-injection following RF irradiation at 1.5 µT (A) and at 6 µT (B). Corresponding pH map (C) obtained upon ratioing the difference ST maps of A and B. Reprinted with permission from Ref. [24], Copyright 2014, American Chemical Society.

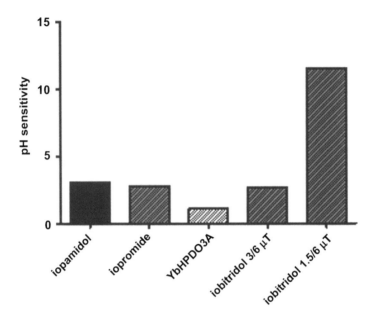

Figure 19.13 Comparison of pH MRI sensitivity of pH-responsive CEST MRI contrast agents. pH sensitivity (ΔR_{pH}) measured as the ratiometric difference between pH levels of 7.4 and 6. (All results were measured on 30 mM solutions in phosphate buffer, $B_0 = 7$ T, 310 K). Reprinted with permission from Ref. [24], Copyright 2014, American Chemical Society.

determination upon irradiating only a single set of mobile protons. In addition, because the chemical shift difference between water and labile protons (e.g., 5.6 ppm for Iobitridol) is much larger than the chemical shift difference between two labile groups (e.g., 1.2 ppm for Iopamidol), the proposed approach has advantages for applications at lower field strength.

The proposed B_1 mismatched ratiometric method offers higher sensitivity than the conventional ratiometric approaches investigated so far. In fact, if we consider the difference in the ratiometric values between the pH values of 6.0 and 7.4 (ΔR_{pH}) to assess pH sensitivity, it corresponds to 3.1, 2.8, and 1.1 for Iopamidol, Iopromide, and YbHPDO3A, respectively, whereas the proposed Iobitridol pH MRI method yielded a ΔR_{pH} of 2.7 (3/6 µT) and a ΔR_{pH} of 11.6 (1.5/6 µT) (Fig. 19.13).

19.5 Conclusion

X-ray contrast agents containing exchangeable protons in their chemical structure act also as CEST agents. They have the advantage of being approved for administration to patients and, because of administration at high doses, they distribute into the regions of interest at concentrations required for acting as effective CEST agents. Moreover, it has been shown that they can be excellent agents for mapping pH. This task has been first addressed by using molecules such as Iopamidol and Iopromide that contain in their structure two chemically different amide functionalities. By exploiting the differences in their pH dependence of the prototropic exchange, it has been possible to set up a method for accurate pH determination that is independent of the local concentration of the agent. Next it has been shown that an analogous ratiometric method for pH assessment is possible by exploiting the dependence of the ST% on the intensity of the applied irradiating field. This is an enormous improvement as it allows also using as responsive agents molecules, which contain only one type of exchangeable proton.

The herein summarized work has shown the great potential of the class of currently used iodinated agents to act as CEST MRI pH reporters, and it has also identified the shortcomings that have to be addressed for their clinical translation. Basically, one needs to improve the SNR in order to reach the level of robustness required by diagnostic clinical tests. This task may be addressed in several ways, either by improving the detection steps or by processing the Z-spectra to remove noise components or to correct for other confounding effects [25–30]. Both routes are currently under intense scrutiny, and we do hope that this will allow enriching the MRI-armory with mapping of pH as an important biomarker of early diagnosis, staging, and therapeutic monitoring of relevant diseases.

References

1. Bourin M, Jolliet P, and Ballereau F. An overview of the clinical pharmacokinetics of X-ray contrast media. *Clin Pharmacokinet*, 1997; 32(3): 180–193.

2. Hasebroock KM and Serkova NJ. Toxicity of MRI and CT contrast agents. *Expert Opin Drug Metab Toxicol*, 2009; 5(4): 403–416.
3. Aime S, Calabi L, Biondi L, et al. Iopamidol: Exploring the potential use of a well-established X-ray contrast agent for MRI. *Magn Reson Med*, 2005; 53(4): 830–834.
4. Longo DL, Dastru W, Digilio G, et al. Iopamidol as a responsive MRI-chemical exchange saturation transfer contrast agent for pH mapping of kidneys: In vivo studies in mice at 7 T. *Magn Reson Med.*, 2011; 65(1): 202–211.
5. Sun PZ, Longo DL, Hu W, Xiao G, and Wu R. Quantification of iopamidol multi-site chemical exchange properties for ratiometric chemical exchange saturation transfer (CEST) imaging of pH. *Phys Med Biol*, 2014; 59(16): 4493–4504.
6. Sun PZ, Wang Y, and Lu J. Sensitivity-enhanced chemical exchange saturation transfer (CEST) MRI with least squares optimization of Carr Purcell Meiboom Gill multi-echo echo planar imaging. *Contrast Media Mol Imaging*, 2014; 9(2): 177–181.
7. Raghunand N, Howison C, Sherry AD, Zhang SR, and Gillies RJ. Renal and systemic pH imaging by contrast-enhanced MRI. *Magn Reson Med*, 2003; 49(2): 249–257.
8. Pereira PC, Miranda DM, Oliveira EA, and Silva AC. Molecular pathophysiology of renal tubular acidosis. *Curr Genomics*, 2009; 10(1): 51–59.
9. Paragas N, Qiu A, Zhang Q, et al. The Ngal reporter mouse detects the response of the kidney to injury in real time. *Nat Med*, 2011; 17(2): 216–222.
10. Longo DL, Busato A, Lanzardo S, Antico F, and Aime S. Imaging the pH evolution of an acute kidney injury model by means of iopamidol, a MRI-CEST pH-responsive contrast agent. *Magn Reson Med*, 2013; 70(3): 859–864.
11. Longo DL, Cutrin JC, Michelotti FC, Lanzardo S, and Aime S. Functional evaluation of normothermic ischemia/reperfusion injury in mice kidney combining CEST-pH mapping and DWI MRI. In *Proceedings of the 8th European Molecular Imaging Meeting (EMIM)*. Torino, Italy, 2013: #99.
12. Jones KM, Randtke EA, Howison CM, et al. Measuring extracellular pH in a lung fibrosis model with acidoCEST MRI. *Mol Imaging Biol*, 2015; 17(2): 177–184.
13. Keupp J, Heijman E, Langereis S, et al. Respiratory triggered chemical exchange saturation transfer MRI for pH mapping in the kidneys at 3 T.

In *Proceedings of the 19th Annual Meeting of ISMRM.* Montreal, Quebec, Canada, 2011: 690.
14. Müller-Lutz A, Khalil N, Schmitt B, et al. Pilot study of Iopamidol-based quantitative pH imaging on a clinical 3T MR scanner. *MAGMA*, 2014; 27(6): 477–485.
15. Chen LQ, Howison CM, Jeffery JJ, Robey IF, Kuo PH, and Pagel MD. Evaluations of extracellular pH within in vivo tumors using acidoCEST MRI. *Magn Reson Med*, 2013; 72(5): 1408–1417.
16. Hingorani DV, Bernstein AS, and Pagel MD. A review of responsive MRI contrast agents: 2005-2014. *Contrast Media Mol Imaging*, 2014; 10(4): 245–265.
17. Gillies RJ, Raghunand N, Garcia-Martin ML, and Gatenby RA. pH imaging. A review of pH measurement methods and applications in cancers. *IEEE Eng Med Biol Mag*, 2004; 23(5): 57–64.
18. Chen LQ, Howison CM, Spier C, et al. Assessment of carbonic anhydrase IX expression and extracellular pH in B-cell lymphoma cell line models. *Leuk Lymphoma*, 2015; 56(5): 1432–1439.
19. Ward KM, Aletras AH, and Balaban RS. A new class of contrast agents for MRI based on proton chemical exchange dependent saturation transfer (CEST). *J Magn Reson*, 2000; 143(1): 79–87.
20. Delli Castelli D, Ferrauto G, Cutrin JC, Terreno E, and Aime S. In vivo maps of extracellular pH in murine melanoma by CEST-MRI. *Magn Reson Med*, 2014; 71(1): 326–332.
21. Sheth VR, Li Y, Chen LQ, Howison CM, Flask CA, and Pagel MD. Measuring in vivo tumor pHe with CEST-FISP MRI. *Magn Reson Med*, 2012; 67(3): 760–768.
22. Delli Castelli D, Terreno E, and Aime S. Yb(III)-HPDO3A: A dual pH- and temperature-responsive CEST agent. *Angew Chem Int Ed Engl*, 2011; 50(8): 1798–1800.
23. Wu Y, Soesbe TC, Kiefer GE, Zhao P, and Sherry AD. A responsive europium(III) chelate that provides a direct readout of pH by MRI. *J Am Chem Soc*, 2010; 132(40): 14002–14003.
24. Longo DL, Sun PZ, Consolino L, Michelotti FC, Uggeri F, and Aime S. A general MRI-CEST ratiometric approach for pH imaging: Demonstration of in vivo pH mapping with Iobitridol. *J Am Chem Soc*, 2014; 136(41): 14333–14336.
25. Sun PZ, Cheung JS, Wang E, Benner T, and Sorensen AG. Fast multislice pH-weighted chemical exchange saturation transfer (CEST) MRI with unevenly segmented RF irradiation. *Magn Reson Med*, 2011; 65(2): 588–594.

26. Sun PZ, Wang E, and Cheung JS. Imaging acute ischemic tissue acidosis with pH-sensitive endogenous amide proton transfer (APT) MRI: Correction of tissue relaxation and concomitant RF irradiation effects toward mapping quantitative cerebral tissue pH. *Neuroimage*, 2012; 60(1): 1–6.
27. Sun PZ, Wang E, Cheung JS, Zhang X, Benner T, and Sorensen AG. Simulation and optimization of pulsed radio frequency irradiation scheme for chemical exchange saturation transfer (CEST) MRI-demonstration of pH-weighted pulsed-amide proton CEST MRI in an animal model of acute cerebral ischemia. *Magn Reson Med*, 2011; 66(4): 1042–1048.
28. Zhu H, Jones CK, van Zijl PC, Barker PB, and Zhou J. Fast 3D chemical exchange saturation transfer (CEST) imaging of the human brain. *Magn Reson Med*, 2010; 64(3): 638–644.
29. Li H, Zu Z, Zaiss M, et al. Imaging of amide proton transfer and nuclear Overhauser enhancement in ischemic stroke with corrections for competing effects. *NMR Biomed*, 2015; 28(2): 200–209.
30. Zaiss M, Xu J, Goerke S, et al. Inverse Z-spectrum analysis for spillover-, MT-, and T1-corrected steady-state pulsed CEST-MRI: Application to pH-weighted MRI of acute stroke. *NMR Biomed*, 2014; 27(3): 240–252.

Index

acid 168, 177, 248, 270, 386
   2-hydroxybenzoic 175
   3-nitrosalicylic 174
   ascorbic 232
   anthranilic 174, 175
   barbituric 168, 336
   carboxylic 380
   deoxyribonucleic 290
   glucuronic 114
   hyaluronic 114, 380, 382
   hydrochloric 162
   lactic 401
   nucleic 29, 162, 177, 284, 286, 290, 291
   perchloric 439
   polyuridilic 29
   ribonucleic 29
   salicylic 173, 174
acute kidney injury (AKI) 451
acute stroke 25, 355
ADC *see* apparent diffusion coefficient
AF *see* annulus fibrosus
AKI *see* acute kidney injury
alcohol 5, 6, 162, 236, 262, 263, 270
amide proton 17, 19, 22, 24, 25, 49, 165, 204, 205, 223, 224, 265, 267, 269, 271, 298, 349–351, 386
amide proton transfer (APT) 25, 28, 33, 69, 348, 353, 354, 358
amine proton 28, 165, 166, 430

annulus fibrosus (AF) 381, 382, 387, 392
apparent diffusion coefficient (ADC) 353–357, 359
APT *see* amide proton transfer
APT effects 26, 349–352, 358
APT imaging 26, 349, 351, 353–355, 359, 361, 362
APT signal 73, 350, 351, 357
APTw hyperintensity 351, 358
APTw images 354, 357, 359, 361, 363
articular cartilage 114, 116, 117, 383, 391
asymmetry analysis 62, 81, 350, 416, 419, 420, 450

bead slurries 124, 125
biological system 211, 260
biomarker 232, 285, 286, 296, 305, 404, 451, 463
   clinical 451
biosensors 122, 125, 144, 206
   transferrin-based 144
   xenon-based 135
Bloch equation 7, 22, 31, 41, 98, 99, 163, 229, 248, 250
Bloch–McConnell equation 140, 173
Bloch simulation 69, 180–183
Bloch theory 227, 238, 252
blood–brain barrier 198, 203, 358, 414

blood urea nitrogen (BUN) 451–453
BUN *see* blood urea nitrogen

cage 122, 135, 136, 138, 144, 339
   biotinylated 138
   molecular 122
   naked 146
calf muscles 433–435, 437
cancer 24, 26, 347, 360, 362, 400–402, 407, 408, 412, 421, 442, 451
   colorectal 408
   human 301
   metastatic 399
   neck 362
cancer metabolism 400, 401, 422
carriers 137
   biocompatible 50
   drug-delivery 325
   liposomal 147
   nanoparticulate 140
cartilage 117, 378, 379, 381–385, 387, 389, 391–393
   damaged 391
   femorotibial 392
cartilage imaging 379, 384
cartilage tissue 116, 378, 379, 383, 386, 392
cavity 122, 313, 318, 321, 326
   inner aqueous 333
   synovial 114, 117
CBF *see* cerebral blood flow
cell lysates 203, 204
cell populations 50, 205
cell 13, 15, 18, 19, 138, 140, 143–146, 195, 204, 206, 272, 330–332, 334, 337–339, 402, 412
   cancer 18, 19, 348, 399–402, 411, 417, 418
   dead 338
   encapsulated 147

   eukaryotic 195
   human aortic endothelial 339
   human brain microvascular endothelial 339
   lanthanide-loaded 331
   malignant 402
   microvascular brain endothelial 140
   pumping 130–132
   rat gliosarcoma 203
   red blood 143, 330, 357
   tumor 24, 400, 401, 418
   white blood 357
central nervous system (CNS) 198, 203
cerebral blood flow (CBF) 26, 353–355, 359, 414, 416
cerebral ischemia 355–357
CERT *see* chemical exchange rotation transfer
CEST *see* chemical exchange saturation transfer
CEST agents 49, 50, 52, 84, 85, 163, 170, 305, 311–313, 324, 325, 327, 330, 336, 407, 408, 448, 456, 463
   heterocyclic 168
   ion-unresponsive 296
   liposome-based 327
   pH-dependent 297
   pH-Responsive 447, 448, 450, 452, 454, 456, 458, 460, 462
   temperature-independent 301
   vesicle-based 52, 328
CEST amplitudes 284, 285, 287, 291, 293, 294, 296, 298–300
   pH-responsive 298
   pH-unresponsive 298
CEST applications 128, 132, 168
CEST contrast 51, 52, 72–74, 78, 86, 165, 166, 170, 172, 176–178, 198–200, 203–206, 235, 323, 324, 326–330, 332–334, 448, 449

CEST effect 81–84, 98, 105–109, 116, 117, 124–126, 140, 141, 267, 268, 270, 271, 285, 287, 294–297, 299, 300, 323, 324, 326, 327, 456, 457
CEST experiment 52, 63, 75, 76, 124, 125, 131–133, 135, 178, 386, 387, 432
CEST images 234, 403, 405–407, 413, 419, 437, 461
CEST imaging 11, 79, 165, 166, 205, 262, 269, 272, 276, 337, 429, 430
CEST intensity 42, 43, 229, 236, 237, 271
CEST maps 328, 329, 336, 433
CEST measurements 84, 105, 117, 131, 166, 406
CEST mechanism 41, 63, 178, 351
CEST MRI 73, 79, 84, 145, 166, 180, 183, 198, 208, 288, 290, 295, 405, 406
CEST peaks 83, 229, 231, 264, 266, 267, 269, 271, 274, 276, 277, 442
CEST pools 108, 109, 134, 136, 141, 333
CEST responses 131, 136, 137, 141, 146, 448
CEST signal 41, 79, 200, 203, 221, 227, 232, 235, 238, 241–244, 267, 284, 287–290, 294–298, 300, 301
  amide-based 228
  enzyme-unresponsive 288, 289
  fake 86
  pH-dependent 298
  pH-unresponsive 299
  water-based 232
CEST spectrum 24, 221, 227, 228, 233, 236, 244, 245, 249–251, 263, 265–270, 272, 275, 276, 284, 292, 293, 300, 457
CEST systems 49, 140, 460

chelate 48, 180, 183, 288, 290–296, 298, 300, 301, 330
  gadolinium-based 197, 198
  ionic 317
  organic 283, 296, 298
chemical exchange rotation transfer (CERT) 70
chemical exchange saturation transfer (CEST) 9–14, 29, 30, 39–42, 49, 58, 61, 62, 73, 74, 78–80, 107–110, 121, 131–134, 177–179, 227–229, 268, 269, 413, 414
chemical shift difference 85, 141, 163, 209, 221, 266, 462
chemical shifts 4–6, 57, 58, 107, 108, 122, 123, 135, 136, 164, 165, 173–175, 222, 243, 248–250, 284–287, 291, 293–296, 299–301, 388
chemical shift selective (CHESS) 18, 19, 433
CHESS *see* chemical shift selective
cholesterol 12, 143, 324, 336
CK *see* creatine kinase
CK reaction 427–432, 435
CNR *see* contrast-to-noise ratio
CNS *see* central nervous system
collagen 114, 116, 286, 380–382
complexes 40–42, 49, 179, 222, 223, 225–228, 236, 238–240, 243, 245, 247, 250, 257–264, 266–269, 271–274, 276
  neutral 319
components 66, 72, 81, 97–100, 104, 106, 222, 223, 383, 430–432
  cellular 387
  exchangeable 393
  polysaccharide 379
  semisolid 97
  signal 81
  trial and error 259

compounds 65, 161–163, 165, 166, 168, 170, 178, 226, 252, 277, 311, 378
  anionic 380
  aromatic 32
  cyclic 168
  diaCEST 34
  hydroxy 5
  millimolar concentration 18
  natural 168
contrast agents 13, 51, 63, 83, 161, 163, 200, 260, 269, 273, 274, 289–291, 301, 311, 313, 383, 384
  diaCEST 178
  exogenous 13
  non-metallic 34
  pH-responsive 448, 456
  polymeric 291
  ratiometric 271
  superparamagnetic MRI 161
  X-ray 448
contrast-to-noise ratio (CNR) 83, 194, 447
creatine 163, 165, 427–430, 441, 442
creatine kinase (CK) 197, 427, 428
cryptophanes 122, 135–137, 140, 141, 146, 208

DDD *see* degenerative disc disease
degenerative disc disease (DDD) 383, 388, 393
dendrimers 22, 29, 31, 177, 285, 312
density functional theory (DFT) 226, 239
detection sensitivity 163, 285, 286, 301, 312, 352, 405, 416
DFT *see* density functional theory
diaCEST agents 83, 132, 163, 164, 168, 177, 178, 235, 242, 324

dipolar contributions 258, 259, 274
dipolar coupling 14, 58
dipolar interactions 100, 130
direct saturation (DS) 28, 62, 64, 76, 79–82, 108, 109, 168, 177, 221, 349, 405
diseases 193, 205, 347, 364, 378, 428, 463
  cardiac 429
  degenerative joint 377
  muscular 428
  neoplastic 400
disorders 362, 377, 383, 429, 441
  liver 406
  metabolism-related 404
  muscular 428
  neurologic 166
  neuromuscular 437
DNA 30, 195, 206, 291, 292
DS *see* direct saturation

enzyme activities 284, 286, 287, 289, 290
enzyme catalysis 286–288
enzymes 194, 196, 197, 200, 202, 205, 206, 210, 284, 286–289
  esterase 290
  glycolytic 401, 414
  protease 288
  tyrosinase 197
  viral 206
exchangeable protons 18–20, 22, 81, 83, 85, 116, 165, 168–170, 219, 220, 262, 264, 267–269, 271, 284, 463
exchangeable sites 389, 393
exchange process 10, 12, 59, 76
exchange rates 7, 63, 64, 66–70, 72, 74, 82, 83, 138, 140, 162, 163, 312, 313, 323, 324, 430, 432, 434, 448, 449
  amide proton 352

backbone NH 176
modulate water 238, 239
prototropic 49
solute-to-water 351
tissue 9

FCD *see* fixed charge density
fixed charge density (FCD) 385, 388, 390
FLEX *see* frequency-labeled exchange
frequency-labeled exchange (FLEX) 22, 65, 71, 126, 164
frequency offsets 50, 51, 72, 78, 97, 99–111, 113–117, 171, 181, 208, 209, 349, 420
frequency positions 98, 102–104, 106, 108, 109

GAG *see* glycosaminoglycan
gagCEST 116, 377–379, 384–386, 389, 391–393
GAG concentrations 383–385, 388, 393
GAG sugars 378–380, 385, 387
gas vesicles (GVs) 140, 142, 208–210
gene 30, 194, 195, 203, 210, 211
  artificial 203
  gas vesicle 209
gene expression 195–197
gene products 195, 197, 198, 200
GFP *see* green fluorescent protein
gliomas 26, 27, 351, 359, 362, 363, 402
glucoCEST 399, 400, 402–410, 412–414, 416–422
glucoCEST signal 408, 412–415, 418, 422
glucose 42, 85, 166, 292, 293, 302, 399–405, 407, 408, 410–412, 414, 416, 420, 422

glucose metabolism 407, 410
glucose uptake 399, 402, 403, 414, 417, 456
glycogen 20, 21, 177, 406, 407, 434
glycolysis 401, 402, 428, 456, 457
  aerobic 400, 401
glycosaminoglycan (GAG) 114, 116, 378–381, 383, 384, 386, 392
gradient-encoded CEST 76–78, 133–135
green fluorescent protein (GFP) 203–205
GVs *see* gas vesicles

HIFU *see* high-intensity focused ultrasound
high-intensity focused ultrasound (HIFU) 362
human scanners 32–34, 73
hydroxyl groups 12, 25, 206, 283, 389, 393, 400, 404, 406, 408
hydroxyl protons 168, 208, 327, 328, 386, 406, 414
hydroxyls 166, 219, 304, 386, 406
hyperCEST 58, 121–123, 125, 126, 136, 140–142, 147, 198, 208, 348
hyperfine shift 258, 259, 266, 274, 284
hyperpolarization 14, 122, 124, 126, 128, 129, 162, 208
hyperpolarized MRI 208, 210
hyperpolarized xenon 129, 198

images 9–11, 13, 68, 69, 73, 74, 110, 111, 233, 234, 269, 351–353, 383, 384, 403, 405, 406, 413, 414, 417–419, 438, 440
  anatomical 433, 434, 439, 461

confocal 47
gagCEST-weighted 391
grayscale 419
histology 363
imaging 10, 17, 18, 20, 22, 24, 26, 28, 30–32, 34, 161, 162, 210, 211, 348, 349, 428, 430, 437, 438
   animal 30
   classical 147
   diffusion 347, 357
   fluorescence 204
   lung 133
   magnetic resonance spectroscopic 162
   multi-color 32
   multiplex 209, 323, 325, 328
   multi-slice 358
   myocardial 439
   perfusion 347
   radionuclide 194
   tomographic 378
   ultrasound 194
imino proton 29, 169, 170, 172, 208
infarction 353, 438, 439
interaction 100, 178, 196, 197, 200, 204, 210, 291, 292, 318
   chemical 314, 316
   electric charge 384
   intramolecular 204
   low-frequency 385
   non-covalent 50
Iobitridol 449, 459–462
Iopamidol 49, 448, 449, 451, 453–456, 459, 462, 463
ischemia 19, 25–28, 353, 355, 357, 451, 453
ischemic strokes 352, 355, 357

kidneys 10–13, 75, 165, 204, 232, 234, 412, 417, 419, 451, 453–455

Krebs cycle 400, 428

labeling pulses 64, 68, 70, 72
label-transfer module (LTM) 65, 66
labile proton 162–166, 168, 173, 176, 385, 404, 416, 462
lanthanide 222, 225, 227, 228, 238, 314, 315, 320, 332
lanthanide complex 40, 41, 235, 238, 246, 314, 315
lanthanide ion 224, 227–229, 234, 238, 239, 246, 257, 301
LF *see* ligand field
ligand field (LF) 223–227
ligand 41, 235, 238, 239, 241, 242, 246, 259, 260, 263, 264, 273, 276, 283, 284, 289, 290, 295, 298, 316, 320
   aliphatic amine 289
   bis-carboxylate 294
   macrocyclic 260
   macrocyclic polyaminocarboxylic 315
   phenolate 300
   phenylbornate 292
   pyrazole 296
   pyridine 294
   quinolinium 296
   tripodal 275
lipoCEST 50, 51, 61, 312, 320, 323, 325, 327, 330, 335
   non-spherical 319, 320
   spherical 319
lipoCEST agents 51, 313, 317, 320, 325–329, 333–335
liposome 51, 140, 277, 285, 312, 314, 316, 318–320, 323–328, 330, 333, 334, 336–339
   arginine-loaded 337
   bulk 320
   non-spherical 51, 318, 325
   spherical 51

length and offset varied saturation (LOVARS) 72
longitudinal magnetization 58, 61, 64, 65, 126, 132
LOVARS *see* length and offset varied saturation
LRP *see* lysine-rich protein
LTM *see* label-transfer module
lysine-rich protein (LRP) 30, 32, 176, 201, 203, 204

macrocycle 260–264
macrocyclic chelate 287, 289–291, 294, 295, 298, 300, 305
macromolecule 12, 14, 50, 97, 176–178, 222, 385
magnetic field inhomogeneity 383
magnetic resonance spectroscopy (MRS) 19, 162, 275, 406, 410, 428–430, 433–437, 439, 441, 442
magnetization 8, 12, 19, 20, 61, 63, 65, 81, 122, 124–126, 129, 131, 132, 134, 141, 162, 229
 non-equilibrium 128
 thermal 128, 132
 transverse 101, 126, 132, 133
magnetization transfer (MT) 5, 7, 12, 14, 76, 82, 84, 97, 105, 108, 198, 276, 386, 460, 462
magnetization transfer contrast (MTC) 12, 13, 20, 27, 65, 69, 72, 79, 81, 82
MCAO *see* middle cerebral artery occlusion
membrane 138, 141–144, 146, 320, 323, 324, 333
 cellular 51, 335
 inhomogeneous 143
 permeable 324
 phospholipid-based 318
 plasma 138

metabolism 400, 401, 404, 407, 408, 410, 413, 414, 417, 441
metabolite 10, 12, 13, 26, 28, 162, 165, 166, 202, 284–286, 292–294, 403, 427–431
 endogenous 172
 low molecular weight 165
 methylphosphate 293
metal ion 178, 179, 197, 259–263, 274, 284, 296
 transition 258, 277
mice 71, 232, 243, 301, 325, 329, 330, 338, 413, 459
 healthy 329, 451
 tumor-bearing 232
mice kidneys 71
middle cerebral artery 355, 356
middle cerebral artery occlusion (MCAO) 26, 352
mitochondria 400, 401, 428
mitochondrial dysfunction 438, 442
mitochondrial malfunction 400, 401
mobile protein 28, 348, 357
mobile proton 48, 49, 312, 317, 338, 448
molecular imaging 193, 285, 311, 348, 362
molecule 6, 10, 13, 20, 22, 52, 57, 163, 165–167, 197, 198, 220, 222–225, 448, 451, 460, 461
 analogue 412
 diaCEST 325
 endogenous 317
 hyperpolarized 48
 inorganic 162
 iodinated 454
MPP *see* mucus-penetrating particles
MRI contrast agents 48, 235, 257, 274, 277, 283, 286, 294
MRI scanners 64, 73, 106, 269, 348, 354, 359, 363, 454, 455

MRS *see* magnetic resonance spectroscopy
MT *see* magnetization transfer
MTC *see* magnetization transfer contrast
MT effects 97–99, 104–107, 111, 113, 114, 116, 170, 276, 277
MT pools 81, 98, 104, 105, 107–109
mucus-penetrating particles (MPP) 172
muscle metabolism 429, 441, 442
muscles 165, 427, 433–437
  skeletal 428
  soleus 434–436
muscle tissue 429, 433, 434, 436

nanocarriers 137, 166, 172, 325
nanovesicles 50, 320, 323, 326
NH proton 22, 168, 170, 248, 262, 269
NMR *see* nuclear magnetic resonance
NMR
  heteronuclear 178
  high-resolution 220
  liquid-crystalline 104
NMR magnet 132, 143
NMR spectrum 20, 39, 57, 58, 61, 101, 102, 122, 123, 135, 220, 263, 265, 267, 269, 411, 430, 431
NMR timescale 22, 135, 169, 180, 258, 264, 386
NOE *see* nuclear overhauser enhancement
NOMAR *see* normalized magnetization ratio
normalized magnetization ratio (NOMAR) 86
nuclear magnetic resonance (NMR) 9, 39, 41, 57, 58, 98, 121, 126, 143, 144, 146, 162, 219, 248, 250, 386, 402
nuclear overhauser enhancement (NOE) 20, 28, 110, 351, 386, 404

oxidation 262, 271, 273, 274, 276
oxygen atoms 239, 240, 246, 261
  acetyl 242
  amide 226
  carboxylate 226
  ketone-type 242
  ligand's 246
  ligating 241

paraCEST 121, 124, 138, 168, 219, 227, 242, 248, 252, 257, 272, 283, 312, 348
paraCEST agents 42, 43, 222, 223, 228, 229, 231, 232, 234, 235, 246, 247, 249, 252, 253, 258, 266–271, 274–277, 283–292, 294–298, 300, 301, 305
patient 33, 166, 168, 354, 355, 358–362, 463
PBS *see* phosphate buffered saline
peptide 24, 26, 32, 162, 176, 177, 194, 200, 204, 206, 210, 287, 288, 338, 349, 357
  arginine-rich 32, 204, 339
  conjugated 288
  lysine-rich 177
perfusion 353, 407, 410, 412
permeability 51, 172, 324, 336, 414
PET *see* positron emission tomography
phantom 81, 233, 269, 270, 274, 328, 393, 431, 432
  agar 31
  iopamidol 450

phosphate buffered saline (PBS) 170, 172, 331, 450
phosphocreatine 427, 430
phospholipid-based bilayer 318, 319, 330
phospholipid membrane 125, 137, 142, 324
pH sensor 42, 49, 243
pixel 80, 85, 86, 110, 275, 392, 419
positron emission tomography (PET) 402, 408, 409, 422
probes 14, 15, 125, 170, 172, 175, 178, 183, 194, 197, 198, 210, 311, 324, 329, 379, 459
    diaCEST 177
    fluorescent 144
    hypoxia 273
    lipoCEST 51
    metal-based 196, 203
    NMR-active 122
    paraCEST 232
    temperature 42
protein backbone 48, 348
proteins 17, 19, 24, 26, 28, 30, 162, 176, 177, 204, 206, 286, 348, 358, 359, 362, 379
    endoplasmic reticulum 348
    ferritin 197
    fluorescence 13
    glycosylated 380
    nuclear 348
    synthetic 30
proteoglycans 380, 381, 385
proton exchange 42, 59, 198, 230, 235, 237, 248, 267, 269, 270
proton MRI 124, 208, 210
proton resonance 220, 264, 266, 275, 386
    aromatic 266
    far-shifted 268
    hyperfine shifted ligand 39, 258
proton 18, 19, 65–71, 107, 108, 128, 173–175, 198, 199, 222, 223, 226, 229–231, 266, 268, 269, 275, 284, 349, 350, 378, 379
    aliphatic 25, 348, 351
    bulk 50, 312, 317, 326
    cytoplasmatic 333
    guanidyl 204, 205
    intraliposomal 319, 320, 324, 327
    intravesicular 313, 330
    macrocyclic CH 264
    non-saturated 62
    pyrazole 271
    semi-solid 28, 76
proton transfer enhancement (PTE) 63, 83, 163
proton transfer ratio (PTR) 27, 60, 61, 63, 71, 72, 82, 163, 351
Provotorov's theory 103, 107
PTE *see* proton transfer enhancement
PTR *see* proton transfer ratio
pulse 18, 19, 61, 64–68, 71–73, 102, 106, 123, 134, 229, 270, 432
    cosine-modulated 104, 107
    frequency-selective 64
    irradiation 60
    phase-modulated 68
    presaturation 220, 229, 258, 268, 269
    rectangular 432
pulse train 24, 68, 78, 270, 432

quantifying exchange using saturation power (QUESP) 31, 170, 199, 289
quantifying exchange using saturation time (QUEST) 31, 68, 135, 199
QUESP *see* quantifying exchange using saturation power
QUEST *see* quantifying exchange using saturation time

radiation necrosis 360, 361, 363
radio frequency (RF) 5, 32, 58, 63, 98, 99, 123, 284, 349
rat brain 24, 27, 350, 352–354, 413, 415
ratiometric approach 14, 285, 286, 300, 448, 460, 461
ratiometric comparison 285, 287, 296, 301
RBCs *see* red blood cells
receptor 139, 196, 210, 286
   cell surface 138
   endothelial 330
   transferrin 138, 197
red blood cells (RBCs) 51, 143, 330–335, 357
reduction 40, 60, 63, 65, 74, 134, 169, 232, 233, 241, 242, 272, 349, 402, 439
   biological 232
   chemical 232
relaxation 25, 58, 73, 97, 126, 130, 132, 162, 358
   exchange-dependent 85
   nuclear 41
   quadrupolar 385
   radiation-free 130
   rotation frame 64
   spin-lattice 100, 385
relaxation rate 7, 162, 163, 228, 250
relaxation time 58, 65, 132, 228, 285, 391, 417, 432, 437
   longitudinal 101, 229, 351
   spin-lattice 414
   transverse 65, 99, 403
reporter gene 30, 194, 195, 197, 198, 200, 204
   CEST MRI 200, 211
   diaCEST 176
   gas vesicle–based 196
   proton-based 211
reporters 196, 198, 199, 203, 208
   fluorescence 144, 146, 204

fluorine-based 211
green fluorescent protein 203
magnetized 198
radionuclide 196
synthetic CEST 30
resonance 39, 40, 48, 50, 123, 125, 179, 180, 220, 233, 248, 249, 264, 327, 328, 338, 430, 431, 448, 449, 460, 461
   biosensor 123
   composite 348, 349
   macrocyclic 263
RF *see* radio frequency
RF irradiation 74, 98, 99, 101–106, 111, 350, 351
   continuous 98
   cosine-modulated 107–110
   multi-frequency 103
RF pulse 14, 32, 67, 72, 73, 84, 101, 305
RF saturation 76, 77, 125, 128, 284, 351

samples 76–78, 131, 132, 140, 144, 182, 270, 320, 328, 388, 392, 420, 431
   aqueous phantom 81
   biological 209
   bovine cartilage 386
   chemical 285
SAR *see* specific absorption rate
saturation 19, 21, 22, 61, 62, 73–79, 84, 87, 97–101, 103–105, 107, 108, 124, 126, 128, 200, 203, 405, 406
   off-resonance 40
   single-frequency 102
   steady-state 18, 74, 75
saturation effects 19, 124, 126, 349, 351
saturation efficiency 61, 70, 74
saturation frequencies 77, 86, 134, 284, 328, 338, 403, 457

saturation power 33, 170, 205, 249, 268, 349, 354, 359, 363, 420, 461
saturation pulse 63, 67–70, 72, 75, 76, 79, 140, 163, 173, 182, 199, 200, 251, 437, 450, 454
  fat 75, 433
  frequency-selective 75
  pre-loop 75
saturation time 31, 125, 170, 230, 231, 237, 244, 350, 354, 359, 363
saturation transfer (ST) 7, 10, 12–14, 49, 52, 68, 70, 71, 124–126, 162, 164, 331, 332, 335, 450, 451, 460, 463
saturation transfer effects 14, 61, 97, 113, 124, 448
sensitivity 43, 50, 126, 129, 140, 146, 231, 232, 243, 285, 286, 290, 311, 312, 330, 338, 432, 433, 439
sensors 42, 124, 140, 146, 206, 232
  biotinylized 124
  cell-associated 147
  genetic–molecular 205, 206
sequences 48, 74, 77, 106, 107, 360
  gradient-echo MRI 355
  MRI 125
  pulse 34, 48, 348
  relaxometry 417
  sodium pulse 392
  spin echo 75
shift 50, 52, 78–80, 109–111, 135, 180, 225, 259, 284, 293, 300, 314, 315, 319, 320, 332–334, 419
  chemical 14, 84, 144, 173, 178, 209, 284, 300
  dipolar 225
  dipolar NMR 222
  hyperfine NMR 42
  hyperfine proton 263
  lanthanide-induced NMR 222
  paramagnetic NMR 222
  paramagnetic proton 258
shift reagents (SRs) 50, 51, 219, 222, 312, 314, 317, 319, 330, 331
specific absorption rate (SAR) 64, 73, 358, 385, 405, 454
SRs *see* shift reagents
ST *see* saturation transfer
stroke 347, 352, 354, 355, 357, 362, 442
sweep imaging with Fourier transfer (SWIFT) 74, 75, 234
SWIFT *see* sweep imaging with Fourier transfer
SWIFT-CEST 74, 75
symmetry 223, 226, 266, 268, 269, 315

TBI *see* traumatic brain injury
therapy 210, 360, 362, 400
  anti-cancer 362
  oncoviral 30
thermal equilibrium 126, 129, 130
threonine 166, 176, 206
thymidine 169, 170, 208
tissue 9, 116, 117, 147, 165, 166, 168, 231, 242, 243, 276, 378, 379, 383–386, 389, 392, 393, 402–404, 421, 422, 439
  biological 165, 178, 198, 348, 349, 380
  brain tumor 442
  damaged 383
  disc 383, 393
  fluid 389
  hypoxic 273
  infarcted 439, 440
  lamb heart 438
  liver 407
  non-cartilage 384

non-infarcted 439
peritumoral 27
semi-solid 20
tumor 73, 400
transfer 12, 18–20, 130, 162, 163, 219, 386
  chemical exchange rotation 70
  high-saturation 456
  spin 219
traumatic brain injury (TBI) 362
tumor models 410, 458
tumors 26–28, 32–34, 172, 203, 329, 330, 332, 336, 351, 352, 357–359, 363, 399–402, 408–410, 412, 413, 417, 418, 456–458
  acidic 457
  active 402
  breast cancer 332
  high-grade 358
  intracerebral 330
  low-grade 32
  malignant 358
  phenotype 408
  wild-type 203

urea 10, 11, 13, 165
urea resonance 11
urea signal 10

van der Waals diameter 135
variable delay multi-pulse (VDMP) 68, 70
variable flip angles (VFAs) 133
VDMP *see* variable delay multi-pulse
vesicles 51, 139, 209, 313, 316, 319, 323, 324, 330, 333, 337
  lysosomal 144
  phospholipid-based 312
  spherical 323
VFAs *see* variable flip angles

Warburg Effect 399–401, 407
WASSR *see* water saturation shift referencing
water exchange 25, 39–43, 220, 222, 235, 236, 239–241, 243, 245, 246, 248, 276
water exchange rates 42, 221, 227, 235, 237, 239, 241, 243, 245–250
water frequency 27, 62, 71, 419
water lifetimes 231, 236–238, 246, 248, 249, 252
water molecules 4, 39, 41, 51, 168, 198, 220, 228, 239, 243, 246, 248, 283, 292, 312
water peak 238, 276, 420, 431
water pool 58, 60, 61, 63, 66, 72, 84, 97, 229
water protons 59, 62, 66, 70, 108, 109, 168, 172, 177, 178, 197, 198, 208, 211, 228–230, 312–317, 328, 349
  bulk 350
  free 437
  intracellular 51
  intraliposomal 314, 321
  metal-coordinated 49
  solvent 313
  unsaturated 349, 350
water resonance 12, 13, 26, 39, 68, 82, 83, 97, 334, 349, 350, 389, 411, 420, 432, 437
water saturation 18, 21, 27, 28, 62, 70, 349, 437
water saturation shift referencing (WASSR) 79, 111, 389
water signal 11, 18, 19, 24, 25, 40, 62, 63, 79, 105–107, 109, 111, 114, 116, 300, 305, 348, 349, 352
water signal intensity 48, 60, 63, 65, 70, 71
water suppression 18, 19

xenon 128, 130, 136, 137, 144, 208
xenon hosts 136, 144
xenon transfer contrast (XTC) 124
XTC *see* xenon transfer contrast

Z-spectroscopy 107, 134
Z-spectrum 21, 26, 27, 62, 67, 68, 76, 80, 81, 99, 107–113, 132, 133, 135, 386, 387, 403–406, 419, 420, 450, 455, 456